Inorganic Chemistry Experiment
无机化学实验

(英汉双语版)

主　编　童国秀

副主编　乔　儒　宫培军　赵玉玲　柏　嵩

科学出版社

北　京

内 容 简 介

本书共精选42个无机化学实验，内容包括无机化学实验的基本操作和常用仪器的使用、无机化学实验基础知识、无机化学中常用平衡常数的测定、元素及其化合物的性质、无机化合物的制备与提纯；并适当加入一些反映无机化学学科发展前沿的综合与设计实验，体现无机化学学科的系统性和科学性；以及部分趣味化学实验，体现无机化学与生活的紧密联系。

本书可作为高等学校化学、化工及相关专业的无机化学实验教材，也可供有关化学专业的工作人员及研究人员参考。

图书在版编目(CIP)数据

无机化学实验=Inorganic Chemistry Experiment：英、汉/童国秀主编.—北京：科学出版社，2019.8
ISBN 978-7-03-061183-3

Ⅰ.①无… Ⅱ.①童… Ⅲ.①无机化学–化学实验–高等学校–教材–英、汉 Ⅳ.①O61-33

中国版本图书馆 CIP 数据核字(2019)第 089829 号

责任编辑：丁　里/责任校对：何艳萍
责任印制：赵　博/封面设计：迷底书装

科 学 出 版 社 出版
北京东黄城根北街16号
邮政编码：100717
http://www.sciencep.com

涿州市般润文化传播有限公司印刷
科学出版社发行　各地新华书店经销
*

2019年8月第 一 版　开本：787×1092　1/16
2024年7月第七次印刷　印张：18 3/4
字数：492 000

定价：69.00元
(如有印装质量问题，我社负责调换)

《无机化学实验》(英汉双语版)
编写委员会

主　编　童国秀

副主编　乔　儒　宫培军　赵玉玲　柏　嵩

编　委(按姓名汉语拼音排序)

　　　　　白晓慧　柏　嵩　宫培军　何亚兵

　　　　　李春霞　吕光磊　乔　儒　童国秀

　　　　　王冬梅　赵　典　赵玉玲

前　言

　　党的二十大报告指出："教育、科技、人才是全面建设社会主义现代化国家的基础性、战略性支撑。"随着"一带一路"倡议的提出,来华留学教育的招生规模不断扩大,我国高校毕业生出国深造的需求也不断上升,教育和人才国际化趋势加强。为了满足21世纪对国际化高素质人才的迫切需求,各高校响应教育部的号召大力推广双语教学或全英文教学。在双语教学、全英文教学的起步时期,选用优秀的双语教材是取得好的教学效果的基础。无机化学是高等学校化学、化工、轻工、应用化学、高分子材料、安全工程、环境工程、生物工程、制药工程、食品科学与工程等专业必修的一门基础课程。无机化学是实验科学,通过无机化学实验教学环节,学生加深对理论知识的理解;掌握无机化学实验的基本方法和操作技能;培养严谨的科学态度、分析问题和解决问题的能力及创新意识,为后续课程的学习以及将来成为21世纪高素质人才打下良好的学习基础。

　　浙江师范大学无机化学教研室紧扣无机化学实验的教学大纲,充分考虑到化学与非化学、师范与非师范专业培养目标和培养要求的差异性,在多年教学实验改革的基础上,从已有的实验内容中精选部分实验,新增了部分综合性、设计性实验以及一些反映无机化学学科发展前沿的实验,进行重新组合,形成了相对独立和完整的无机化学实验新体系。在基础知识与基本操作部分中突出实验仪器和操作知识的新颖性和时代性,增加综合性、设计性实验的比例,达到提高学生创新能力的目的。同时,修订、整合目前所用教材元素化学实验部分设计繁杂、不符合绿色化学要求的实验,突出安全性、系统性和趣味性。

　　本书内容分为六部分,共42个实验。第1章绪论介绍了化学实验的重要意义、教学目的、学习要求和方法;第2章实验室基础知识;第3章基础无机化学实验,通过14个实验加深对无机化学反应原理的理解和掌握;第4章元素无机化学实验,通过9个实验了解常见的s、p、d区元素及其化合物的性质;第5章综合与设计实验,通过13个实验训练提高学生无机化学实验基本操作的能力、分析问题和解决问题的能力;第6章生活中的化学,选编了紧密联系生活的6个实验,目的是激发学生的科学探究兴趣和培养创新意识。

　　参加本书编译的所有人员均是长期从事无机化学及实验的教学工作,具有深厚的化学专业知识背景,同时大部分人员具有博士学历或出国留学背景,具有较强的英文读写能力,这为双语教材的编写提供了保证。

　　本书编译成员为:童国秀(第1、2章和实验1、5、9、21、30~33、42的编译及全文统稿)、乔儒(实验15~16、18~20、22~23的编译及第4章校对)、宫培军(实验24~28、34~35的编译及第5章部分实验的校对)、赵玉玲(实验12、17、29、36的编译及第3、5章部分实验的校对)、柏嵩(实验7~8、37~41的编译及第3章部分和6章实验的校对)、王冬梅(实验11、13~14的编译)、赵典(13个实验的电子课件整理)、吕光磊(7个实验的电子课件整理)、白晓慧(全书英文校对)、李春霞(实验2~4的编译)、何亚兵(实验6、10的编译)。在此对李良超、林秋月、林建军、王晓娟、吕天喜、冯洁及其他参与编译工作的研究生表示感谢。

　　由于编者水平有限,书中的疏漏和不妥之处在所难免,恳请读者批评指正。

编　者
2023年6月

目 录

前言

Chapter 1　Introduction ·· 1
　1.1　Significance of Chemical Experiments ··· 1
　1.2　Teaching Objectives of Inorganic Chemistry Experiment Course ············· 1
　1.3　Requirements and Methods of Experiment Course ····························· 2
Chapter 2　Basic Knowledge of the Lab ··· 6
　2.1　Laboratory Safety Knowledge, Discipline and Precautions ···················· 6
　2.2　The Use, Cleaning and Drying of Inorganic Chemical Apparatuses ········· 11
　2.3　Introduction and Use of Commonly Used Heating Equipment ··············· 21
　2.4　General Knowledge of Chemical Reagents ····································· 23
Chapter 3　Basic Inorganic Chemistry Experiments ································ 28
　Exp. 1　Receiving, Washing and Drying the Apparatuses ························· 28
　Exp. 2　Transferring Chemical Reagents and Operating Test Tubes ·············· 29
　Exp. 3　Preparation of Solutions ··· 32
　Exp. 4　Determination of the Molecular Weight of CO_2 ························· 36
　Exp. 5　Determination of Crystal Water Content in Blue Vitriol ················· 39
　Exp. 6　Purification and Qualitative Test of Crude Salt ··························· 42
　Exp. 7　Determination of the Degree of Dissociation and Dissociation Constant of Acetic Acid ·· 45
　Exp. 8　Qualitative Identification of Some Essential Elements in Organisms ······ 48
　Exp. 9　Determination of the Solubility Product of PbI_2 ························· 50
　Exp. 10　Determination of Chemical Reaction Rate and Activation Energy ········ 53
　Exp. 11　Preparation of Potassium Permanganate by Solid Alkali Fusion Oxidation Method ·· 58
　Exp. 12　Preparation of Ammonium Ferrous Sulfate Crystals ····················· 60
　Exp. 13　Purification and Solubility Determination of Potassium Nitrate ········· 63
　Exp. 14　Oxidation-reduction Reaction and Equilibrium ·························· 66
Chapter 4　Elemental Inorganic Chemistry Experiments ························· 71
　Exp. 15　s-Block Elements (Alkali Metals and Alkaline Earth Metals) ··········· 71
　Exp. 16　p-Block Nonmetallic Elements (Ⅰ) (Halogen, O, S) ···················· 74
　Exp. 17　p-Block Nonmetallic Elements (Ⅱ) (N, P, C, Si, B) ···················· 76
　Exp. 18　p-Block Metallic Elements (Al, Sn, Pb, Sb, Bi) ························· 79
　Exp. 19　Properties of Element Compounds in the d-Block (Ⅰ) ·················· 82
　Exp. 20　Properties of Element Compounds in the d-Block (Ⅱ) ·················· 85

Exp. 21　ds-Block Metals (Cu, Ag, Zn, Cd, Hg) ……………………………………… 87
Exp. 22　Separation and Identification of Unknown Cations ……………………… 91
Exp. 23　Separation and Identification of Unknown Anions ……………………… 97

Chapter 5　Comprehensive and Design Experiments …………………………………… 105
Exp. 24　Preparation and Composition Measurement of a Cobalt (Ⅲ) Complex ………… 105
Exp. 25　Preparation and Component Analysis of the Large Crystal of Cuprammonium Sulfate (Ⅱ) ……………………………………………………………… 108
Exp. 26　Preparation and Quality Analysis of Zinc Gluconate …………………… 111
Exp. 27　Preparation and Photocatalytic Performance of Titanium Dioxide Nanoparticles ……………………………………………………………… 114
Exp. 28　Preparation of Zinc Vitriol ………………………………………………… 118
Exp. 29　Preparation and Composition Analysis of Potassium Bis (oxalato) Copper (Ⅱ) Dihydrate ………………………………………………………………… 121
Exp. 30　Synthesis and Composition Analysis of Potassium Trioxalatoferrate (Ⅲ) ……… 124
Exp. 31　Preparation and Content Analysis of Calcium Peroxide ………………… 127
Exp. 32　Preparation of Basic Copper Carbonate ………………………………… 128
Exp. 33　Determination of the Composition and Stability Constant of Titanium (Ⅳ) Peroxide Complex ……………………………………………………… 130
Exp. 34　Preparation and Content Analysis of Cobalt Ferrite $CoFe_2O_4$ ………… 136
Exp. 35　Preparation, Optical Absorption Properties and Stability of Gold (Silver) Colloids …………………………………………………………………… 138
Exp. 36　Interaction of Metal Ions and Serum Albumin …………………………… 142

Chapter 6　Chemistry in Life …………………………………………………………… 148
Exp. 37　Preparation and Application of $Na_2S_2O_3$ ……………………………… 148
Exp. 38　Home-made Plant Acid-base Indicator …………………………………… 150
Exp. 39　Qualitative Identification of Iodine in Kelp and Salt ……………………… 152
Exp. 40　Interesting Chemical Experiment (Ⅰ) …………………………………… 153
Exp. 41　Interesting Chemical Experiment (Ⅱ) …………………………………… 157
Exp. 42　Identification of Salt and Sodium Nitrite ………………………………… 160

第1章　绪论 …………………………………………………………………………… 163
1.1　化学实验的重要意义 …………………………………………………………… 163
1.2　无机化学实验课的教学目的 …………………………………………………… 163
1.3　实验课的学习要求和方法 ……………………………………………………… 163

第2章　实验室基础知识 ……………………………………………………………… 168
2.1　实验室安全知识、纪律及注意事项 …………………………………………… 168
2.2　无机化学常用仪器的使用、洗涤与干燥 ……………………………………… 171
2.3　常用加热器具的介绍与使用 …………………………………………………… 178
2.4　化学试剂的一般知识 …………………………………………………………… 180

第3章　基础无机化学实验 …………………………………………………………… 184

实验 1	仪器的认领、洗涤与干燥	184
实验 2	试剂的取用和试管操作	185
实验 3	溶液的配制	187
实验 4	二氧化碳相对分子质量的测定	190
实验 5	胆矾结晶水的测定	193
实验 6	粗盐的提纯与定性检验	195
实验 7	乙酸解离度和解离常数的测定	198
实验 8	生物体中几种必需元素的定性鉴定	200
实验 9	碘化铅溶度积的测定	202
实验 10	化学反应速率与活化能的测定	204
实验 11	固体碱熔氧化法制备高锰酸钾	208
实验 12	硫酸亚铁铵晶体的制备	210
实验 13	硝酸钾提纯和溶解度测定	212
实验 14	氧化还原反应及平衡	215

第 4 章 元素无机化学实验 219

实验 15	s 区元素(碱金属、碱土金属)	219
实验 16	p 区非金属元素(一)(卤素、氧、硫)	221
实验 17	p 区非金属元素(二)(氮、磷、碳、硅、硼)	223
实验 18	p 区金属元素(铝、锡、铅、锑、铋)	225
实验 19	d 区元素化合物的性质(一)	227
实验 20	d 区元素化合物的性质(二)	229
实验 21	ds 区金属(铜、银、锌、镉、汞)	232
实验 22	未知阳离子的分离与鉴定	234
实验 23	未知阴离子的分离与鉴定	239

第 5 章 综合与设计实验 245

实验 24	一种钴(Ⅲ)配合物的制备与组成测定	245
实验 25	硫酸四氨合铜(Ⅱ)大晶体的制备及组成分析	247
实验 26	葡萄糖酸锌的制备与质量分析	249
实验 27	纳米二氧化钛的制备和光催化性能	251
实验 28	七水合硫酸锌的制备	255
实验 29	二水合二草酸根合铜(Ⅱ)酸钾的制备与组成分析	256
实验 30	三草酸根合铁(Ⅲ)酸钾的合成及组成分析	259
实验 31	过氧化钙的制备及含量分析	261
实验 32	碱式碳酸铜的制备	262
实验 33	过氧化氢合钛(Ⅳ)配合物的组成和稳定常数的测定	263
实验 34	钴铁氧体 $CoFe_2O_4$ 的制备及含量分析	268
实验 35	金(银)胶体的制备、光学吸收性质和稳定性	270
实验 36	金属离子与血清白蛋白的相互作用	273

第6章 生活中的化学 ·· 277
  实验 37 硫代硫酸钠的制备和应用 ·································· 277
  实验 38 自制植物酸碱指示剂 ·· 279
  实验 39 海带和食盐中碘的定性鉴定 ································ 280
  实验 40 化学趣味实验(一) ·· 281
  实验 41 化学趣味实验(二) ·· 284
  实验 42 食盐和亚硝酸钠的鉴别 ···································· 286

Chapter 1 Introduction

1.1 Significance of Chemical Experiments

Chemistry is a discipline that creates new substances. At present chemistry is not only a core discipline but also a science with strong practicality because it is at the center of a multilateral relationship, whether in terms of the classification of natural sciences, the level of research objects, or the cross and integration of emerging disciplines. Chemistry provides a great deal of material base for human production and life and continuously creates surprising results through experimental researches. Chemists have already not only discovered and synthesized millions of natural and non-natural compounds, but also been able to synthesize new compounds by molecular design. Although modern chemistry has entered the stage of paying equal attention to the theory and practice, it is still based on the experimental data because the formation of chemical theories and rules is based on the experiment which is the only criterion to verify the theory. Experiments are the bridge connecting chemistry with productivity development. Chemical experiments promote the rapid development of chemistry and lead human beings into a new material world.

1.2 Teaching Objectives of Inorganic Chemistry Experiment Course

If there is no strict training in laboratory work and independent ability in experiment operations, it is impossible for someone, who is about to engage in chemistry and its related majors, to truly master and understand the chemical knowledge, let alone having explorations, inventions and creations through experiments. Inorganic chemistry experiment is the first major professional basic course for students who major in chemistry, applied chemistry or other chemistry related discipline in colleges and universities. It allows students to understand the facts and laws of inorganic chemistry through the observation of the experimental phenomena and property changes, thereby reaching greater understanding about the related basic principles and knowledge. In addition, it helps students be proficient in the basic operating skills of inorganic chemistry experiments, master the general methods of preparing inorganic compounds and initially possess the ability to synthetical design and experiment exploration through practical training. An experiment course is a teaching activity that is organized and guided by a teacher and completed independently by students. It is committed on improving students' ability to observe, analyze, solve problems and write papers; training students to possess the scientific attitude and innovation ability of seeking truth from facts, independent thinking, tenacity and persistent; allowing students to develop good working habits and scientific literacy. Therefore, experiment teaching has a special objective and role that cannot be replaced by theoretical teaching.

1.3 Requirements and Methods of Experiment Course

This teaching material includes the following experiment modules: basic knowledge and basic operation experiments, inorganic chemistry principle experiments, properties experiments of inorganic compounds, preparation experiments of inorganic compounds, comprehensive and design experiments. Each experiment includes objectives, principles, contents, basic operations, matters needing attention and other requirements. In order to achieve the learning objectives and requirements of the experiment course, one must have correct learning methods. Lu Jiaxi, a famous chemist in China, had a brilliant exposition on how to do chemical experiments well: clear head, clever hands and clean habits—that is, in more details, being rich in basic knowledge of chemical theory, being adept at using theory to guide experiments, analyze and solve problems, being solid in standard experimental operation skills, and forming good laboratory working habits. The learning method of inorganic chemistry experiments can be roughly divided into the following three steps.

1.3.1 Preview

In order to obtain good results from experiments, catch problems and correct mistakes in time and avoid the security risks of laboratory, students are required to have good preparation before each experiment. Through reading experiment materials, textbooks and related references, students should have a clear understanding of experiment objectives, experiment principles, experiment contents, operation points, matters needing attention, etc. On this basis, students should break down the experiment contents into several steps, refine and tease apart these steps with clear, concise and well-organized words and symbols, and then complete a good preparation note. Gap-filling is a good way to write experiment contents, including the name, concentration and amount of reagents, the apparatus used and the operation names in each step. Besides, please pay attention to leaving the blank space after each step for recording the experimental phenomena or data. In this way, a notebook for preparing notes is also a notebook for experiment records. Whereas if students have no idea of the experiment contents and do the experiment with confusion, the progress of the experiment will be affected, some contents may be omitted, some reagents or apparatus may be confused, and what's worse, some accidents may occur. Therefore, each student is required to have a notebook for preview and record, and someone who doesn't preview before the experiment is not allowed to enter the laboratory.

1.3.2 Experiment

Do the experiment according to the methods, procedures and the amount of reagents in the experiment materials, and do the following:

(1) Operate the equipment seriously, observe experiments carefully, and record experimental phenomena and data timely, detailedly and truthfully.

(2) Follow the experimental operating procedures and equipment operating specifications

strictly.

(3) Abide by the rules of the laboratory strictly, keep the table and equipment clean and tidy, and make an overall arrangement of the schedules.

(4) Think, analyze and solve problems diligently and independently. Supposing that the experimental phenomena and results are inconsistent with the theory, students should respect the facts, carefully analyze and check the operational problems. If there is any problem in the operation, please correct it without delay. Also, students can examine problems through doing the control experiments, blank experiments, or the self-designing experiments with the approval of teachers. If a time-consuming experiment is to be repeated, students should ask teachers for permission in advance.

1.3.3 Lab Report

After the experiment, students should explain and discuss the phenomena, process and calculate the data, get the results and discuss them with classmates, finish the lab report, and submit the report to the teacher. The lab report is required to be completed, professional in words and symbols, standard and neat in handwriting, accurate in phenomena and results, reasonable in interpretation and discussion, well organized and tidy.

A lab report should include the following sections:

(1) The title, objective and principle of the experiment.

(2) Experimental contents: steps (the name, concentration and amount of reagents, the apparatus used, the operation method, etc.), phenomena (the state and color of the material) and data (in table form).

(3) Experimental phenomena and explanations: use chemical reaction formulas or chemical principles to explain and summarize the experimental phenomena.

(4) Experimental data and results: use chemical formulas or adopt the drawing method to analyze and calculate the data, and get the results.

(5) Problems and discussions: have a brief description of the new operation or the key points in the experimental operation, or have an analysis and discussion of the problems and experiences encountered in the experiment, or answer the questions in the teaching material.

Several types of lab report formats are listed below for reference.

Lab Report: Inorganic Chemistry Property Experiments

Title_____
Name_____   Major_____   Class_____
No._____   Advisor_____   Date_____

1. Objectives:

2. Contents:

Experimental procedures	Experimental phenomena	Explanations and reaction equations

3. Problems and Discussions:

Lab Report: Inorganic Chemistry Determination Experiments

Title_____

Name_____ Major_____ Class_____

No._____ Advisor_____ Date_____

1. Objectives:

2. Principles:

3. Procedures:

4. Data Recording and Processing:

5. Problems and Discussions:

Lab Report: Inorganic Chemistry Preparation Experiments

Title_____

Name_____ Major_____ Class_____

No._____ Advisor_____ Date_____

1. Objectives:

2. Principles:

3. Procedures:

4. Phenomena and Reaction Equations:

5. Results (Colour, State, Productivity and Yield):

6. Problems and Discussions:

Chapter 2 Basic Knowledge of the Lab

2.1 Laboratory Safety Knowledge, Discipline and Precautions

An experimental lesson is a teaching activity organized and instructed by a teacher and completed independently by students. It undertakes important functions of observing experimental phenomena, consolidating theoretical knowledge, learning and mastering experimental skills and methods, developing scientific rigor and good laboratory working habits, cultivating innovative awareness and improving scientific literacy.

Fresh students know little about the lab work, and inorganic chemistry experiment is the first basic chemistry experiment course after they enter the university. In order to make students familiar with the experimental teaching, standardize the teaching order and master some laboratory work knowledge as soon as possible, relevant rules and regulations must be formulated, and at the same time, related education and training should be actively carried out. However, it cannot be accomplished overnight but runs through the chemistry experiment teaching in various disciplines.

2.1.1 Basic Layout and Common Equipment of Lab

1. Basic Layout

Inorganic chemistry laboratory must have water, electricity, ventilation and other systems, and the basic layout is as follows:

(1) Student lab benches and cabinets; tables and cabinets for placing common reagents, materials and apparatuses; the passage between benches is both an accessible way and an escape way, so there should not be stools or other things.

(2) Both sides of experimental benches and the inside of fume hoods are equipped with a faucet, under which are sinks that have access to the waste water basin.

(3) The experimental bench, wall and fume hood are equipped with a variety of outlets and an electricity brake.

(4) The common hoods are installed in a corner of the laboratory, and the exhaust vents which have a straight channel to the building roof are installed over student lab benches. Reagents such as volatile reagents and organic reagents are placed in a fume hood.

2. Common Instruments and Equipment

Common instruments and equipment include balances, analytical balances, barometers, centrifuges, vacuum pumps, electric stoves, constant temperature water baths, electrically heated drying ovens. Other equipment and materials are placed in the designated places by category: iron stands and tripods are placed in the designated cabinet; utility clamps, iron rings, rubber stoppers and rubber tubes are placed in the designated drawer; deionized water, fuel alcohol and a variety of public reagents are placed on a public platform; a small amount of commonly used reagents are dispensed in the reagent bottles or dropping bottles which are placed on the student lab bench neatly and orderly, and promptly returned to the original place after use.

2.1.2 Laboratory Rules

Experiments are a kind of practical activity that people apply or explore scientific knowledge and laws, and they have their own characteristics and laws. Doing experiments should follow a certain work procedure and rules, which is different from attending lectures or reading books. In the long-term laboratory work, researchers sum up laboratory rules, safety rules and school-related rules and regulations, which is an important prerequisite and guarantee to maintain the good working order and environment, prevent accidents and do experiments well. Therefore, everyone must keep the rules.

(1) Preview the experiment carefully: prepare for the experiment carefully, understand the experimental principles and requirements and determine the experimental procedure before entering the lab to do the experiment, and the additional experiment should be approved by the teacher.

(2) Comply with the operating procedures: comply with the equipment operating specifications and experimental operating procedures seriously and strictly. Observe carefully, think and work independently, make a reasonable overall arrangement and record the experimental phenomena and data promptly and accurately during the experiment. If there are mistakes, the experiment should be repeated, but a time-consuming experiment should be agreed by the teacher first. Analyze the problem, process the data and write a lab report seriously after the experiment.

(3) Comply with the laboratory discipline: no shouting, casual walking, calling or texting in the laboratory. Do not be late, leave early, or be absent without the teacher's permission, and the absentee should catch up on all the experiments.

(4) Take good care of the state property: be careful with the instrument and equipment during the experiment, and the precision instrument should be promptly restored and registered after use. If the instrument or equipment is abnormal or damaged, stop using it and report to the teacher immediately. Public equipment should be wiped up and put back to the original place after use. Do not use other people's glassware at will.

(5) Practise strict economy: save water, electricity, reagents and materials under the premise of not affecting the experimental effect. Take reagents according to the specified grade and dosage. Wash or rinse the apparatuses many times with a little water at each. Apparatuses, reagents and products may not be brought out of the laboratory, and good experimental habits should be developed.

(6) Reduce pollution: reduce the environmental pollution caused by the toxic gas or the waste. Comply with the strict management and use system when using highly toxic chemicals, including registration before use, recycling or destruction after use, lab cleaning and hand washing. Experiments producing toxic gases are carried out in a fume hood. Prepare two containers, one is for solid waste such as waste paper and matchsticks, and the other is for liquid waste. Pour the waste into the public trash can and waste liquid can separately after the experiment. But do not pour the solid waste and acidic waste into the sink directly to prevent the clogging and corrosion of sewers.

(7) Keep the laboratory clean and orderly: place the reagents neatly in the public bench and put them back promptly after use. Keep the bench dry and tidy at any time, take the needed apparatuses and place them in the front of the lab bench neatly. After the experiment, clean the apparatuses, put them back to the cabinet from high to low and from inside to outside. The lab bench and sink are wiped clean with a damp cloth.

(8) Student on duty system: after each experiment, the students on duty are responsible for cleaning the laboratory, including checking the reagent bottles, apparatuses and equipment, turning off the tap and electrical equipment, mopping the floor, wiping the public equipment and common benches clean, emptying the waste can, closing windows and doors, and finally passing the teacher's examination.

(9) Report immediately if an accident occurs: if an accident occurs, keep calm, report to the teacher immediately and take urgent action.

2.1.3 Laboratory Safety Rules

The chemistry lab has water, electricity, equipment and reagents, and many of the reagents are flammable, explosive, corrosive and toxic. Therefore, it is necessary to attach great importance to safety. The safety problems and precautions in the experiment must be fully understood before each experiment, and the laboratory safety regulations, equipment operating rules and laboratory operation specifications must be observed strictly during the experiment. At the same time, it is necessary to learn some security and emergency medical knowledge. In the event of an accident, immediate emergency treatment should be made. Specifically, the following rules should be obeyed:

(1) Do not contact the power supply with wet items or hands. Turn off the tap, electrical equipment, and alcohol burner immediately after use. Extinguish the lighted match immediately after use and do not throw it anywhere.

(2) Experiments involving flammable and explosive materials should be carried out away from the fire and heat.

(3) Experiments involving toxic and irritating gases should be carried out in the fume hood, and never sniff the gas directly from the mouth of bottles or pipes.

(4) Do not look down from the top of the heated container to avoid the splashing of reactants to the face. Do not face the mouth of the tube containing liquid when it is being heated. Keep stirring when a solid or concentrated liquid is heated to avoid the splashing of the solid, and goggles can be used.

(5) Be careful with strong acid and concentrated alkali which are both highly corrosive, and keep them from the skin, clothes, and more importantly, eyes. Inject the concentrated sulfuric acid into the water to dilute it. Hold the labeled side of the reagent bottle rather than just grabbing the bottle cap with fingers to transfer the reagent bottle.

(6) Do not take food or drink into the laboratory. Lab coats must be worn and long hair must be tied up. No long fingernails, rings or bracelets, vests, shorts or slippers in the lab. Do not look down, bend over or lean on the bench while doing experiments.

(7) Chemical reagents are prohibited to have direct contact with the skin or mouth. The remaining waste may not be poured into the sink but should be recycled for centralized processing after the experiment.

(8) Wash your hands after the experiment before leaving the lab.

2.1.4 Keeping the Commonly Used Reagents

Pay attention to preventing fire, water, volatilization, exposure and deterioration when keeping the chemical reagents. Improper storage will bring damage or threat to the property of the state, the lab environment and people health. Chemical reagents should be kept by different ways according to their different characteristics such as flammability, lability, toxicity, causticity and deliquescence. Dangerous chemicals refer to the chemicals which may cause a combustion explosion influenced by external factors such as light, heat, air, water and impact, or the chemicals which are highly toxic and corrosive.

Elementary and inorganic salt solids with stable properties should be kept in the reagent cabinet, and the inorganic reagents and organic reagents should be stored separately. Moreover, dangerous reagents must be regulated strictly and stored by category.

(1) Flammable reagents:

(i) Flammable liquids are mainly organic reagents, which are volatile and can burn in the open fire. Therefore, they should be kept alone in a cool and well-ventilated place away from the fire and heat.

(ii) Flammable solids include sulphur, red phosphorus, magnesium powder and aluminum powder. They all have low burning point, and they should be stored in a cool, dry and well-ventilated place away from the fire and heat. White phosphorus can spontaneously ignite in the air, and it should be kept in water away from light.

(iii) Some solids are flammable in water, such as lithium, sodium, potassium, calcium carbide and zinc powder. They can react with water to emit flammable gases. They must be stored in a cool and dry place. What's more, lithium should be sealed with paraffin wax, and sodium and potassium should be kept in kerosene.

(2) Explosives: a reagent that is easy to decompose in the event of heat, friction or impact, releasing a large amount of gas and producing an explosion, such as ammonium nitrate, bitter acid and trinitrotoluene. They should be stored in a shady and cool place, and they must be handled gently.

(3) Strong oxidizing reagents: strong oxidizing reagents such as potassium chlorate, potassium nitrate, hydrogen peroxide, sodium peroxide, permanganate and dichromate are easy to decompose, and an explosion may occur when they react with acid or reducing substances, or suffer the heat or impact. Therefore, such substances must not be stored together with combustible materials or reducing substances, but should be placed in a cool and ventilated place.

(4) Highly toxic chemicals: cyanide, arsenic compounds, mercury and its compounds are all highly toxic chemicals, which should be locked in a fixed iron cabinet and kept by two specially-assigned people. Soluble chromium, cadmium, barium, lead, antimony, thallium salt are also toxic chemicals, which should be safekept as well. Receive the chemical when you need it and return the leftovers promptly after use, and the registration system should be adopted.

(5) Corrosive reagents: strong acid, hydrogen fluoride, strong alkali, bromine, phenol and other highly corrosive reagents should not be placed together with strong oxidizing, flammable and explosive reagents.

(6) Other reagents:

(i) Deliquescent reagents: highly oxidized metal chlorides such as antimony trichloride and stannic (stannous) chloride, sulphide, sodium hydroxide, potassium hydroxide and other deliquescent reagents must be placed in a dry place.

(ii) Reagents that are susceptible to decomposition or metamorphism: nitric acid, nitrate, iodide, hydrogen peroxide, ferrous salts and nitrite should be stored in a brown bottle to avoid light.

2.1.5 Emergency Treatment of Laboratory Accidents

If there is an accident during the experiment, the following emergency measures can be taken.

(1) Incised wound: clean the wound with sterilized cotton, pick out the foreign matter, and then put some iodine on the cut or bind up the wound with a "hemostatic paster", and go to the hospital if necessary.

(2) Scald: the wound may not be washed with water, but should be coated with scald ointment or "Wanhua Oil".

(3) Chemical burn:

(i) Acid or alkali corrosion: wipe off the chemical reagent on the skin with a clean cloth or absorbent paper and rinse the wound with plenty of water. Saturated sodium bicarbonate solution or dilute ammonia water can be used to rinse the wound caused by acid corrosion, while 3%~5% acetic acid or 3% boric acid solution can be used for the wound caused by alkali corrosion, and finally the wound is washed with water, and go to the hospital if necessary.

(ii) Splashing of acid or alkali into the eyes: rinse the eyes immediately with plenty of running water, then 3%~5% sodium bicarbonate or 3% boric acid solution, and go to the hospital.

(4) Inhaling irritant or poisonous gas: the mixed vapours of alcohol and ethyl ether can be inhaled to treat chlorine or hydrogen chloride intoxication; go outside immediately for fresh air if you feel unwell when inhaling hydrogen sulfide or carbon monoxide.

(5) Eating a poison by mistake: take orally a cup of warm water containing 5~10 mL of dilute copper sulfate solution, and reach into the throat with your fingers to vomit the poison, then go to the hospital immediately.

(6) Electric shock: cut off the power immediately, and if necessary, administer artificial respiration and send the wounded to the hospital.

(7) Outfire: the combustion of matter requires air and temperature, so the way to put out the fire is to isolate the air or lower the temperature. If a fire breaks out during the experiment accidentally, put out the fire immediately and then take some measures such as cutting off the power immediately, removing the inflammables and stopping the ventilation to prevent fire spreading. Fire extinguishing methods and fire-extinguishing equipment should be selected according to the cause of the fire.

(i) As for general fire, wet rags, asbestos cloth and sand can be used to put out a small fire while water and foam extinguishers can be used to put out a big fire.

(ii) When clothes are on fire, don't panic or run away because the wind may feed the fire. The correct approach is to remove the clothes immediately or use a damp cloth or asbestos cloth to cover the fire. If the burning area is larger, you can roll on the ground to extinguish the fire.

(iii) The fire caused by active metals such as Na, K, Mg and Al may not be put out by water or foam extinguishers so as to avoid the generation of combustible gases which will feed the fire, and it should be put out by sand or dry powder extinguishers.

(iv) The fire caused by organic solvents may not be put out by water or foam extinguishers so as to avoid the fire spreading, and it should be put out by carbon dioxide extinguishers, 1211 fire extinguishers, special fireproof cloth, sand or dry powder fire extinguishers.

(v) When electric appliances catch fire, turn off the power first and then use fireproof cloth and dry powder extinguishers to put out the fire. Do not use water or foam fire extinguishers to avoid electric shock.

2.2 The Use, Cleaning and Drying of Inorganic Chemical Apparatuses

2.2.1 Commonly Used Apparatuses

Commonly used apparatuses are shown in Fig. 2.1.

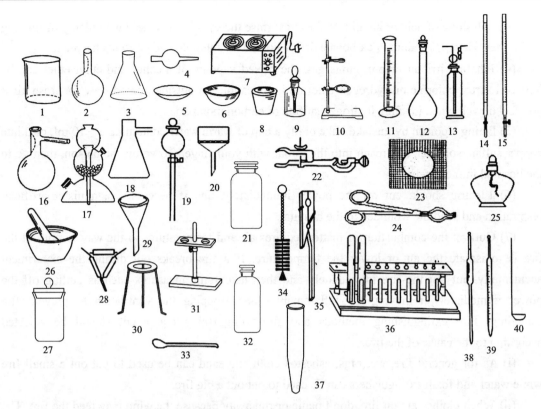

Fig. 2.1 Commonly used apparatuses

1. beaker; 2. flask; 3. Erlenmeyer flask; 4. drying tube; 5. watch glass; 6. evaporating dish; 7. water bath; 8. crucible; 9. dropping bottle; 10. iron stand and iron ring; 11. measuring cylinder; 12. volumetric flask; 13. gas-washing bottle; 14. base burette; 15. acid burette; 16. distillation flask; 17. Kipp's apparatus; 18. suction flask; 19. separating funnel; 20. Büchner funnel; 21. wide-mouth bottle; 22. iron clamp; 23. asbestos net; 24. crucible tong; 25. alcohol burner; 26. mortar; 27. weighing bottle; 28. pipeclay triangle; 29. funnel; 30. tripod; 31. funnel stand; 32. narrow-mouth bottle; 33. medicine spoon; 34. test tube brush; 35. test tube clamp; 36. test tube rack and test tube; 37. centrifuge tube; 38. transfer pipette; 39. measuring pipette; 40. burning spoon

2.2.2 The Use and Precautions of Commonly Used Apparatuses

1. Test Tubes, Centrifuge Tubes

Specifications: test tubes are divided into hard or soft ones, ones with or without graduation, ones with or without a branch pipe, etc. The capacity (mL) of test tubes includes 5, 10, 15, 20, 25, 50 and so forth. The test tubes without graduation are classified according to tube diameter × length (mm), such as 15×75.

Main use: they can be used as a reaction container for a small amount of reagent at room temperature or heating conditions, which is convenient for operation and observation. They can also be used as a collection container for a small amount of gas. Test tubes with a branch pipe can be used to test the gaseous product. Centrifuge tubes can be used for precipitation separation.

Usage and precautions: the reaction liquid should be no more than 1/2, and it should be no more than 1/3 when heated to prevent the liquid splashing. The outer wall should be dried to prevent the uneven heating that causes tube breaking and scalding. When heating the liquid, tilt the tube at 45 degree to avoid bumping, and the mouth of the tube may not directly face people. When heating the solid, slightly tilt the mouth of the tube down to increase the heated surface and avoid the reflux of the condensate water. Centrifuge tubes may not be directly heated.

2. Beakers

Specifications: there are hard beakers and soft beakers, and they are classified according to the capacity (mL) generally: 25, 50, 100, 150, 200, 250 and so forth.

Main use: they can be used as a reaction container for a large amount of reagent at room temperature or heating conditions, so the reactants are easy to be mixed evenly. They can also be used to prepare the solutions.

Usage and precautions: the reaction liquid should be no more than 2/3 to prevent the liquid splashing, and it should be no more than 1/2 when heated. The outer wall should be dried, and the bottom of the beaker should be cushioned with asbestos net to prevent uneven heating.

3. Erlenmeyer Flasks

Specifications: there are many kinds of Erlenmeyer flasks, such as hard or soft ones, ones with graduation or without graduation, wide-mouth or narrow-mouth ones and miniature ones. The capacity (mL) of Erlenmeyer flasks includes 50, 100, 150, 200 and so forth.

Main use: they can be used as a reaction container, which is convenient for oscillation and suitable for titration.

Usage and precautions: do not use it to hold too much liquid to prevent the liquid splashing. It should be cushioned with asbestos net or placed in a sink in the experiment to prevent uneven heating.

4. Dropping Bottles

Specifications: they are made of glass, and they are divided into brown ones or colorless ones. Each dropper has a rubber head. The capacity (mL) of dropping bottles includes 15, 30, 60, 125 and so forth.

Main use: they can be used to hold a small amount of liquid reagent or solution for easy access.

Usage and precautions: brown bottles are used to store substances that are easy to decompose or unstable when exposed to light. The dropper must not be filled too much or inversed so as to prevent the corrosion of the rubber head. The dropper is not allowed to be messed up to prevent the reagent pollution.

5. Funnels

Specifications: they are divided into long-necked ones and short-necked ones. Their specifications are represented by the caliber (mm), including 60, 100 and so forth. The hot funnel is used for the thermal filtration.

Main use: they can be used to filter or pour the liquid. Long-necked funnels are often equipped with gas generators and used to add liquid.

Usage and precautions: do not heat them directly. The tip of the funnel neck must be close to the wall of the container that holds the filtrate during the filtration. When a long-necked funnel is used to add liquid, the neck should be inserted into the liquid.

6. Weighing Bottles

Specifications: their specifications are represented by the capacity (mL). The high type includes 10, 20 and so forth, and the short type includes 5, 10, 15 and so forth.

Main use: they can be used to accurately weigh a certain amount of solids.

Usage and precautions: do not heat them. The matching lid must not be lost or messed up, and it should be cleaned when it is not used. Put a piece of paper on the grinding mouth to prevent adhesion.

7. Measuring Cylinders

Specifications: their specifications are represented by the capacity (mL), including 5, 10, 20, 25, 50, 100, 200 and so forth. Something with a large upper part and a small lower part is called a measuring cup.

Main use: they can be used to take a certain volume of liquid.

Usage and precautions: when reading the scale, make sure your sight line tangent to the meniscus of the liquid. They may not be heated or used as an experimental container (such as dissolution and dilution), or used to measure the hot solution or liquid.

8. Pipettes

Specifications: they are divided into graduated tube pipettes and single graduated pipettes with a fat belly, as well as automatic pipettes. According to the maximum scale (mL), there are 1, 2, 5, 10, 25 and so forth. The scale of micropipettes includes 0.1, 0.2, 0.25 and so forth.

Main use: they can be used to precisely transfer a certain volume of liquid.

Usage and precautions: when the surface of the pipetted liquid is above the scale, hold the pipette mouth to gently release the air so that the liquid surface is lowered to the scale. Press the pipette mouth with your index finger, move to the designated container and release your finger to inject the liquid. Use the target solution to wash the pipette three times before use. The last drop must not be blown out (unless there is a "blow" word on the pipette).

9. Volumetric Flasks

Specifications: the capacity (mL) below the scale includes 25, 50, 100, 250, 500, 1000 and so forth. There are also volumetric flasks with a plastic stopper.

Main use: they can be used to prepare the solutions with accurate concentrations.

Usage and precautions: dissolve the solute in a beaker first, and after the solution is cooled, transfer the solution into a volumetric flask. Do not heat the volumetric flask, otherwise the accuracy may be affected. Moreover, volumetric flasks cannot replace the reagent bottles to store the solutions.

10. Suction Flasks and Büchner Funnels

Specifications: Büchner funnels are made of pottery and porcelain, and their specifications are represented by the caliber. Suction flasks are made of glass, and their specifications are represented by the capacity (mL).

Main use: they can be used as a set for the vacuum filtration of crystals or precipitates in the inorganic preparation experiments.

Usage and precautions: do not heat them directly. The filter paper should be slightly smaller than the inner diameter of the funnel, and all the holes should be covered to avoid the leakage. Pump the air before the filtration, and after the filtration, release the air and then turn off the pump.

11. Evaporating Dishes

Specifications: they are divided into flat bottom ones and round bottom ones. Their specifications are represented by the upper caliber (mm).

Main use: they can be used to evaporate and concentrate the solutions, and different evaporating dishes of different materials are selected according to the property of the liquid.

Usage and precautions: heat them directly but do not cool them rapidly.

12. Crucibles

Specifications: they are made of porcelain, quartz or iron. Their specifications are represented by the caliber (mm).

Main use: they can be used to calcinate the solid in hot conditions, and different crucibles of different materials are selected according to the property of the solid.

Usage and precautions: they can be placed on a pipeclay triangle directly for strong heat or calcination. When the heating or reaction is completed, the crucible tong should be preheated and should be placed on the asbestos net.

13. Narrow-mouth Bottles

Specifications: they are made of glass, and they can be divided into ones with or without grinding mouth, colorless ones, brown ones and blue ones. According to the capacity (mL), there are 100, 125, 250, 500, 1000 and so forth. Narrow-mouth bottles are also called reagent bottles.

Main use: they can be used to store solutions and liquid chemicals.

Usage and precautions: do not heat them directly. The stopper may not be soiled or messed up. A rubber stopper should be used if the bottle is for containing lye. When the narrow-mouth bottle with a grinding stopper is not in use, it should be washed and a piece of paper should be placed on the grinding mouth. A colored bottle is used to hold the solution or liquid that is easy to decompose in light or less stable.

14. Wide-mouth Bottles

Specifications: they are made of glass, and they can be divided into colorless ones, brown ones, and ones with or without a grinding mouth. The grinding mouth of a wide-mouth bottle has a stopper, while the grinding mouth of a gas bottle doesn't have. According to the capacity (mL), there are 30, 60, 125, 250, 500 and so forth.

Main use: they are used to store solid chemicals or collect gases.

Usage and precautions: do not heat them directly. They cannot contain lye, and their stoppers may not be soiled or messed up. A little sand or water should be put at the bottom of the bottle in a gas combustion experiment. A piece of frosted glass should be used to cover the bottle after collecting the gas.

15. Acid Burettes / Basic Burettes

Specifications: burettes are made of glass, and they are divided into acid ones (with a glass stopper) and basic ones (with a glass nozzle connected with the latex tube). According to the maximum scale (mL), there are 25, 50, 100 and so forth. The scale of microburettes includes 1, 2, 3, 4, 5, 10, etc.

Main use: they are used for titration or measuring the volume of the liquid.

Usage and precautions: wash them before use and rinse them with the target solution three times before loading. When an acid burette is in use, open its stopper with your left hand. When a basic burette is in use, use your left hand to gently pinch the glass bead in the latex tube with your left hand, and the solution can be released. The bubbles in basic burettes should be removed, and the stoppers of acid burettes should be greased with Vaseline, and the lower end of basic burettes may not be washed with washing liquid. Note that acid burettes and basic burettes may not be exchanged for use.

16. Drying Tubes

Specifications: they are made of glass, and they have different shapes, which are graded according to the size.

Main use: they are used for gas drying.

Usage and precautions: the desiccant granules should be of moderate size. The tightness of drying tubes should be moderate when they are used. Both ends of them should be stopped with cotton. The desiccant should be renewed immediately when they get damp, and the drying tubes should be cleaned after use. Do not mix up the two ends of the drying tube (the larger end is the inlet while the smaller end is the outlet), and the drying tube should be fixed on an iron stand for use.

17. Gas-washing Bottles

Specifications: they are made of glass, and they have a variety of shapes. According to the capacity (mL), there are 125, 250, 500, 1000 and so forth.

Main use: they are used for gas purification, and they can also be used as safety bottles (or buffer bottles).

Usage and precautions: pay attention to the correct connection method (the inlet pipe is immersed in the liquid), and the injected washing liquid should be above 1/3 of the container, which must not exceed 1/2.

18. Iron Stands

Specifications: iron stands are made of iron, and some clamps are made of aluminum now. There are circular and rectangular iron stands.

Main use: they are used for fixing or placing the reaction vessels. Iron rings can also be used in place of funnel holders.

Usage and precautions: when apparatus is fixed on an iron stand, the barycentre of the apparatus and the iron stand should fall in the middle of the chassis of the iron stand. It is unadvisable to hold the apparatus with an iron clamp too tightly or too loosely, just keep the apparatus from turning. The heated iron ring cannot be hit or dropped to the ground.

19. Brushes

Specifications: their specifications are expressed by their sizes or purposes, such as test tube brushes, burette brushes.

Main use: they are used to scrub the glassware.

Usage and precautions: hold a brush in place when it is used to wash something. Pay attention to the completeness of the bristles on the top of the brush.

20. Mortars

Specifications: they are made of porcelain, and they can also be made of glass, agate or iron. They are graded according to the size of the caliber.

Main use: they are used to grind or mix solid materials. Different mortars are selected according to the nature and hardness of the solid.

Usage and precautions: bulk materials can only be crushed, not be pounded. The amount of the solid should not exceed 1/3 of the mortar volume. Explosive substances can only be gently crushed, not be ground.

21. Test Tube Racks

Specifications: they are made of wood or aluminum, and they have different shapes and sizes.

Main use: they are used to place test tubes.

Usage and precautions: the heated test tube should be hold by a test tube clamp and hung on a rack.

22. Test Tube Clamps

Specifications: they are made of wood, bamboo or metal wire (steel or copper), and they have different shapes.

Main use: they are used for clamping test tubes.

Usage and precautions: hold the upper end of a test tube with a clamp. Do not put your thumb on the movable part of the clamp, and be sure to cover and remove the test tube clamp from the bottom of the test tube.

23. Tripods

Specifications: they are made of iron, and they have different sizes and heights. They are relatively firm.

Main use: they are used to place larger or heavier heating containers.

Usage and precautions: asbestos net should be placed before placing a heating container (except the water bath) on a tripod. The position of the heating flame should be appropriate, and an oxidizing flame is generally used for heating.

24. Combustion Spoons

Specifications: the head of the spoon is made of copper, and the rest of the spoon is made of iron.

Main use: they are used to test the flammability of some solid materials, and they can be used in the combustion reactions of solids and gases.

Usage and precautions: put the combustion spoon into the gas bottle slowly from top to bottom, and do not touch the bottle wall. In the combustion experiments of sulfur, potassium and sodium, a little asbestos or sand should be put in the bottom of the spoon. The head of the spoon should be washed and dried immediately after use.

25. Pipeclay Triangles

Specifications: they are twisted by iron wire and covered with porcelain tubes. They have different sizes.

Main use: they are used for placing the crucible which is set on fire.

Usage and precautions: check whether the wire is broken or not before use. The crucible should be placed sideways on one of the three porcelain tubes. After use, remove the crucible carefully.

26. Medicine Spoons

Specifications: they are made of ox horns, porcelain or plastic. Most of them are plastic now.

Main use: they are used to transfer solid chemicals. A medicine spoon has two ends, one is larger and the other is smaller. The two ends of the medicine spoon are selected for use according to the amount of chemicals.

Usage and precautions: after transferring a chemical, the medicine spoon must be washed and dried with filter paper, and then it can be used for next transferring.

27. Asbestos Net

Specifications: it is an iron net which is woven with iron wire, and the middle is smeared with asbestos. It has different sizes.

Main use: asbestos is a kind of poor heat conductor, which can make the object evenly heated.

Usage and precautions: it must be checked first before use, and if the asbestos has come off, the asbestos net cannot be used. Do not make it in contact with water, and do not roll up or fold it.

28. Crucible Tongs

Specifications: they are made of iron, and they vary in the size and length (do not open and close the tongs too tightly or too loosely).

Main use: they are used to hold the crucibles which are being heated, and they can also be used to put or take the crucibles (or hot evaporating dishes) in a high temperature electric furnace (Muffle furnace).

Usage and precautions: the crucible tongs must be clean when it is in use. After use, place them flat on the test bench, with their tips up (if the temperature is too high, asbestos net can be used). After the experiment, the crucible tongs should be wiped clean, dried and placed in the laboratory cabinet.

2.2.3 Washing the Commonly Used Glassware

There are many washing methods for glassware, which should be taken according to the requirements of the experiment, the nature of the dirt and the degree of contamination. The commonly used washing methods are as follows:

(1) Water: use water and a brush to remove the dust and impurities from the glassware.

(2) Cleaning powder, soap or detergent: they are used to remove the oil and organic matter, and hot alkaline liquid is usable if the oil and organic matter cannot be washed away.

(3) Chromic acid lotion: it is needed when the requirements for the clean glassware are very high in the quantitative experiments, and the washing method is as follows:

(i) Wash the glassware with water or washing powder, and try to remove the water from the container so as not to dilute the lotion.

(ii) Pour the lotion into the glassware and soak the inner wall in the lotion repeatedly to dissolve the dirt.

(iii) Pour the lotion back into the original bottle for reuse.

(iv) Plug the lotion bottle with a stopper to prevent the lotion from absorbing water and losing efficacy.

Chromic acid lotion has strong acidity, oxidization and detergency, and it is suitable for washing the oil and organic matter. It is very corrosive, so don't splash it on clothes or skin. When the color of the lotion turns from dark brown to green—that is, potassium dichromate is reduced to chromium sulfate, the lotion should be prepared renewedly. Do not use a brush or chromic acid lotion to wash the cuvette.

(4) Concentrated HCl: it can be used to wash away the oxidants such as manganese dioxide that attach to the wall of the glassware, and it can also be used to wash away most insoluble inorganic matter. For example, the porcelain crucible that has burned the precipitation can be washed with hot HCl (1 : 1) first, and then with lotion.

(5) Potassium permanganate lotion of sodium hydroxide: it can be used to wash away the oil and organic matter, and the manganese dioxide precipitation left on the wall after washing can be washed away with hydrochloric acid.

(6) Other washing methods: appropriate reagents may be selected according to the property of the dirt, such as washing AgCl precipitation with ammonia water and washing the sulfide precipitation with nitric acid and hydrochloric acid.

After taking all the above methods, Ca^{2+}, Mg^{2+}, Cl^- and other ions are often left on the glassware. If these ions are not allowed in the experiment, they should be washed away with distilled water. The purpose of using distilled water is to wash the tap water away, so try to use it in accordance with the principle of low intensity (less each time) and high frequency (normally 3 times).

Basic requirements for washing glassware:

(1) No insoluble substances or oil in the wall of the washed glassware. The checking method is to make the glassware completely wet with water, and then turn it upside down to observe whether there is a thin and well-distributed water film on the wall, and if so, the glassware has been cleaned.

(2) Qualitative and quantitative experiments have higher requirements for the clean degree of the glassware because impurities may influence the accuracy of experiments. However, some experiments such as ordinary inorganic preparation and properties experiments have no strict requirement for the clean degree of the glassware, and the only requirement is to scrub the glassware clean. Therefore, the degree of washing should be determined according to the actual situation.

(3) The washed glassware may not be wiped by cloth or paper because cloth and paper fibers may remain on the wall.

2.2.4 Drying the Apparatuses

Commonly used apparatuses can be dried in the following ways (Fig. 2.2):

(1) Air dry: apparatuses which are not in urgent need can be hung upside down in a clean laboratory cabinet or rack to air dry.

(2) Drying oven: pour out the water in the washed apparatus, then put the apparatus in an oven and make its mouth downward, and place an enamel plate in the bottom of the oven to hold the

water from the apparatus to avoid the water dripping onto and damaging the heating wire.

(3) Heating: some commonly used beakers and evaporating dishes can be dried with small fire on the asbestos net. A test tube can be moved back and forth with a clamp over the flame until it is dry, but the mouth of the tube must be downward until there is no water so as to prevent the backflow of the water and the tube breaking.

(4) Airflow dryer: suitable for test tubes, measuring cylinders and other apparatuses.

(5) Electric dryer.

(6) Organic solvents: take away water vapors through the evaporation.

(a) rack   (b) dryer   (c) electric dryer   (d) airflow dryer   (e) oven

Fig. 2.2 Apparatus drying

2.3 Introduction and Use of Commonly Used Heating Equipment

2.3.1 Alcohol Burner and Alcohol Blast Burner

1. Alcohol Burner

The alcohol burner (heating temperature is usually 400~500℃) is made up of a lamp shade, a lamp wick and a lamp kettle, as shown in Fig. 2.3 (a).

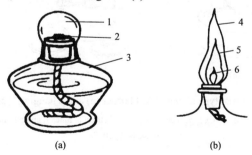

Fig. 2.3 Construction of an alcohol burner (a) and a diagram of the lamp flame (b)

1. lamp shade; 2. lamp wick; 3. lamp kettle; 4. outer flame; 5. inner flame; 6. flame core

When using an alcohol burner, check the wick first. If the top of the wick is uneven or charred, you need to cut a little to make it even. Then check whether there is alcohol in the lamp. When adding alcohols to the lamp, extinguish the flame first, and then take out the wick and add the alcohol with a funnel. The volume of alcohol in the lamp should be between 1/2 and 2/3 of the total volume. Light the alcohol burner with a match. It is absolutely forbidden to use an alcohol burner to light another alcohol burner. When extinguishing the flame, use a lamp shade to cover it rather than blowing it out with your mouth. The diagram of the correct use is shown in Fig. 2.4.

Fig. 2.4　A diagram of the use of an alcohol burner

Safety requirements: alcohol is flammable and should be used strictly in accordance with standard operations. Do not upset the alcohol burner or spill the alcohol outside the container to avoid a fire. Do not panic if the spilled alcohol burns on the table, and use a damp rag to cover it immediately.

2. Alcohol Blast Burner

The alcohol blast burner (heating temperature usually is 800~900℃) also uses alcohol as its fuel, but it gasifies the alcohol and mixes the gaseous alcohol with air. The flame temperature of the alcohol blast burner is high and stable, which can reach 800~900℃. There are two kinds of alcohol blast burner, seat-type one and hanging-type one, as shown in Fig. 2.5.

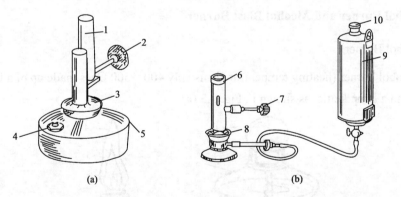

Fig. 2.5　Construction of a seat type alcohol blast burner (a) and a hanging type alcohol blast burner (b)

1. tube; 2. air conditioner; 3. preheating dish; 4. copper cap; 5. alcoholic kettles; 6. tube; 7. air conditioner; 8. preheating dish; 9. alcoholic storage tank; 10. lid

The usages of seat-type blast burners and hanging-type blast burners are basically the same.

(1) Adding alcohol: add alcohol in an alcoholic kettle or a storage tank, and do not add it during use to avoid a fire.

(2) Warming up: heat the preheating dish filled with the alcohol. As for the hanging-type blast burner, turn on the switch under the alcohol storage tank, then turn off the switch after alcohol comes out of the mouth of the tube and turn it on before next ignition; as for the seat-type blast burner, turn off its air conditioner, adjust it slowly after preheating for a while to prevent the spouting of alcohol.

(3) Lighting: when the tube is warmed up, turn on the air conditioner and ignite the light with a match.

(4) Outfire: close the air conditioner or cover the tube with the asbestos net to extinguish the flame.

When the hanging-type blast burner is not used, turn off the switch under the alcoholic storage tank, and the alcohol should be poured out if the blast burner is not used for a long time.

Safety requirements: Alcohol is flammable and should be used strictly in accordance with standard operations. The tube must be hot before igniting the alcohol blast burner, otherwise the liquid alcohol may spout out and cause a fire. The seat-type blast burner is used for up to 30 min, and the hanging-type blast burner should not be used for too long. If you need to continue to use the blast burner, you can reignite it after adding alcohol.

2.3.2 Electric Heating Equipment

Electric heating equipment includes electric furnaces, electric hot plates, electric jackets and high temperature furnaces, as shown in Fig. 2.6.

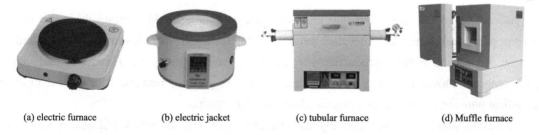

(a) electric furnace   (b) electric jacket   (c) tubular furnace   (d) Muffle furnace

Fig. 2.6   Frequently used electric heating equipment

(1) The temperature of the electric furnace and electric jacket can be adjusted by the external transformer, and the closed electric furnace is generally used for safety reasons. A piece of asbestos net should be used to separate the furnace from the container (e.g. a breaker) for even heating. The heating area of the electric heating plate is larger than that of the electric furnace, which can be used to add larger sample or a larger number of samples.

(2) The commonly used high temperature furnaces include tubular furnaces and Muffle furnaces, which are mainly used for the high temperature burning or reaction. The temperature is controlled by the thermocouple. If the heating element is a piece of heating wire, the highest temperature can reach 950℃ or so; if the silicon carbide rod is used, the highest temperature can reach 1300℃ or so.

2.4   General Knowledge of Chemical Reagents

2.4.1   Classification and Storage of Chemical Reagents

There are many kinds of chemical reagents, and their classification standards are not the same. The standard of chemical reagents in China include national standard, chemical industry standard and enterprise standard. According to the content of impurities in the reagent, chemical reagents (general reagents) produced in our country is basically divided into four grades. The grade and applicable scope are shown in Table 2.1.

Table 2.1 The grade and applicable scope of the chemical reagent

Grade	Name	Abbreviation	Tab color	Applicable scope
first grade	superior pure (guaranteed reagent)	GR	green	precision analysis and research work
second grade	analytical pure (analytical reagent)	AR	red	analysis experiment
third grade	chemical pure	CP	blue	primary chemical experiment
fourth grade	laboratory reagent	LR	yellow	industrial or chemical preparation
biochemical reagents	biochemical reagent (biological stain)	BR	brown or rose red	biochemical experiment

According to the regulation, the label of reagent bottles should indicate the name of the reagent, chemical formula, molecular weight, grade, technical specification, product standard number, production license, production batch number, manufacturer, etc. Hazardous and highly toxic reagents should have the corresponding marks. In general, reagents of chemical pure or analytical pure are often used in inorganic chemistry experiments.

Reagents should be kept separately and convenient for use, and safety and quality should be paid attention to during storage. General chemical reagents should be sealed and stored by category in low temperature, dry and well-ventilated places. Special chemical reagents have special preservation methods. White phosphorus should be preserved in water, and metal potassium and sodium are usually kept in kerosene. When chemical reagents are divided into different bottles, the solid reagent should be stored in a wide-mouth bottle, the liquid reagent or prepared solution should be stored in a narrow-mouth bottle or a dropping bottle with a dropper, the light-sensitive reagent or solution such as silver nitrate should be stored in a brown bottle, and each bottle should be labelled with the name, date and specification or concentration.

2.4.2 Transferring Chemical Reagents

Read the label of the reagent bottle before transferring the reagent, and then remove the bottle stopper and be careful to put it top side down on the experiment bench to avoid polluting the bench, and take the reagent as many as you need. After that, return the stopper to its original bottle promptly and then put the reagent bottle back into its original place. Everyone should form a good experimental habit.

1. Opening methods of Reagent Bottles

(1) Corks: when you want to remove the cork on the solid reagent bottle, you can hold the

bottle and make the bottle slanted on the bench, and then use an awl to insert into the cork and pull it out. The scraps attached to the mouth of the bottle will not fall into the bottle, and they can be wiped off with toilet paper gently.

(2) Plastic stoppers: hydrochloric acid, sulfuric acid and nitric acid and other liquid reagent bottles are mostly equipped with a plastic stopper. If the bottle cannot be opened, you can wrap the head of the stopper with a rag which has been soaked in hot water, and then give the stopper a hard twist. In this way, the stopper can be removed once it is loose.

(3) Glass stoppers: the stopper can be removed gently by being turned in the horizontal direction or shaken in the transverse direction. If it fails, hold the top of the bottle tightly and tap the stopper with a wooden handle or mallet, or tap the stopper with the edge of the table. Note: do not use the hammer.

2. Methods of Transferring Solid Reagents

(1) Take the reagent with a clean and dry medicine spoon. The larger end of the medicine spoon is for a large amount of the solid while the smaller end of the medicine spoon is for a small amount of the solid. Each medicine spoon is for special use, and the used spoon must be washed and dried before next use.

(2) When a certain amount of solid is required, the solid can be weighed on a piece of dry weighing paper, and the corrosive or deliquescent solid should be weighed on a watch glass or in a glass container.

(3) When adding a solid reagent into a test tube (especially a wet test tube), the reagent can be put on a folded piece of paper and put in the test tube about 2/3 [Fig. 2.7 (a) and (b)]. When adding the blocky solid, the tube should be tilted so that the solid slides down along the wall of the test tube slowly [Fig. 2.7 (c)] so as not to break the test tube.

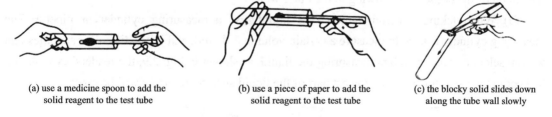

(a) use a medicine spoon to add the solid reagent to the test tube

(b) use a piece of paper to add the solid reagent to the test tube

(c) the blocky solid slides down along the tube wall slowly

Fig. 2.7 Methods of transferring solid reagents

Notes

(1) Do not exceed the required dosage. The reagent cannot be returned to the original bottle and can be placed in the designated container for someone else.

(2) When solid particles are larger, they can be crushed in a clean and dry mortar. The amount of the solid in the mortar shall not exceed 1/3 of the total volume.

(3) The toxic reagent is to be used under the guidance of the teacher.

3. Methods of Transferring Liquid Reagents

(1) When taking a liquid reagent from a dropping bottle, use the dropper in the dropping bottle. The dropper must not be in the experimental container to avoid the contamination of the reagent [Fig. 2.8 (a)]. If a small amount of liquid is taken from the reagent bottle, it is required to use the special dropper attached to the reagent bottle. The dropper with reagent may not be inclined or the mouth of the dropper may not be tilted upward to prevent the liquid from flowing into the latex head of the dropper.

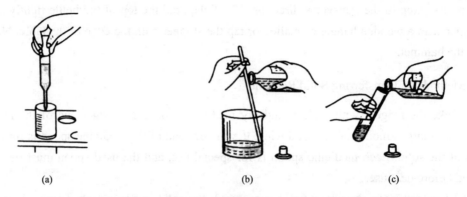

Fig. 2.8 Dropping the liquid into a test tube (a) and the tilt-pour process (b and c)

(2) When taking a liquid reagent from a narrow-mouth bottle, the tilt-pour process is applied. Remove the stopper first and put it top side down on the bench, then hold the labeled side of the reagent bottle in the hand and slowly tilt the bottle to let the reagent flow into a test tube along the tube wall or flow into a beaker along a glass rod [Fig. 2.8 (b) and (c)]. After the required amount is reached, lean the bottle against the container and gradually raise the bottle to prevent the droplets at the mouth of the bottle from flowing to the outer wall.

(3) When taking a certain amount of liquid, use a measuring cylinder or pipette. The measuring cylinder is used to measure a certain volume of liquid, and it has different capacities that we can select as needed. When measuring the liquid, as shown in Fig. 2.9, the readers' eyes should be kept in the same line with the lowest part of the liquid surface, or errors will be made.

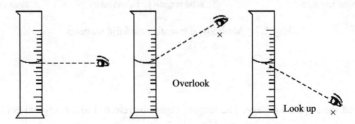

Fig. 2.9 Read the liquid volume in a measuring cylinder

Notes

(1) Take the liquid reagent as needed, and the extra reagent cannot be poured back into the original reagent bottle.

(2) Learn to estimate the amount of liquid when using a dropper to transfer the reagent. How many drops are there in 1 mL? How much does 5 mL take up in a test tube? Build a concept of "a small amount".

(3) The amount of solution poured into the test tube should not exceed 1/3 of the total volume during the experiment.

References

Zhao X H. 2014. Inorganic Chemistry Experiment. 4th ed. Beijing: Higher Education Press

Chapter 3 Basic Inorganic Chemistry Experiments

Exp. 1 Receiving, Washing and Drying the Apparatuses

Objectives

(1) To be familiar with the rules and requirements of inorganic chemistry laboratory.

(2) To receive commonly used apparatuses in inorganic chemistry experiments, and be familiar with the names, specifications and precautions.

(3) To learn and practice methods of washing and drying commonly used apparatuses.

Procedures

(1) Receive the commonly used apparatuses in inorganic chemistry experiments according to the number of the apparatus, specification and model one by one, as shown in Table 3.1.

Table 3.1 Commonly used apparatuses in inorganic chemistry experiments

Name	Specification	Number	Remark	Name	Specification	Number	Remark
test tube rack		1		porcelain crucible	with a cover	1	
test tube clamp		1		spot plate	white	1	
test tube brush		1	for public	glass sheet		1	
test tube	10 mm×150 mm	10		alcohol burner		1	
centrifuge tube		6		forcep		1	
beaker	50 mL	2		asbestos net		1	
beaker	100 mL	2		pipeclay triangle		1	
beaker	250 mL	2		rubber suction bulb		1	
beaker	≥400 mL	2		combustion spoon		1	
measuring cylinder	10~20 mL	1		crucible tong		1	
measuring cylinder	50 mL	1		evaporating dish		1	
gas bottle		1		plastic squeeze bottle		1	
filter flask		1		Büchner funnel		1	
flask		1		glass funnel		1	
Erlenmeyer flask	150 mL	2		watch glass		1	
Erlenmeyer flask	250 mL	3					
volumetric flask	50 mL, with a cover	1					
volumetric flask	100 mL, with a cover	1					

(2) Washing: take correct washing methods to wash different apparatuses, hand two of washed apparatuses over to the teacher for check, and place all washed apparatuses in a cabinet reasonably and orderly.

(3) Drying.

Notes

(1) The waste liquid in the test tube must be first poured into the waste liquid cylinder, and then the test tube should be washed many times with a little water at each.

(2) Multiple test tubes must not be held in one hand at the same time. The apparatuses should be washed with a brush gently. The apparatuses in the cabinet should be placed for easy access.

(3) Aqua regia should be newly prepared if it is used to wash something (aqua regia is unstable).

(4) Chromic acid lotion is often placed in a fume hood, and it will be retrieved after use.

(5) Most of inorganic chemistry experiments belong to microchemistry. All the reagents involved in the experiment should be available in the second-grade reagent bottles in the chemical box. If not, they are placed on the public chemical bench. The second-grade reagents are only used in the experiment on the same day, so it is unnecessary to take much. Do not mix the reagent bottles or replace the caps.

Questions

(1) Why does the tube tip slightly down when drying the test tube?

(2) What kind of apparatus must not be dried by heating, and why?

(3) Draw a simple diagram of the centrifugal tube, multi-purpose dropper, beaker, measuring cylinder and volumetric flask, and explain their main uses and matters needing attention.

Exp. 2　Transferring Chemical Reagents and Operating Test Tubes

Objectives

(1) To understand the grade standards and the applicable scope of the purity of chemical reagents as well as types of bottled reagents.

(2) To learn and master how to weigh the solid samples and measure the liquid reagents as well as how to use them.

(3) To practice and master the operation of test tubes and the method of heating solid or liquid samples.

(4) To practice some basic operations such as stir, dissolution, grinding and the cleaning of apparatus.

(5) To know the chemical principles of experiments.

Principles

1. "Tri-color Cup" Experiment

Non-polar I_2 molecules are easily soluble in many organic solvents and present different colors. Iodine is brown in ethanol or isoamyl alcohol due to the formation of the solvates while it is purple in solvents with smaller dielectric constant such as carbon disulfide and carbon tetrachloride due to the presence of iodine molecules. Although the solubility of iodine is small in an aqueous solution, its solubility increases obviously in the potassium iodide solution. Furthermore, the greater the concentration of iodized salt is and the more iodine dissolves, the deeper color the solution presents. I_3^- ion is formed in the solution: $I_2 + I^- \rightleftharpoons I_3^-$. From this balance, the aqueous solution of iodine presents the color due to the presence of iodine molecules.

2. "Copper Sulfate Discoloration" Experiment

Copper sulfate pentahydrate ($CuSO_4 \cdot 5H_2O$) is known as blue vitriol which is the most common copper salt. It is a kind of blue orthorhombic crystal which can dehydrate gradually at different temperatures:

$$CuSO_4 \cdot 5H_2O \xrightarrow{375\ K} CuSO_4 \cdot 3H_2O \xrightarrow{423\ K} CuSO_4 \cdot H_2O \xrightarrow{523\ K} CuSO_4$$

When heated to above 873 K, the white $CuSO_4$ will decompose into CuO, SO_2, SO_3 and O_2. Obviously, the binding force of each water molecule is not exactly the same. Experiments show that four water molecules are bound to Cu^{2+} by coordination bond, and the fifth water molecule is bound to two coordinated water molecules and SO_4^{2-} by hydrogen bond. The simple planar structure of $CuSO_4 \cdot 5H_2O$ is as follows:

$$\begin{array}{c} H_2O \diagdown \quad \diagup OH_2 \quad \diagdown H \cdots O \diagdown \quad \diagup O \\ Cu \qquad\qquad\qquad\qquad\qquad S \\ H_2O \diagup \quad \diagdown OH_2 \quad \diagup H \cdots O \diagup \quad \diagdown O \end{array}$$

When dehydration occurs by heating, the two water molecules on the left of Cu^{2+} are first lost, then the two on the right side are lost, and finally the one conjugated to SO_4^{2-} by hydrogen bond is lost.

3. "Five-color Tube" Experiment

In an aqueous solution, Ni^{2+} and ethylenediamine(en) can form coordination complexes with the coordination ratios of 1 : 1, 1 : 2 and 1 : 3, which are different in stability and colors: $[Ni(H_2O)_4(en)]^{2+}$ (light blue), $[Ni(H_2O)_2(en)_2]^{2+}$ (blue) and $[Ni(en)_3]^{2+}$ (purple). Ni^{2+} and dimethylglyoxime (dmg) can form a kind of chelate nickel(II) dimethylglyoxime $[Ni(dmg)_2]$, in an aqueous solution, which is a bright red precipitate.

$$[Ni(H_2O)_6]^{2+}(green) + en \longrightarrow [Ni(H_2O)_4(en)]^{2+}(light\text{-}blue) + 2H_2O$$

$$[Ni(H_2O)_6]^{2+}(green) + 2en \longrightarrow [Ni(H_2O)_2(en)_2]^{2+}(blue) + 4H_2O$$

$$[Ni(H_2O)_6]^{2+} (green) + 3en \longrightarrow [Ni(en)_3]^{2+} (purple) + 6H_2O$$

$$[Ni(H_2O)_6]^{2+} (green) + 2dmg \longrightarrow Ni(dmg)_2 (red) + 6H_2O + 2H^+$$

4. The Volume of Oxygen in the Air

Red phosphorus is one of allotropes of elementary phosphorus. Different allotropes of phosphorus can be transformed into each other under different conditions.

$$\text{black phosphorus} \xleftarrow{\text{high temperature}}_{\text{high pressure}} \text{white phosphorus} \xrightarrow{\text{isolate air at 400℃}} \text{red phosphorus}$$

The chemical properties of red phosphorus are more stable than others. It can be burned at over 400℃ and cannot be dissolved in organic solvents. Red phosphorus is burned in the air to produce phosphorus oxides which are all white solids and easily soluble in water.

$$4P \xrightarrow{O_2 \text{(insufficient)}} P_4O_6 \xrightarrow[\triangle]{O_2} P_4O_{10}$$

Experimental Items

Apparatuses and materials: test tube, test tube clamp, beaker, wash bottle, measuring cylinder, mortar, medicine spoon, platform scale, alcohol burner, iron stand, iron clamp, glass rod, dropper, rubber stopper.

Reagents: iodine, potassium iodide, red phosphorus, copper sulfate, carbon tetrachloride, isoamyl alcohol, nickel sulfate (0.1 mol/L), ethylenediamine (25%, *V/V*), dimethylglyoxime (1%).

Procedures

1. "Tri-color Cup" Experiment

Take a measuring cup (or measuring cylinder) of 10 mL and add 2 mL of carbon tetrachloride along the wall of the cup, followed by the injection of 5 mL of water and 2 mL of isoamyl alcohol. Afterwards, put a small spoon of solid KI and iodine (the solid iodine has been porphyrized in a mortar) respectively on the surface of a watch glass and mix them evenly. Then stick the mixture with a wetted glass rod which is then inserted into the measuring cup containing the above solutions, stir the mixed solution gently and observe the colors of its three layers.

2. "Discoloration of Copper Sulfate" Experiment

Add several grains of $CuSO_4 \cdot 5H_2O$ crystal into a test tube, and carry out the experiment according to the method of heating solid reagents. Stop heating when all the crystal becomes white, and add 3 to 5 drops of water when the test tube cools to the room temperature. Observe the color change, and try to touch the wall of test tube.

3. "Five-color Tube" Experiment

Take five test tubes and add 10 drops of 1 mol/L $NiSO_4$ solution into each test tube. Add 1~2 drops of 25% ethylenediamine (en) solution into the first test tube, 3~4 drops of 25% en solution into the second, 5~6 drops of 25% en solution into the third, 9~10 drops of 1% dimethylglyoxime (dmg) solution into the fourth, and the fifth test tube is used for contrasting the color without the addition of en and dmg. After fully oscillation of the test tubes, observe and compare different colors of the coordination complexes in five test tubes.

4. The Volume of Oxygen in the Air

Take a dry and clean hard-quality test tube, put about 0.2 g of red phosphorus at the bottom of the tube and cover the test tube tightly with a rubber stopper. Heat the test tube on the alcohol burner (caution!) and cool to room temperature when the flame of red phosphorus goes out. Invert the test tube in a big beaker filled with water, pull out the rubber stopper underwater (note: never make the mouth of the tube above water), and observe the phenomenon of the liquid surface in the test tube. When the water doesn't rise any longer, cover the test tube with the rubber stopper underwater. Then take out the test tube and observe how much volume of water in the test tube accounts for the total volume to infer the volume of oxygen in the air.

Notes

(1) In the "tri-color cup" experiment, the glass rod cannot be used to stir the solution up and down in order to avoid the mixing of two organic layers.

(2) As the organic solvent is highly volatile and highly toxic, the waste liquid should be treated in time after the "tri-color cup" experiment to avoid environmental pollution.

Questions

What should we pay attention to when heating liquid or solid reagents in a test tube?

References

Teaching and Research Section of Inorganic Chemistry of Beijing Normal University etc. 1991. Inorganic Chemistry Experiment. 2nd ed. Beijing: Higher Education Press

Exp. 3   Preparation of Solutions

Objectives

(1) To master the basic preparation methods of the mass fraction, molality, amount-of-substance concentration, volume by volume concentration of solutions, etc.

(2) To learn the usages of hydrometers, pipettes, volumetric flasks, platform scales, analytic balances, etc.

(3) To understand the preparation of special solutions.

Principles

1. Several Expressions of the Solution Concentration

1) Mass Fraction

It represents the mass fraction (w) of solute dissolved in the solution, which is typically called the mass percent concentration.

$$w = \frac{m_{solute}}{m_{solution}}$$

So

$$m_{solute} = \frac{w\rho_{solvent}V_{solvent}}{1-w}$$

In the formula, w is the mass fraction of the solute; m_{solute} is the mass of the solid reagent (g); $m_{solution}$ is the mass of the solution (g); $\rho_{solvent}$ is the density of the solvent (g/mL); $V_{solvent}$ is the volume of the solvent (mL).

2) Molality

Molality (b) refers to the amount of substance of the solute dissolved in 1000 grams of solvent (mol/kg). In general, water is used as the solvent.

$$b = \frac{n}{1000 \text{ g solvent}}$$

So

$$x_{solute} = \frac{Mbm_{solvent}}{1000} = \frac{Mb\rho_{solvent}V_{solvent}}{1000}$$

In the formula, b is the molality (mol/kg); x_{solute} is the mass of solute (g); M is the molar mass of the solid (g/mol); $m_{solvent}$ is the mass of the solvent (g); and the descriptions of other symbols are the same as before.

3) Amount-of-substance Concentration

The amount-of-substance concentration (c) indicates the amount of substance of the solute dissolved in 1 L of solution.

$$c = \frac{m_{solute}}{VM}$$

So

$$m_{solute} = cVM \qquad c_1V_1 = c_2V_2$$

In the formula, c is the amount-of-substance concentration (mol/L); and the descriptions of other symbols are the same as before.

4) Volume by Volume Concentration

It means that a certain volume of solute is dissolved in a certain volume of solvent (V/V).

2. Basic Methods of Solution Preparation

In inorganic chemistry experiments, the prepared solutions usually include general solutions, standard solutions and special solutions.

1) Preparing General Solutions

The general solution refers to the solution whose accuracy of concentration is undemanding. In general, some apparatuses with low accuracy can meet the requirements, such as the platform scale, measuring cylinder and beaker with calibration. This preparation process is also called rough preparation.

The following methods are commonly used to prepare general solutions.

a. Direct Water Dissolution Method

This method is suitable for solid reagents which can be dissolved easily in water without hydrolysis, such as NaOH, NaCl and KNO_3. When preparing the solution, weigh a certain amount of the solid with a platform balance first and add it into a beaker. Then add an appropriate amount of distilled water to dissolve the solid by stirring and dilute it to the required volume. Finally, transfer the solution into the reagent bottle.

b. Dilution Method

When preparing dilute solutions of some liquid reagents such as strong acid, strong alkali and a certain concentration of electrolyte solution, measure the required concentrated solution with a measuring cylinder and dilute it with an appropriate amount of distilled water. It's important to note that concentrated sulfuric acid should be slowly poured into water under continuous stirring when preparing its dilute solution, and this order should not be reversed.

2) Preparing Standard Solutions

The standard solution refers to the solution whose accurate concentration is known, that is, the experiment has high requirement on the accuracy of the solution concentration. In general, some apparatuses with high accuracy are used to prepare the standard solution, such as analytic balance, pipette and volumetric flask. This preparation process is also known as the accurate preparation. There are two methods to prepare the standard solution.

a. Direct Method

Weigh accurately a certain amount of standard reagent with an analytic balance and put it into a beaker, then add an appropriate amount of pure water (ion exchange water or secondary distilled water). After the reagent is dissolved, transfer it into a volumetric flask, dilute it with water to the required scale and shake well. The accurate concentration can be calculated from the weighing data and the dilution volume.

b. Calibration Method

For non-standard reagents, the standard solution cannot be prepared by direct method, but it can be prepared to the approximately desired concentration and then calibrated with a standard reagent or a standard solution of known exact concentration.

When adopting the dilution method to prepare a dilute standard solution, the pipette can be used to transfer an appropriate volume of concentrated solution into a volumetric flask of a certain volume.

3) Preparing Special Solutions

The special solution refers to the solution in which some solutes are prone to hydrolyze, break down when exposed to light or carry out the redox reaction during the dissolution process. Therefore, some substances must be added or certain measures must be taken during the preparation to prevent the solution from losing effectiveness during the preservation.

When preparing $FeCl_3$, $SnCl_2$, $SbCl_3$ or $BiCl_3$ solutions, add a certain amount of diluted hydrochloric acid to dissolve it, then dilute it to the required volume with distilled water, and finally transfer it into the reagent bottle for preservation. As for substances which are easy to hydrolyze, such as $SnCl_4$, they can be dissolved in concentrated hydrochloric acid first and then diluted with water. As for solutions which are prone to carry out the redox reaction, measures must be taken to prevent them from oxidization and losing effectiveness during the preservation. For instance, some Sn particles and Fe chips should be added in Sn^{2+} and Fe^{2+} solutions, respectively. Solutions like $AgNO_3$, $KMnO_4$ and KI, which are easy to decompose when exposed to the light, should be stored in clean brown bottles. Solutions susceptible to chemical corrosion should be stored in suitable containers.

Some substances are less soluble in water, such as iodine. We can prepare I_2 water by the following methods: dissolve I_2 with KI solution first, then dilute it with water to the required volume, shake well and finally transfer the solution into the reagent bottle.

Experimental Items

Apparatuses and materials: beaker, pipette, measuring pipette, volumetric flask, dropping bottle, measuring cylinder, reagent bottle, weighing bottle, platform scale, analytic balance, etc.

Reagents: NaCl, KCl, $CaCl_2$, $NaHCO_3$, $SnCl_2$, tin grain, concentrated sulfuric acid, concentrated hydrochloric acid, HAc solution (2.00 mol/L), Na_2CO_3(AR), oxalic acid dihydrate (AR).

Procedures

1. Preparing General Solutions

(1) Prepare 50 mL of 3 mol/L H_2SO_4 solution roughly[1].

(2) Prepare 20 mL of 0.2 mol/L Na_2CO_3 solution roughly.

(3) Prepare 50 mL of 0.200 mol/L HAc solution from HAc solution with known concentration of 2.00 mol/L.

(4) Prepare 100 mL of saline solution with mass fraction of 0.90% accurately according to the mass ratio (NaCl : KCl : $CaCl_2$: $NaHCO_3$ = 45 : 2 : 1.2 : 1).

2. Preparing $H_2C_2O_4 \cdot 2H_2O$ Standard Solution

Weigh 1.5000~1.6000 g of $H_2C_2O_4 \cdot 2H_2O$[2] solid in a beaker accurately, add a small amount of pure water (secondary distilled water or deionized water) to dissolve the crystal completely, and transfer the solution to a volumetric flask. Then use a small amount of water to rinse the beaker and glass rod several times and transfer all the rinsing liquid into a 250 mL volumetric flask. Finally dilute the solution with pure water to the required scale and shake it well. Calculate the exact concentration of the solution.

3. Preparing Special Solution

Prepare 100 mL of 0.1 mol/L $SnCl_2$ solution (pay attention to preventing hydrolysis during the preparation and oxidation during the storage).

Notes

[1] The comparison table of the relative density and mass fraction of concentrated H_2SO_4

d_4^{20}	1.8144	1.8195	1.8240	1.8279	1.8312	1.8337	1.8355	1.8364	1.8361
x/%	90	91	92	93	94	95	96	97	98

[2] The standard substance oxalic acid dihydrate ($H_2C_2O_4 \cdot 2H_2O$) has the molecular weight of 126.03, and it contains two H^+ that can react with OH^-, which is used to calibrate the concentration of alkali.

Questions

(1) How to prevent the oxidation and hydrolysis when preparing $SnCl_2$ solution?

(2) Does the volumetric flask need to be dried when preparing the solution accurately? Does the volumetric flask need to be rinsed with the diluent when preparing the dilute solution from the concentrated solution?

(3) Does the clean pipette need to be rinsed with the sucked solution, and why?

Exp. 4　Determination of the Molecular Weight of CO_2

Objectives

(1) To master the principles and methods of determining the molecular weight of gases by the relative density.

(2) To master the use of Kipp's apparatus and some basic operations, such as gas washing and drying.

(3) To consolidate the use of analytic balances.

Principles

According to Avogadro's law, the mass ratio of two ideal gases with the same volume under the conditions of the same temperature and pressure is equal to the ratio of their molecular weight.

$$\frac{m_A}{m_B} = \frac{M_A}{M_B}$$

In the formula, m_A and m_B represent the mass of gas A and gas B, respectively; M_A and M_B

represent the molecular weight of gas A and gas B, respectively. Therefore, as long as we determine the mass of two gases of the same volume under the same temperature and pressure and the molecular weight of one gas is known, the molecular weight of the other can be determined.

It is well known that the average molecular weight of air is 29.0, so the molecular weight of CO_2 can be determined as long as the mass of CO_2 and air under the same conditions (temperature, pressure and volume) is measured, that is:

$$M_{CO_2} = 29.0\, m_{CO_2}/m_{air}$$

The mass of CO_2 can be obtained directly by weighing twice.

For the first weighing, the mass of the container filled with air is:

$$m_1 = m_{air} + m_{bottle} + m_{stopper}$$

For the second weighing, the mass of the container filled with CO_2 is:

$$m_2 = m_{CO_2} + m_{bottle} + m_{stopper}$$

$$m_2 - m_1 = m_{CO_2} - m_{air}$$

$$m_{CO_2} = m_2 - m_1 + m_{air}$$

According to the atmospheric pressure (p) and thermodynamic temperature (T), the air mass m_{air} of the same volume can be calculated by the ideal gas state equation.

$$m_{air} = \frac{pV}{RT} \times 29.0$$

The volume V in the formula can be calculated by weighing the mass of the container filled with water.

$$m_3 = m_{H_2O} + m_{bottle} + m_{stopper}$$

$$m_3 - m_1 = m_{H_2O} - m_{air} \approx m_{H_2O}$$

$$V = m_{H_2O}/\rho_{H_2O}$$

Experimental Items

Apparatuses and materials: analytical balance, platform scale, Kipp's apparatus, gas-washing bottle, drying tube, Erlenmeyer flask, glass tube, rubber tube, thermometer, air pressure gauge.

Reagents: limestone, HCl (6 mol/L), $NaHCO_3$ (1 mol/L), concentrated sulfuric acid.

Procedures

1. Preparing CO_2

The experimental device for preparing CO_2 is shown in Fig. 3.1. The CO_2 produced from the Kipp's apparatus is washed by $NaHCO_3$ solution in the gas-washing bottle 2 to remove a small amount of H_2S, HCl and other acidic gas impurities, and further washed by the concentrated sulfuric acid in the gas-washing bottle 3 and water is removed, thereby making the derived CO_2 gas dry and pure.

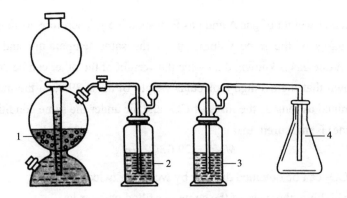

Fig. 3.1 The experimental device for the preparation, purification and collection of CO_2
1. Kipp's apparatus; 2. gas-washing bottle (1 mol/L $NaHCO_3$ solution); 3. gas-washing bottle (concentrated H_2SO_4); 4. Erlenmeyer flask

2. Weighing

(1) Weighing the air-filled flask and the stopper: take a dry and clean Erlenmeyer flask, stuff it with a suitable rubber stopper and make a mark on the rubber stopper to fix the position in which the stopper is inserted into the flask. Then obtain the total mass (air, flask and stopper) by an analytical balance, which is denoted as m_1.

(2) Weighing the CO_2-filled flask and the stopper: after the CO_2 produced from the Kipp's apparatus is purified and dehydrated, it is introduced into the bottom of the Erlenmeyer flask. When the Erlenmeyer flask is full of CO_2, take out the guide pipe slowly and insert the rubber stopper to the original mark position, and then obtain the total mass (CO_2, flask and stopper) by the same analytical balance, which is denoted as m_2. Repeat the operations until the mass difference between the first time and second time is only 1~2 mg.

(3) Weighing the water-filled flask and the stopper: fill the flask with water and insert the rubber stopper to the original mark position. Then dry the outer wall of the flask with absorbent paper and obtain the total mass (water, flask and stopper) by a platform scale, which is denoted as m_3. Moreover, record the temperature T and atmospheric pressure p when the experiment is conducted.

Data Recording and Processing

Room temperature: $t =$ ____ ℃; atmospheric pressure: $p =$ _____ mmHg
The total mass of (air + flask + stopper) $m_1 =$ _____ g
For the first time, the total mass of (CO_2 + flask + stopper) = _____ g
For the second time, the total mass of (CO_2 + flask + stopper) = _____ g
The total mass of (CO_2 + flask + stopper) $m_2 =$ _____ g
The total mass of (water + flask + stopper) $m_3 =$ _____ g
The volume of the flask $V = (m_3 - m_1) / 1.00 =$ _____ mL
The mass of the air in the flask $m_{air} =$ _____ g

The mass of CO_2 $m_{CO_2} = m_2 - m_1 + m_{air}$ = _____ g

The molecular weight of CO_2 M_{CO_2} = _____

Error _____

Questions

(1) Complete data recording and result processing, and analyze the cause of error.

(2) Point out the functions of each part in the diagram of the experimental device and write down the relevant reaction equations.

(3) How to test the airtightness of the device?

(4) How to test whether the flask is full of CO_2?

(5) Why should the total mass of (CO_2+ flask + stopper) and (air + flask + stopper) be obtained by the same analytical balance, while the total mass of (H_2O + flask + stopper) can be obtained by a platform scale?

References

Institute of Inorganic Chemistry, Beijing Normal University. 2001. Inorganic Chemistry Experiment. 3rd ed. Beijing: Higher Education Press

Institute of Inorganic Chemistry, Sun Yat-Sen University. 1992. Inorganic Chemistry Experiment. 3rd ed. Beijing: Higher Education Press

Exp. 5   Determination of Crystal Water Content in Blue Vitriol

Objectives

(1) To learn the principles and methods of determining the crystal water content in crystalline hydrates.

(2) To master the use of mortars, dryer, etc. and the basic operations of sand bath, constant weight, etc.

(3) To get more familiar with the use of the analytical balance.

Principles

When crystals are precipitated by many substances in aqueous solution, they often contain a certain number of water molecules called crystal water. When the crystalline hydrate is heated to a certain temperature, a part or all of the crystal water can be removed. Therefore, if we are going to determine the crystal water in the crystalline hydrate (when heated, the crystal water can be removed and the crystalline hydrate will not break down), we usually place a certain amount of crystalline hydrate in a crucible which has been burnt to a constant weight, heat it to a higher temperature for dehydration (not to exceed the decomposition temperature of the substance to be determined), then transfer the crucible into the dryer, and finally weigh it on an analytical balance after it cools to room temperature. From the weight loss value of the crystalline hydrate heated at high temperature, the mass fraction of the crystal water contained in the crystalline hydrate and the amount of crystal water in per mole of the salt can be calculated, thus the chemical formula of the crystalline hydrate can be determined.

Copper sulfate pentahydrate ($CuSO_4 \cdot 5H_2O$), commonly known as blue vitriol, is a sky-blue crystal. The experiment results show that four water molecules in $CuSO_4 \cdot 5H_2O$ are bound with Cu^{2+} by coordination bond, and the fifth one is bound with two coordination water molecules and SO_4^{2-} by hydrogen bond. The simple structure diagram of $CuSO_4 \cdot 5H_2O$ is shown in Fig. 3.2.

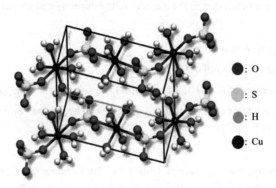

Fig. 3.2 The structure diagram of $CuSO_4 \cdot 5H_2O$

When heated to lose water, the two water molecules on the left of Cu^{2+} are first lost, then the two on the right side are lost, and finally the one conjugated to SO_4^{2-} by hydrogen bond is lost. If continue to be heated above 600℃, the white $CuSO_4$ will decompose into CuO, SO_2, SO_3 and O_2. Therefore, $CuSO_4 \cdot 5H_2O$ is gradually dehydrated according to the following reactions (note: dehydration temperature is related to the pressure, particle size, heating rate and other factors):

$$CuSO_4 \cdot 5H_2O \xrightarrow{48℃} CuSO_4 \cdot 3H_2O + 2H_2O$$

$$CuSO_4 \cdot 3H_2O \xrightarrow{99℃} CuSO_4 \cdot H_2O + 2H_2O$$

$$CuSO_4 \cdot H_2O \xrightarrow{218℃} CuSO_4 + H_2O$$

In this experiment, $CuSO_4 \cdot 5H_2O$ with known mass crystal is heated and all the crystal water is removed and weighed, thus the number of crystal water in the chemical formula can be calculated.

Experimental Items

Apparatuses and materials: analytical balance, dryer, crucible, pipeclay triangle, crucible tongs, iron stand, iron ring, sand bath (evaporating dish, sand), alcohol burner, thermometer, filter paper.

Reagents: blue vitriol.

Procedures

1. Constant Weight of the Crucible

Place a washed crucible and crucible cover on the pipeclay triangle, dry it by a small fire, and burn it with an oxidizing flame to red heat. After it cools to slightly above room temperature, transfer the crucible into the dryer with clean crucible tongs and let it cool to room temperature (note: after the hot crucible is placed in the dryer, the dryer lid should be opened 1 or 2 times in a short time to avoid it being difficult to open owing to the decreased pressure). Take it out and weigh it on an analytical balance. Reheat it to above dehydration temperature, cool it, and weigh it until the weight is constant.

2. Weighing Hydrated Copper Sulfate and Preparing the Sand Bath

Add 1.0~1.2 g of fine blue vitriol crystal into a crucible with constant weight to form a uniform layer, and then obtain the accurate total mass of the crucible and blue vitriol by the same analytical balance, and the mass of hydrated copper sulfate m_1 can be obtained by subtracting the mass of the crucible. Place the crucible in a sand bath and bury three quarters of its volume in the sand, then insert a thermometer (the measuring range is 300 ℃) in the sand close to the crucible and its end should be at the same level with the bottom of the crucible roughly.

3. Dehydration of Hydrated Copper Sulfate

Heat the sand bath to about 210 ℃, then slowly heat up to 280 ℃ and control the temperature at the range of 260~280 ℃. Stop heating when the powder inside the crucible turns from blue to off-white (15~20 min). Then transfer the crucible into the dryer with clean crucible tongs for cooling to room temperature. After the outer wall of the crucible is wiped off with filter paper, weigh the total mass of the crucible and anhydrous copper sulfate on the same analytical balance. The mass of anhydrous cupric sulfate m_2 can be obtained by subtracting the mass of the crucible. Repeat the above operations (the sand bath heating, cooling and weighing) until constant weight is reached (this experiment requires the difference between two weighing is ≤1 mg). After the experiment, pour anhydrous copper sulfate into the recovery bottle.

Data Recording and Processing

Data recording and processing of determining the crystal water content in blue vitriol are shown in Table 3.2.

Table 3.2 Data recording and processing of determining the crystal water content in blue vitriol

Mass of empty crucible/g			Mass of empty crucible and blue vitriol/g	After heating, mass of the crucible and blue vitriol/g		
m_1	m_2	Average value		m_1	m_2	Average value

The mass of blue vitriol m_1 = _____ g

The mass of $CuSO_4$ m_2 = _____ g

The amount of substance for $CuSO_4$ $n_2 = m_2/159.6$ g/mol = _____ mol

The mass of crystal water m_3 = _____ g

The amount of substance for crystal water $n_3 = m_3/18.0$ g/mol = _____ mol

The crystal water of 1 mol $CuSO_4 = n_3/n_2 =$ _____

The chemical formula of blue vitriol _____

Questions

(1) Why do we use the sand bath to heat and control the temperature at 280℃?

(2) Why do we have to repeat the burning operation? What is its role?

(3) Why do we weigh the heated crucible after it cools to room temperature? Why do we place the heated crucible in the dryer to cool down?

References

Institute of Inorganic Chemistry, Beijing Normal University. 2001. Inorganic Chemistry Experiments. 3rd ed. Beijing: Higher Education Press

Institute of Inorganic Chemistry, Sun Yat-sen University. 1992. Inorganic Chemistry Experiments. 3rd ed. Beijing: Higher Education Press

Exp. 6 Purification and Qualitative Test of Crude Salt

Objectives

(1) To learn the principles and methods of preparing reagent grade sodium chloride from crude salt.

(2) To master weighing, measuring, dissolution, filtration, evaporation, concentration, crystallization, drying and other basic operations.

(3) To learn the methods of qualitatively identifying SO_4^{2-}, Ca^{2+} and Mg^{2+}.

Principles

Insoluble impurities (such as sediment) in crude salt can be removed by dissolution and filtration. The soluble impurities in crude salt are mainly Ca^{2+}, Mg^{2+}, K^+, SO_4^{2-}, etc., and appropriate reagents are added to make them form insoluble compound precipitates and be removed. However, the selected precipitator should comply with the principle that no new impurities are introduced or the introduced one can be removed in the next step.

(1) Removal of SO_4^{2-}: add excessive $BaCl_2$ solution to crude salt, and then remove insoluble substances and $BaSO_4$ precipitate by filtration.

$$Ba^{2+} + SO_4^{2-} == BaSO_4\downarrow$$

(2) Removal of Ca^{2+}, Mg^{2+} and excessive Ba^{2+}: add excessive NaOH and Na_2CO_3 solutions to

the filtrate, and then remove the produced precipitate by filtration.

$$Mg^{2+} + 2OH^- = Mg(OH)_2\downarrow$$
$$Ca^{2+} + CO_3^{2-} = CaCO_3\downarrow$$
$$Ba^{2+} + CO_3^{2-} = BaCO_3\downarrow$$

(3) Removal of excessive mixed alkali in the solution by neutralization with hydrochloric acid.

$$2H^+ + CO_3^{2-} = H_2O + CO_2\uparrow$$
$$H^+ + OH^- = H_2O$$

(4) Neither K^+ in crude salt nor above precipitants have any effect, and since the solubility of KCl is greater than that of NaCl and the content is less, NaCl crystallizes first during evaporation and concentration while KCl is left in the solution.

Experimental Items

Apparatuses and materials: platform scale, beaker, measuring cylinder, iron stand, iron ring, glass funnel, Büchner funnel, suction bottle, circulating water pump, evaporating dish, asbestos net, alcohol burner, medicine spoon, filter paper, pH test paper.

Reagents: crude salt, $BaCl_2$ (1 mol/L), NaOH (6 mol/L), Na_2CO_3 (saturated), HCl (6 mol/L), HAc (6 mol/L), $(NH_4)_2C_2O_4$ (saturated), magneson, the mixture of NaOH (2 mol/L) and Na_2CO_3 (saturated, 50%).

Procedures

1. Purification of Crude Salt

(1) Weigh 10.0 g of crude salt on the platform scale and put it into a 100 mL beaker, dissolve the sample with 40 mL of water, and heat and stir the solution. When the solution is boiling, add 1 mol/L $BaCl_2$ solution (about 2 mL) dropwise while stirring until the precipitation is complete. Continue heating it for 5 min to make the $BaSO_4$ particles grow and easy to precipitate and filter. In order to check whether the precipitation is complete, the beaker can be removed from the asbestos net. After the precipitation is settled, tilt the beaker and add a few drops of 1 mol/L $BaCl_2$ solution along the wall of the test tube. Then pour the remaining clear liquid by filtration at the normal pressure or decantation.

(2) Add 1 mL of 6 mol/L NaOH solution and 2 mL of saturated Na_2CO_3 solution (or about 4 mL of NaOH-Na_2CO_3 mixed solution, with a pH of about 11) to the above filtrate, and then heat the solution to boiling. In order to check whether the precipitation is complete, the beaker can be removed from the asbestos net. After the precipitation is settled, tilt the beaker and add a few drops of Na_2CO_3 solution along the wall of the test tube to check for the precipitation formation. If the precipitation is no longer produced, filter the solution with a Büchner funnel under diminished pressure and retain the filtrate.

(3) Add 6 mol/L HCl solution to the filtrate dropwise until the solution is slightly acidic (pH is

2~3).

(4) Pour the filtrate into an evaporating dish, heat it with a small fire and concentrate the filtrate until it becomes thick liquor like gruel. Do not dry out the solution.

(5) After cooling, try to drain the product by vacuum filtration. Put the product back into the evaporating dish, heat and dry it over a small fire until there is no steam rising from it.

(6) Weigh the product after it cools to room temperature. Finally, put the product back into the designated container and calculate the yield.

2. Testing of Product Purity

Weigh 1 g of crude salt and the product obtained from the experiment, respectively, and then dissolve each of them in 5 mL of distilled water. After filtering the crude salt solution, add the two clear solutions into three test tubes and form three groups respectively. Then test their purity by the control study.

(1) Testing of SO_4^{2-}: add 2 drops of 6 mol/L HCl to the solutions in the first group, which makes the solutions acidic. Then add 3 to 5 drops of 1 mol/L $BaCl_2$ solution, and if there is a white precipitate, the existence of SO_4^{2-} is proved. Record the results and have a comparison.

(2) Testing of Ca^{2+}: add 2 drops of 6 mol/L HAc to the solutions in the second group, which makes the solutions acidic. Then add 3~5 drops of saturated $(NH_4)_2C_2O_4$ solution, and if the white precipitate is produced, the existence of Ca^{2+} is proved. Record the results and have a comparison.

(3) Testing of Mg^{2+}: add 3~5 drops of 6 mol/L NaOH to the solutions in the third group, which makes the solutions alkaline. Then add 1 drop of magneson [magneson is a kind of organic dye which is red or purple in alkaline solution, but sky blue after being adsorbed by $Mg(OH)_2$]. If there is a blue precipitate, the existence of Mg^{2+} is proved. Record the results and have a comparison.

Data Recording and Processing

    (1) Purification of crude salt.

    Product appearance:_____

    Product mass: _____ g; Yield percentage: _____ %

(2) Testing of product purity (Table 3.3).

Table 3.3 Experimental phenomena and conclusions for testing product purity

Number	Experiment contents	Experimental phenomena		Conclusions and equations
		Crude salt	Refined salt	
(1)	add $BaCl_2$ solution			
(2)	add $(NH_4)_2C_2O_4$ solution			
(3)	add NaOH solution and magneson			

Notes

(1) When using the platform scale, pay attention to keeping it clean.

(2) When the salt solution is concentrated, it must not be dried.

(3) The difference of devices and operations between normal filtration and vacuum filtration.

(4) The use of pH test paper.

(5) The addition of $BaCl_2$, NaOH, Na_2CO_3, HCl and other reagents cannot be overdone, and the order of removing impurity cannot be changed.

Questions

(1) What is the basis for adding 40 mL of water to dissolve 10.0 g of salt? What is the effect of too much or too little water?

(2) How to remove the excessive precipitant $BaCl_2$, NaOH and Na_2CO_3 added during the experiment?

(3) Why can't the purified experimental solution be dried when it is concentrated?

(4) Why hydrochloric acid solution is added when testing SO_4^{2-}?

(5) In the purification of crude salt, can the solutions in step (1) and (2) be combined to be filtered?

(6) Why do solutions need to be heated to boiling after the addition of $BaCl_2$ and Na_2CO_3 during the purification process?

(7) What factors contribute to the overhigh yield of the product?

References

Institute of Inorganic Chemistry, Beijing Normal University. 2001. Inorganic Chemistry Experiment. 3rd ed. Beijing: Higher Education Press

School of Medicine and Chemical Engineering, Zhejiang Taizhou College. 2011. Basic Experiment I (Inorganic Chemistry Experiment). Hangzhou: Zhejiang University Press

Exp. 7 Determination of the Degree of Dissociation and Dissociation Constant of Acetic Acid

Objectives

(1) To understand the principle of determining the dissociation constant of a weak acid by pH method, and deepen the understanding of dissociation degree and dissociation constant.

(2) To master the titrimetric principle, titrimetric operation and judgement of titrimetric end point.

(3) To learn how to use the pH meter.

(4) To consolidate the operation and use of the pipette and volumetric flask.

Principles

A weak electrolyte is an electrolyte that is only partially ionized in an aqueous solution. When the weak electrolyte reaches the ionization equilibrium in the aqueous solution, the concentration of each substance in the solution can be determined and the dissociation equilibrium constant K_a can be obtained. As a monoprotic weak acid, acetic acid (CH_3COOH, abbreviated as HAc) has the following equilibrium in an aqueous solution:

$$HAc(aq) \rightleftharpoons H^+(aq) + Ac^-(aq)$$

Initial concentration (mol/L)	c	0	0
Equilibrium concentration (mol/L)	$c(1-\alpha)$	$c\alpha$	$c\alpha$

The expression of dissociation equilibrium constant K_a is:

$$K_a = \frac{[H^+][Ac^-]}{[HAc]} = \frac{[H^+]^2}{c-[H^+]} \quad (3.1)$$

Ionization degree:

$$\alpha = \frac{[H^+]}{c} \times 100\% \quad (3.2)$$

In the formula, c is the initial concentration of HAc solution; $[H^+]$, $[Ac^-]$ and $[HAc]$ are the equilibrium concentrations of H^+, Ac^- and HAc, respectively.

At a certain temperature, if the total concentration of HAc is known (the solution with unknown concentration can be titrated with NaOH standard solution, thus its concentration can be determined), we can determine the pH value of the solution by pH meter and calculate the concentration of H^+, then K_a and α of HAc at that temperature can be calculated with the above formulas. By determining K_a values of a series of HAc solution with different concentrations, the obtained mean value is the dissociation constant of HAc at that temperature.

In addition, the dissociation constant of HAc can also be determined by half-neutralization method. When NaOH is used to neutralize 50% of the acid in HAc solution, the system composition becomes a buffer solution composed of HAc-NaAc with an equal concentration. According to the formula (3.3) of the buffer solution, the pH value of the solution now is equal to pK_a.

$$pH = -\lg[H^+] \quad (3.3)$$

Experimental Items

Apparatuses and materials: pH meter, air dry oven, volumetric flask, transfer pipette, measuring pipette, beaker, thermometer, rubber suction bulb, filter paper.

Reagents: HAc standard solution (about 0.2 mol/L, accurate to 4 significant figures), NaOH standard solution (about 0.2 mol/L, accurate to 4 significant figures).

Procedures

1. Determination of the HAc Solution Concentration

Take phenolphthalein as an indicator, calibrate the accurate concentration of HAc solution by titration with NaOH standard solution of known concentration, and record the results in the Table 3.4.

Table 3.4   Data recording and processing of the HAc solution concentration   Room temperature____ ℃

Experiment Number		I	II	III
Concentration of NaOH solution/(mol/L)				
Volume of HAc solution/mL				
Volume of NaOH solution/mL				
Concentration of HAc solution/(mol/L)	Measured value			
	Average value			

2. Direct Determination of the pH Value of HAc Solution

1) Preparation of HAc Solution with Various Concentrations

Pipette 2.50 mL, 5.00 mL and 25.00 mL of HAc standard solution into 50 mL volumetric flasks respectively, dilute them with distilled water to the scale and shake well. Afterwards, calculate the concentration of each HAc solution. Then pour the above solutions and the HAc standard solution into 4 dry 100 mL beakers (numbered 1, 2, 3 and 4) respectively, which are used to test their pH values.

2) Determination of the pH Value of HAc Solution

Determine the pH values of the above HAc solutions by pH meter according to the sequence from the most dilute to the most concentrated, fill in the Table 3.5, calculate the K_a and α, and record the room temperature at that time.

Table 3.5   Data recording and processing of the pH value of HAc solution   Room temperature ____ ℃

Number	c_{HAc}/(mol/L)	pH	[H$^+$]/(mol/L)	α	Dissociation constant K_a	
					Measured value	Average value
1						
2						
3						
4						

Notes

(1) The pH meter is a precise instrument which should be used correctly. Please refer to the method of using

acidity meter for specific methods and matters needing attention.

(2) When using the standard buffer solution to adjust the pH meter, please avoid introducing water or impurities and recycle it after use.

(3) The K_a determined in this experiment is qualified within the range of $1.0\times10^{-5}\sim2.0\times10^{-5}$ (at 25℃, the literature value is 1.76×10^{-5}).

Questions

(1) According to the results of the experiment and calculation, explain how the dissociation degree of a weak electrolyte is affected by the concentration and ion composition.

(2) Why should the pH value of HAc solutions with different concentrations be determined according to the sequence from the most dilute to the most concentrated? Why should the room temperature be measured?

(3) Compare the measured value with the standard dissociation constant of HAc, and analyze the reason of the deviation.

(4) The given pH value has only 2 significant figures, so how many significant figures of $[H^+]$, K_a and α in the table should be reserved?

Exp. 8 Qualitative Identification of Some Essential Elements in Organisms

Objectives

To study how to qualitatively identify Ca, Fe, P and other elements in plants and how to process the plant samples.

Principles

An organism is a complicated organic system, which is formed by different types of compounds composed of multiple elements. It is generally believed that essential elements of life include more than 20 elements, such as carbon, hydrogen, oxygen, nitrogen, sulfur, phosphorus, chlorine, silicon, iodine, sodium, potassium, calcium, magnesium, iron and copper. The necessity, distribution and content of each element in the biological body vary greatly with the species, location, as well as organs or tissues of biology. Boron, for example, is a necessary element for the growth of some green plants and algae, but not for mammals. Calcium and phosphorus, the constituent element of hydroxyapatite in bone, are very high in animal bones, and at the same time, phosphorus is also an important constituent element of the nucleic acid. As a trace element, iron exists in hemoglobin, myoglobin, nitrogenase, chlorophyll and other biological macromolecules, and it is involved in a variety of important metabolic activities in the organism. The lack of these essential elements or trace elements will result in affecting and restricting the development and biological function of the organism, such as cramps and osteoporosis caused by calcium deficiency, anemia caused by iron deficiency and yellowing of plant leaves.

This experiment aims to qualitatively identify the calcium, iron and phosphorus elements in plant samples (such as the blade). The calcium, iron and phosphorus in the plant samples are converted into Ca^{2+}, Fe^{3+} and phosphate forms respectively by processing the plant samples, and then the characteristic reaction is used for identification. The mechanisms of reactions for identification are as follows:

$$Ca^{2+} + C_2O_4^{2-} = CaC_2O_4\downarrow \text{ (white)}$$
$$Fe^{3+} + [Fe(CN)_6]^{4-} = Fe_4[Fe(CN)_6]_3\downarrow \text{ (blue)}$$
$$Fe^{3+} + nSCN^- = [Fe(SCN)_n]^{(3-n)+} \text{ (red)}$$
$$HPO_4^{2-} + 12MoO_4^{2-} + 3NH_4^+ + 23H^+ = (NH_4)_3[P(Mo_{12}O_{40})] \cdot 6H_2O\downarrow \text{ (yellow)} + 6H_2O$$

The processing method of raw material in this experiment can also be applied to test whether the materials, such as bones and egg yolks, contain calcium, iron and phosphorus elements.

Experimental Items

Apparatuses and materials: porcelain crucible, crucible tongs, pipeclay triangle, tripod, alcohol burner, combustion spoon, funnel, beaker, test tube, tweezers, leaves (half green and half yellow), cotton, pH test paper.

Reagents: HNO_3 (concentrated, 6 mol/L, 0.1 mol/L), ammonia water (2 mol/L), KSCN solution, $(NH_4)_2MoO_4$ solution, $K_4[Fe(CN)_6]$ solution, $(NH_4)_2C_2O_4$ solution, red phosphorus, limestone.

Procedures

1. Raw-material Processing

Take 6~10 g of fresh leaves, and then heat and burn the leaves hold by tweezers or crucible tongs on the alcohol burner. Afterwards, crush the black ashes (carbide), collect them in a porcelain crucible and continue to heat the crucible until the ashing is complete (gray). Then weigh 0.3~0.5 g of grey powder, transfer it into a test tube and add 1 mL of concentrated HNO_3 to nitrate the sample. 3 min later, add 3 mL of water, filter it, and wash the precipitate with 1 mL of water. Finally, merge the filtrate for further use.

2. Qualitative Test

Divide the filtrate into four equivalent portions, add $(NH_4)_2MoO_4$ solution, KSCN solution and $K_4[Fe(CN)_6]$ solution to the former three portions respectively. In the fourth portion, the pH is adjusted to weakly alkaline with 2 mol/L $NH_3 \cdot H_2O$, then $(NH_4)_2C_2O_4$ solution is added. Observe and record the phenomena.

3. Controlled Trials

(1) Take a small amount of red phosphorus with a combustion spoon, heat the red phosphorus and then burn it in the gas bottle. Afterwards, add 2 mL of water to the gas bottle and shake the bottle gently. After white smoke being absorbed by water, transfer the solution to a test tube and

boil it. Then add 5 drops of 6 mol/L HNO_3 solution and $(NH_4)_2MoO_4$ solution, and observe the phenomenon. Use a glass rod to rub the inside wall of the test tube to promote the precipitation of yellow precipitate $((NH_4)_3[P(Mo_{12}O_{40})]\cdot 6H_2O)$.

(2) Place a small piece of cotton in a test tube, add 1 mL of 6 mol/L HNO_3 solution, and heat the test tube (avoid boiling and decomposition of nitric acid). Then add 3 mL of water, filter it, divide the filtrate into two parts, add KSCN solution and $K_4[Fe(CN)_6]$ solution respectively, and observe the phenomenon.

(3) Take a piece of $CaCO_3$ with the size of mung bean, add 2 mL of 0.1 mol/L HNO_3 solution, and adjust the solution to be weakly alkaline with $NH_3 \cdot H_2O$. Then add $(NH_4)_2C_2O_4$ solution, and observe the phenomenon.

Compare the above experimental phenomena and determine whether the sample contains calcium, iron and phosphorus elements or not.

Notes

(1) The ashing of leaves must be complete.

(2) The extent of heating should be controlled and the mouth of the test tube should not face people when cotton is heated and nitrated, which avoid the eruption of cotton from the test tube caused by boiling nitric acid to hurt others.

Questions

Some students didn't find the significant experimental phenomena. Some have even found the phenomenon of controlled trials were not as significant as those of leaf samples. Please explain the reasons.

Exp. 9   Determination of the Solubility Product of PbI_2

Objectives

(1) To understand how to use ion-exchange method to determine the solubility product of insoluble electrolytes.

(2) To learn the general use of ion exchange resins and practice basic operations of the acid-base titration.

Principles

At a certain temperature, insoluble strong electrolytes (such as PbI_2) have dissociation equilibrium in saturated solutions:

$$PbI_2(s) \rightleftharpoons Pb^{2+}(aq) + 2I^-(aq)$$

The solubility product constant of PbI_2 is expressed as: $K_{sp}=[Pb^{2+}][I^-]^2$, which varies with the temperature. The K_{sp} can be calculated by determining $[Pb^{2+}]$ or $[I^-]$ ($[Pb^{2+}]:[I^-]=1:2$) in PbI_2 saturated solution at a certain temperature.

Ion exchange resins are a kind of organic polymer compound and they contain exchangeable ions. They can be divided into cation exchange resins and anion exchange resins. The cation exchange resin contains acidic groups, for example, $R-SO_3^-H^+$ contains H^+ which can exchange with the cation in the solution. While the anion exchange resin contains basic groups, for example, $R-NH_3^+OH^-$ contains OH^- which can exchange with the anion in the solution. A strong acid-type cation exchange resin is used in this experiment. When a certain volume of saturated PbI_2 solution is exchanged through the resin, the H^+ on the resin is exchanged with Pb^{2+}. The exchange reaction can be expressed by the following formula:

$$2R\text{-}H^+ + Pb^{2+} \rightleftharpoons R_2\text{-}Pb^{2+} + 2H^+$$

Collect the effluent liquid after exchange, titrate the content of H^+ with standard alkali solution to obtain the concentration of PbI_2 saturated solution, and then calculate the solubility product constant.

The resin with Pb^{2+} can be washed by dilute nitric acid solution without Cl^-, and then the resin can be retransformed into the acid type (namely resin regeneration).

Experimental Items

Apparatuses and materials: chromatographic column, base burette, transfer pipette, measuring cylinder, small beaker, Erlenmeyer flask, funnel, funnel stand, iron stand, burette clamp, glass rod, thermometer, quantitative filter paper, pH test paper.

Reagents: strong acid-type cation exchange resin, sodium hydroxide standard solution (about 0.005 mol/L), nitric acid (1 mol/L), PbI_2 saturated solution, brominated thymol blue indicator.

Procedures

1. Transformation and Packing

Soak the cationic exchange resin in dilute acid solution (1 mol/L) for 24~48 h to transform it into hydrogen type completely. Take an appropriate volume of preactivated cation exchange resin with a beaker, remove the soaked acid, wash it 3~5 times with pure water, inject the resin dispersed in pure water into the column. The total height of the packing is about 20 cm. The amount of water in the column can be controlled by adjusting the cock. Add a small amount of pure water to wash the resin for several times until the pH of effluent liquid is equal to that of water, and then turn off the cock. In the above operations and the subsequent ion exchange process, the resin should always be immersed in the solution, in case the air bubble in the resin will affect the exchange efficiency of ions. If the resin bed appears bubbles, add distilled water to 2~3 cm higher than the resin surface and gently stir the resin with a glass rod to remove air bubbles.

2. Exchange and Washing

Filter 50~60 mL of PbI_2 saturated solution, absorb 25.00 mL of filtrate by a pipette and put it into the ion exchange column. Control the velocity of the fluid to 30~40 drops/min (if the velocity of the fluid is too fast, the iron exchange will be not complete). Use a clean Erlenmeyer flask to hold the effluent. When the PbI_2 solution on the upper surface of the resin column is close to flow out, use pure water to clean the exchange resin in batches until the pH of effluent liquid is equal to the pH of water. Collect all effluent liquid and cleaning mixture in the Erlenmeyer flask. The used resin is recovered to the designated container, and it can be reused after being washed with pure water, nitric acid and pure water.

3. Titration

Add 2 drops of brominated thymol blue indicator to the solution in the Erlenmeyer flask, titrate with 0.0050 mol/L NaOH standard solution to the end point (the color of the solution turns bright blue from yellow). Write down the volume of consumed NaOH standard solution and calculate the solubility product of PbI_2.

Data Recording and Processing

Temperature_____ °C
Volume of PbI_2 saturated solution_____ mL
Concentration of NaOH standard solution_____ mol/L
Location of NaOH standard solution in burette: before titration____ mL, after titration____ mL
Volume of NaOH standard solution_____ mL
Concentration of $[Pb^{2+}]$ in PbI_2 saturated solution_____ mol/L
K_{sp} of PbI_2_____

Notes

(1) The funnel and beaker used for filtering and transferring the PbI_2 saturated solution must be dried.

(2) The measured value of K_{sp} in this experiment is qualified within the range of $10^{-9} \sim 10^{-8}$.

Questions

(1) Why should we control the flow rate of the liquid during the ion exchange process to ensure that it is not too fast?

(2) Why should the liquid level always be kept higher than the ion exchange resin layer?

(3) Please analyse the reason of errors when comparing with $K_{sp}(PbI_2) = 7.1 \times 10^{-9}$ (298 K) in the literature.

(4) What effect does incomplete resin transformation have on the experimental results?

(5) What will be the impact on the experimental results if there is a small loss of the effluent liquid during the process of exchange and washing?

Exp. 10 Determination of Chemical Reaction Rate and Activation Energy

Objectives

(1) To understand the effects of the concentration, temperature and catalyst on chemical reaction rates.

(2) To determine the reaction rate of $(NH_4)_2S_2O_8$ reacting with KI, and calculate the reaction order, reaction rate constant and activation energy.

(3) To be familiar with the use of measuring cylinders and stopwatches.

Principles

$(NH_4)_2S_2O_8$ can react with KI in an aqueous solution:

$$S_2O_8^{2-} + 3I^- = 2SO_4^{2-} + I_3^- (aq) \tag{3.4}$$

The rate equation can be expressed as:

$$v = k c_{S_2O_8^{2-}}^m c_{I^-}^n$$

In the formula, v is the instantaneous speed rate; $c_{S_2O_8^{2-}}$ and c_{I^-} are the initial concentrations of $S_2O_8^{2-}$ and I^-, respectively; k is the reaction rate constant; m and n are the reaction orders of reactant $S_2O_8^{2-}$ and I^-, respectively; $(m + n)$ is the overall reaction order.

Only the average reaction rate during a period time can be determined in the experiment:

$$\bar{v} = \frac{-\Delta c_{S_2O_8^{2-}}}{\Delta t}$$

That is to say, it is the change of $S_2O_8^{2-}$ concentration during the time Δt.

In order to measure the change of $S_2O_8^{2-}$ concentration during the time Δt, it is necessary to add a certain volume of $Na_2S_2O_3$ solution and starch solution with known concentrations simultaneously to the mixed solution of $(NH_4)_2S_2O_8$ and KI solutions. Thus, there is another reaction during the reaction (3.4):

$$2S_2O_3^{2-} + I_3^- = S_4O_6^{2-} + 3I^- \tag{3.5}$$

The reaction (3.5) almost completes instantaneously, which is much faster than reaction (3.4). Thus, I_3^- generated in reaction (3.4) immediately reacts with $S_2O_3^{2-}$ to produce colorless $S_4O_6^{2-}$ and I^-, and no characteristic blue color of iodine and starch is observed. When $S_2O_3^{2-}$ is consumed completely, reaction (3.5) does not occur and reaction (3.4) is still ongoing, then the I_3^- generated in reaction (3.4) will appear blue when reacting with the starch.

From the beginning of the reaction to the appearing time of blue (Δt), the change of $S_2O_3^{2-}$ concentration is:

$$\Delta c = -[c_{S_2O_3^{2-}(\text{end})} - c_{S_2O_3^{2-}(\text{beginning})}] = c_{S_2O_3^{2-}(\text{beginning})}$$

According to the reaction (3.4) and (3.5), we can get:

$$-\Delta c_{S_2O_8^{2-}} = \frac{c_{S_2O_3^{2-}\text{(beginning)}}}{2}$$

By changing the initial concentrations of $S_2O_8^{2-}$ and I^-, the different time intervals of consuming a certain amount of $S_2O_8^{2-}$ ($-\Delta c_{S_2O_8^{2-}}$) are determined. The initial rate of reactants with different initial concentrations is calculated, and the rate equation and reaction rate constant are determined.

Experimental Items

Apparatuses and materials: thermostatic water bath, beaker, measuring cylinder, stopwatch, thermometer, glass rod or electromagnetic stirrer, coordinate paper.

Reagents: $(NH_4)_2S_2O_8$ (0.20 mol/L), KI (0.20 mol/L), $Na_2S_2O_3$ (0.01 mol/L), KNO_3 (0.20 mol/L), $(NH_4)_2SO_4$ (0.20 mol/L), $Cu(NO_3)_2$ (0.02 mol/L), starch solution (5 g/L).

Procedures

1. Effects of Concentration on the Chemical Reaction Rate

The experiment of number I in Table 3.6 is carried out at room temperature. Measure 20.0 mL of KI solution, 8.0 mL of $Na_2S_2O_3$ solution and 2.0 mL of starch solution with three measuring cylinders, add them to a beaker, and mix them well. Then use another measuring cylinder to take 20.0 mL of $(NH_4)_2S_2O_8$ solution and pour it quickly into the above mixture. At the same time, start the stopwatch, stir the solution continuously and observe the phenomenon carefully. Stop the stopwatch immediately when the solution just appears blue, and then record the reaction time and room temperature.

Use the same method to do number II, III, IV and V experiments in Table 3.6.

Table 3.6　Effects of concentration on the chemical reaction rate　Room temperature＿＿＿　℃

	Experiment number	I	II	III	IV	V
Reagent dosage/mL	0.20 mol/L $(NH_4)_2S_2O_8$	20.0	10.0	5.0	20.0	20.0
	0.20 mol/L KI	20.0	20.0	20.0	10.0	5.0
	0.01 mol/L $Na_2S_2O_3$	8.0	8.0	8.0	8.0	8.0
	5 g/L starch solution	2.0	2.0	2.0	2.0	2.0
	0.20 mol/L KNO_3	0	0	0	10.0	15.0
	0.20 mol/L $(NH_4)_2SO_4$	0	10.0	15.0	0	0
Initial concentration/(mol/L)	$(NH_4)_2S_2O_8$					
	KI					
	$Na_2S_2O_3$					
Reaction time Δt / s						
Concentration change of $S_2O_8^{2-}$ /(mol/L)						
Reaction speed rate v /[mol/(L·s)]						

2. Effects of Temperature on the Chemical Reaction Rate

According to experiment IV in Table 3.7, place a beaker containing a mixed solution of KI, $Na_2S_2O_3$, KNO_3 and starch and a small beaker containing $(NH_4)_2S_2O_8$ solution in an ice-water bath for cooling. When their temperature is 10 ℃ lower than room temperature, add $(NH_4)_2S_2O_8$ solution to the mixed solution rapidly, start the stopwatch and keep stirring. When the solution just appears blue, record the reaction time. This experiment is recorded as number VI.

Use the same method to do another experiment, but place the beaker in a hot water bath and heat it to 10 ℃ higher than room temperature. This experiment is recorded as number VII.

The data of experiment VI, VII and IV are recorded in Table 3.7 for comparison.

Table 3.7 Effects of temperature on the chemical reaction rate

Experiment number	VI	IV	VII
Reaction temperature t /℃			
Reaction time Δt / s			
Reaction rate v /[mol/(L·s)]			

3. Effects of Catalysts on the Chemical Reaction Rate

Add KI, $Na_2S_2O_3$, KNO_3 and starch solutions into a 150 mL beaker according to the experiment IV in Table 3.7, and then add 2 drops of $Cu(NO_3)_2$ solution. After stirring the mixed solution well, quickly add $(NH_4)_2S_2O_8$ solution, stir and start the stopwatch. What conclusions can be drawn from the qualitative comparison of the reaction rate of this experiment with that of experiment IV in Table 3.7?

Data Recording and Processing

1. Calculation of the Reaction Order and Reaction Rate Constant

$$v = k c_{S_2O_8^{2-}}^m c_{I^-}^n$$

Take a logarithm on both sides:

$$\lg v = m \lg c_{S_2O_8^{2-}} + n \lg c_{I^-} + \lg k$$

When c_{I^-} is invariant (experiment I, II and III), a straight line is drawn according to $\lg v$ and $\lg c_{S_2O_8^{2-}}$, and the slope m can be obtained. Similarly, when $c_{S_2O_8^{2-}}$ is invariant (experiment I, IV and V), a straight line is drawn according to $\lg v$ and $\lg c_{I^-}$, and the slope n can be obtained. The reaction order is ($m + n$). The reaction rate constant k can be obtained according to Table 3.8.

Table 3.8 Calculation of the reaction rate constant

Experiment number	I	II	III	IV	V
lg v					
lg $c_{S_2O_8^{2-}}$					
lg c_{I^-}					
m					
n					
Reaction rate constant k					

2. Calculation of the Reaction Activation Energy

The relationship between the reaction rate constant k and the reaction temperature T is generally as follows:

$$\lg k = A - \frac{E_a}{2.30RT}$$

In the formula, E_a is the activation energy of the reaction; R is the molar gas constant; and T is the thermodynamic temperature. Measure the k value at different temperatures, draw a straight line according to lg k and $1/T$, and the slope is $-\dfrac{E_a}{2.30R}$, then the activation energy E_a can be obtained. Fill the data in Table 3.9.

Table 3.9 Calculation of the activation energy

Experiment number	VI	VII	IV
Reaction rate constant k			
lg k			
$1/T$			
Activation energy E_a			

Notes

(1) KI, $Na_2S_2O_3$, starch, KNO_3 and $(NH_4)_2SO_4$ solutions can be measured by the same measuring cylinder, but $(NH_4)_2S_2O_8$ solution must be measured by a separate measuring cylinder.

(2) After mixing KI, $Na_2S_2O_3$, starch, KNO_3 and $(NH_4)_2S_2O_8$ solutions well, pour $(NH_4)_2S_2O_8$ solution quickly into the above mixture, and at the same time, start the stopwatch and stir the solution.

(3) Stop the stopwatch immediately when the solution just appears blue.

(4) In the experiments of heating or cooling, the mixed solution of KI, $Na_2S_2O_3$, starch, KNO_3 and $(NH_4)_2SO_4$, and the solution of $(NH_4)_2S_2O_8$ should be heated or cooled separately.

(5) The thermometers must be used separately.

(6) Keep the temperature constant.

(7) The KI solution should be colorless and transparent, rather than being light yellow which contains precipitated I_2. The $(NH_4)_2S_2O_8$ solution should be freshly prepared because it's easy to break down over time. If the pH of freshly prepared $(NH_4)_2S_2O_8$ solution is less than 3, it indicates that the reagent has been decomposed and it is not suitable for the experiment. If the reagent contains a small amount of impurities such as Cu^{2+} and Fe^{3+}, these impurities will have catalytic effects on the reaction, and add a few drops of 0.10 mol/L EDTA solution if necessary.

(8) In the experiment of "effects of temperature on the chemical reaction rate", if the room temperature is below 10℃, the temperature conditions can be changed to three cases: room temperature, 10℃ higher than room temperature and 20℃ higher than room temperature.

(9) Experimental guidance:

(i) The use of a stopwatch.

(ii) The order and speed of adding $(NH_4)_2S_2O_8$ solution and KI solution.

(iii) The timing operation in the experiment.

(iv) Students can obtain the reaction order by drawing with the guidance of teachers so as to calculate the reaction rate constant and activation energy.

(10) The error of the measured value of activation energy in this experiment is no more than 10% (literature value: 51.8 kJ/mol).

Questions

(1) What effect do the following operations have on the experiment?

(i) The measuring cylinders are not separate.

(ii) $(NH_4)_2S_2O_8$ solution is added first, and finally KI solution is added.

(iii) $(NH_4)_2S_2O_8$ solution is slowly added to the mixed solution of KI and other solutions.

(2) Why do we add KNO_3 solution or $(NH_4)_2SO_4$ solution in the experiment II, III, IV and V, respectively?

(3) What should be paid attention to in the timing operation of each experiment?

(4) If the reaction rate is expressed by the concentration change of I^- or I_3^- instead of $S_2O_8^{2-}$, is the reaction rate constant k the same?

(5) How to determine the reaction order of chemical reactions? Explain it with the result of this experiment.

(6) Calculate the reaction activation energy by Arrhenius formula. Compare it with the value obtained by drawing.

(7) It is known that A (g) ⟶ B (l) is a second-order reaction, and its data are as follows. Try to calculate the reaction rate constant k.

p_A / kPa	40	26.6	19.1	13.3
t / s	0	250	500	1000

Exp. 11 Preparation of Potassium Permanganate by Solid Alkali Fusion Oxidation Method

Objectives

(1) To learn the basic principles and operations of preparing potassium permanganate from manganese dioxide by alkali fusion method.

(2) To be familiar with basic operations such as melting and leaching.

(3) To consolidate the basic operations such as filtration, crystallization and recrystallization.

(4) To grasp the transformation relationships among various oxidation states of manganese.

Principles

Manganese dioxide, the main component of pyrolusite, can be oxidized to potassium manganite when it co-melts with alkali in the presence of strong oxidants (such as potassium chlorate):

$$3MnO_2 + KClO_3 + 6KOH \xrightarrow{melting} 3K_2MnO_4 + KCl + 3H_2O$$

After soaking the frit with water, MnO_4^{2-} in the aqueous solution is unstable to have disproportionated reaction with the decrease of alkalinity of the solution. Generally, the tendency of disproportionation and the reaction rate is relatively smaller in weakly basic or nearly neutral media. However, in weakly acid media, the disproportionation of MnO_4^{2-} is easy to happen, and MnO_4^- and MnO_2 are produced. If CO_2 gas is injected into potassium manganite solution, the following reaction may take place:

$$3K_2MnO_4 + 2CO_2 = 2KMnO_4 + MnO_2\downarrow + 2K_2CO_3$$

After manganese dioxide is removed by vacuum filtration, the solution is concentrated and the acicular dark purple potassium permanganate crystal can be precipitated.

Experimental Items

Apparatuses and materials: iron crucible, Kipp's apparatus, crucible tongs, pipeclay triangle, Büchner funnel, oven, evaporating dish, beaker, watch glass, instruments for titration analysis, number-eight iron wire.

Reagents: manganese dioxide, potassium hydroxide, potassium chlorate, calcium carbonate, sodium sulfite, oxalic acid, industrial hydrochloric acid, analytical sulfuric acid.

Procedures

1. Melting and Oxidation of Manganese Dioxide

Weigh 2.5 g of potassium chlorate solid and 5.2 g of potassium hydroxide solid, add them into an iron crucible and mix the material evenly with an iron bar. Place the iron crucible on the pipeclay triangle, clamp it with crucible tongs and heat it with a small fire while stirring it with an iron bar. When the mixture is melted, add 3.0 g of manganese dioxide solid into the iron crucible carefully by several times to prevent the splashing of sparks. Stir it vigorously to prevent it from caking or sticking to the crucible wall with the increasing viscosity. After drying the reactants, raise the temperature and heat it with strong heat for 5 min to obtain the dark green manganese acid potassium melt, and then mash it as much as you can with an iron bar.

2. Leaching

After the iron crucible containing the melt cools down, mash the frit with an iron bar as much as possible, place the crucible sidelong into a 250 mL beaker filled with 100 mL of distilled water, heat it over a small flame until the melt is completely dissolved, and then take out the crucible with crucible tongs carefully.

3. Disproportionation of Potassium Manganate

Carbon dioxide gas is injected into the leaching liquid under high temperature until the disproportionation of potassium manganate is complete (the solution can be dipped on the filter paper with a glass rod. If the filter paper has only purple red trace and no green trace, it indicates that the disproportionation of potassium manganate is complete. At that time pH is between 10 and 11), and then let it stand for a moment and filter it.

4. Evaporation and Crystallization of Filtrate

Pour the filtrate into the evaporating dish, evaporate and concentrate it until the potassium permanganate crystal film starts to appear on the surface, cool the crystal naturally, then drain the potassium permanganate crystal by vacuum filtration.

5. Drying the Potassium Permanganate Crystal

Transfer the crystal to the watch glass with known quality and separate it with a glass rod. Put it into the oven (80℃ is advisable, and it should not exceed 240℃) to dry for 0.5 h, weigh it after cooling and calculate the yield.

6. Purity Analysis

The laboratory is equipped with primary standards of oxalic acid and sulfuric acid. Design an analysis scheme to determine the content of potassium permanganate in the prepared product.

7. Transformation among Various Oxidation States of Manganese (optional)

Use the self-made potassium permanganate crystal to design the experiment as shown in the Fig. 3.3 to realize the transformation among various oxidation states of manganese. Write down the experimental steps and related ionic equations.

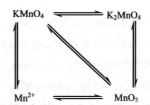

Fig. 3.3 Transformation among various oxidation states of manganese

Notes

(1) Reference data (Table 3.10).

Table 3.10 The solubility of some compounds at different temperatures Unit: g/100 g H_2O

Compounds	t/°C										
	0	10	20	30	40	50	60	70	80	90	100
KCl	27.6	31.0	34.0	37.0	40.0	42.6	45.5	48.3	51.1	54.0	56.7
$K_2CO_3 \cdot 2H_2O$	51.3	52	52.5	53.2	53.9	54.8	55.9	57.1	58.3	59.6	60.9
$KMnO_4$	2.83	4.4	6.4	9.0	12.56	16.89	22.2	—	—	—	—

(2) The excess CO_2 will result in lower pH and a large amount of $KHCO_3$. The solubility of $KHCO_3$ is much lower than that of K_2CO_3, and $KHCO_3$ will be precipitated together with $KMnO_4$ during the concentration process.

(3) The installation and debugging of Kipp's apparatus.

(4) The dissolution, filtration and crystallization of solids.

Questions

(1) Why do we use an iron crucible instead of a porcelain crucible to prepare potassium manganate?

(2) Why do we use an iron bar instead of a glass rod to stir in the experiment?

(3) In the third procedure, the solution should be stirred with a glass rod instead of an iron bar. Why?

(4) Summarize the construction and application method of Kipp's apparatus.

(5) Can HCl instead of CO_2 be used to realize K_2MnO_4 disproportionation?

(6) What is the maximum conversion rate of potassium permanganate obtained from the disproportionation of potassium manganate in an acidic medium? What other experimental methods can be adopted to increase the conversion rate of potassium manganate?

Exp. 12 Preparation of Ammonium Ferrous Sulfate Crystals

Objectives

(1) To consolidate the experiment operations such as heating in the water bath, dissolution, crystallization and vacuum filtration.

(2) To learn to use the difference of the solubility to prepare ammonium ferrous sulfate.

(3) To master the properties of ferrous sulfate and ammonium ferrous sulfate.

(4) To understand how to use visual colorimetry to test the product purity.

Principles

Ammonium ferrous sulfate is also known as Mohr's salt [$(NH_4)_2SO_4 \cdot FeSO_4 \cdot 6H_2O$]. It is a light blue-green crystal or powder, which is soluble in water but not in ethanol. Mohr's salt is not easily oxidized in air and is more stable than ferrous sulfate, so it is often used to prepare the standard solution of ferrous ions in the quantitative analysis. Ammonium ferrous sulfate is also an important chemical raw material with extensive uses.

In this experiment, excess iron is dissolved in dilute sulfuric acid to get ferrous sulfate:

$$Fe + H_2SO_4 \Longrightarrow FeSO_4 + H_2\uparrow$$

Ferrous sulfate reacts with the same mole of ammonium sulfate to obtain ammonium ferrous sulfate:

$$FeSO_4 + (NH_4)_2SO_4 + 6H_2O \Longrightarrow (NH_4)_2SO_4 \cdot FeSO_4 \cdot 6H_2O$$

The solubility of ammonium ferrous sulfate in the water is lower than that of its each component at 0~60℃, according to Table 3.11. Therefore, the crystallized Mohr's salt can be prepared from heating, concentrating and cooling the mixed solution of concentrated ferrous sulfate and ammonium sulfate.

Table 3.11 The solubility of the three salts   Unit: g/100 g H_2O

Temperature/℃	$FeSO_4 \cdot 7H_2O$	$(NH_4)_2SO_4$	$(NH_4)_2SO_4 \cdot FeSO_4 \cdot 6H_2O$
10	40.0	73.0	18.12
20	48.0	75.4	21.2
30	60.0	78.0	24.5
40	73.3	81.0	27.9
50	—	84.5	31.3
70	—	91.9	38.5

Experimental Items

Apparatuses and materials: platform scale, Erlenmeyer flask, Büchner funnel, suction flask, colorimetric tube, comparator block, filter paper.

Reagents: iron powder, $(NH_4)_2SO_4(s)$, $(NH_4)_2SO_4 \cdot FeSO_4 \cdot 6H_2O(s)$, H_2SO_4 (3 mol/L, concentrated), HCl (3 mol/L), KSCN (w is 0.25), ethanol (95%).

Procedures

1. Preparation of Ammonium Ferrous Sulfate

1) Preparation of Ferrous Sulfate

Weigh 2.0 g of iron powder with a platform scale, put it into a 100 mL Erlenmeyer flask, then add 10 mL of 3 mol/L H_2SO_4. The Erlenmeyer flask is heated in a water bath (about 60℃) in the fume hood for about 30 min until air bubbles do not appear any more. Add a little water appropriately to keep the original volume. Add 1 mL of H_2SO_4 (3 mol/L) after the reaction. Then filter the solution under reduced pressure while it is hot, wash the precipitate with a little hot water and transfer the filtrate into an evaporating dish.

2) Preparation of the Saturated Solution of Ammonium Sulfate

According to the quantity of $FeSO_4$ in the solution and the relation n [$(NH_4)_2SO_4$] : n ($FeSO_4$) = 1:1, weigh the required $(NH_4)_2SO_4$ solid to prepare $(NH_4)_2SO_4$ saturated solution at the corresponding temperature.

3) Preparation of Ammonium Ferrous Sulfate

Add $(NH_4)_2SO_4$ saturated solution to the $FeSO_4$ solution (the pH should be close to 1. If the pH is high, add a few drops of dilute H_2SO_4 to adjust), evaporate the solution in a water bath and concentrate it until the crystal film appears on the surface. Afterwards, put it aside for slow cooling to obtain the ammonium ferrous sulfate crystal. Then remove the mother liquid by vacuum filtration, wash the crystal with 95% ethanol and remove the liquid as much as possible. Finally, transfer the crystal to the watch glass to dry for a while, observe its color and shape, weigh it and calculate the yield.

2. Quantitative Analysis of Fe (III)

1) Preparation of Fe (III) Standard Solution (prepared by the preparation room)

Weigh 0.5030 g of $Fe_2(SO_4)_3·9H_2O$, dissolve it in a little water, add 2.5 mL of concentrated H_2SO_4, transfer the solution into a 1000 mL volumetric flask and add water to the scale. The concentration of Fe^{3+} in this solution is 0.1000 g/L (0.1000 mg/mL).

2) Preparation of the Standard Level

Take 0.50 mL of Fe (III) standard solution into a 25 mL colorimetric tube, add 2 mL of 3 mol/L HCl solution and 1 mL of KSCN solution (w is 0.25). Then add oxygen-free water to the scale to obtain a standard solution which is equivalent to the first grade reagent (the concentration of Fe^{3+} is 0.05 mg/g, that is, w is 0.005%).

Also, take 1.00 mL and 2.00 mL of Fe (III) standard solution respectively to obtain a standard solution which is equivalent to the second grade and third grade reagent (the concentration of Fe^{3+} is 0.10 mg/g and 0.20 mg/g respectively, that is, w is 0.01% and 0.02% respectively).

3) Determination of the Product Level

Weigh 1.0 g of product and dissolve it with 15 mL of oxygen-free water in a 25 mL colorimetric tube. After the product is dissolved completely, add 2 mL of 3 mol/L HCl solution and 1 mL of KSCN solution (w is 0.25), dilute the solution to scale with oxygen-free water and shake it well. Then compare the color with the standard level to determine the product level.

Notes

(1) In the process of water bath heating, the Erlenmeyer flask should be oscillated continuously to prevent

caking.

(2) The reaction time should not be too long.

(3) Patiently stirring is necessary during the preparation of ammonium sulfate saturated solution.

(4) The ammonium ferrous sulfate crystal should be washed by ethanol after the suction filtration.

Questions

(1) Why should iron powder be excessive during the preparation of ferrous sulfate?

(2) Can the final product $(NH_4)_2SO_4 \cdot FeSO_4 \cdot 6H_2O$ be directly heated and dried in the evaporating dish, and why?

(3) The yield of ammonium ferrous sulfate production should be calculated on the base of the amount of H_2SO_4. Why?

(4) Why must the solution be acidic during the preparation of ammonium ferrous sulfate crystals? Is stirring necessary for evaporation and concentration?

(5) How to prevent ferrous ions from being hydrolyzed and oxidized?

Exp. 13 Purification and Solubility Determination of Potassium Nitrate

Objectives

(1) To learn how to determine the solubility of potassium nitrate roughly and draw the solubility curve.

(2) To understand the relationship between the solubility of potassium nitrate and temperature.

(3) To use the solubility curve for purification of crude potassium nitrate.

(4) To learn how to qualitatively test chloride ions.

Principles

The solubility of salts in water refers to their concentrations in a saturated aqueous solution at a given temperature, generally expressed as the mass (g) of the dissolved salt per 100 g water. In general, the solubility of potassium nitrate is determined by adding a certain amount of potassium nitrate to a certain amount of water. Heat the mixed solution to dissolve completely, and then cool the solution while stirring continuously until the crystal just appears. The solution concentration is the solubility at this temperature, which is expressed as g/100 g H_2O.

The solubility of different compounds is affected by different temperatures: ①the solubility of the most compounds is greatly affected by temperature, such as KNO_3, and the solubility increases with the increasing temperature; ②the solubility of some compounds changes a little with temperature, such as NaCl; ③the solubility of some compounds decreases with the increasing temperature, such as $Ca(OH)_2$. Therefore, the difference of solubility can be used to separate one compound from others through changing the temperature. For example, heat the mixed solution of KNO_3 and NaCl to a certain temperature, and then cool it to purify KNO_3 through the solubility variation with the temperature.

The solubility of KNO_3 is measured mainly by crystallization and solute mass. The crystallization method includes the rising temperature method and the falling temperature method. The falling temperature method is as follows: dissolve a certain amount of solute in a certain amount of water to make an unsaturated solution. Measure the solution temperature at the moment of crystal appearing after the solution cools down slowly, and calculate the solubility. This method needs to keep the mass conservation of solute and add solvent to reach the saturation state in the cooling process. The experimental equipment is shown in Fig. 3.4.

The solute mass method is to weigh the KNO_3 saturated solution in an evaporating dish with known exact mass at a given temperature. Heat the saturated solution, evaporate all the moisture, and then weigh the evaporating dish, so as to calculate the solubility of KNO_3 at this temperature.

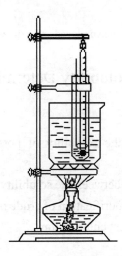

Fig. 3.4 The experimental equipment of measuring the solubility of KNO_3 with crystallization method

Experimental Items

Apparatuses and materials: platform scale, Büchner funnel, suction flask, beaker, large test tube, thermometer, evaporating dish, tripod, pipeclay triangle, glass rod, alcohol burner, filter paper.

Reagents: KNO_3 (AR, s), crude KNO_3 (mixed with NaCl, s), HNO_3 (5 mol/L), $AgNO_3$ (0.1 mol/L).

Procedures

1. Purification of Potassium Nitrate

(1) Dissolution and evaporation.

Weigh 10.0 g of crude KNO_3 into a hard test tube, add 35 mL of water and heat the test tube in a glycerine bath. After the salt is dissolved completely, continue to heat and evaporate the solution to obtain 2/3 of its original volume, and then filter the solution while it is hot, the crystal is_____. Then filter it with a hot funnel, place the filtrate into a small beaker for natural cooling and the precipitated crystal is_____. KNO_3 crystal is obtained by vacuum filtration and dried in a water bath.

(2) Recrystallization of crude products.

0.1~0.2 g of crude product is reserved for purity test and the remaining crude product is dissolved in distilled water according to the mass ratio (crude product : distilled water=2 : 1). Heat and stir the solution until the crystal is dissolved completely. After the solution cools to room temperature, filter it and dry it in a water bath to obtain KNO_3 crystal with high purity.

(3) Purity test.

2. Testing the Purity of KNO_3 Qualitatively

Add 0.1 g of crude product and the primary recrystallization product into 2 small test tubes respectively, add 2 mL of distilled water to dissolve them and add 1 drop of 5 mol/L HNO_3 solution for acidification, then add 2 drops of 0.1 mol/L $AgNO_3$ solution. Test the total chlorine content in the sample according to the reagent grade standard.

Take 1.0 g of sample (accurate to 0.01 g), heat it to 400℃ for decomposition and calcinate it for 15 min at 700℃. After cooling, dissolve it in distilled water and dilute it to 25 mL, add 2 mL of 5 mol/L HNO_3 and 0.1 mol/L $AgNO_3$ solutions, shake them well, and place the solution for 10 min. The turbidity shall not be greater than the standard.

3. Determining the Solubility of KNO_3 by the Rising Temperature Method

(1) Weigh 3.5 g, 1.5 g, 1.5 g, 2.0 g and 2.5 g of KNO_3 respectively with a platform scale.

(2) Add 10 mL of distilled water and 3.5 g of KNO_3 into a large test tube, and then heat and stir the solution to dissolve it completely in a water bath.

(3) Take the test tube out of the water bath, insert a clean thermometer, and use a glass rod to stir gently and rub the tube wall. At the same time, observe the thermometer reading and record the reading as soon as the crystal appears.

(4) Put the test tube into the water bath and heat it again to dissolve the crystal completely. Then repeat the operations of (3) and measure the temperature at which the crystal appears.

(5) Add 1.5 g of KNO_3 into the test tube [the total amount of KNO_3 in the test tube is 3.5 + 1.5 = 5.0 (g)], and repeat the above operations of (3) and (4).

(6) Repeat the same operations of (5), and measure the temperatures (the thermometer does not need to be washed) at which the crystal appears when adding 1.5 g, 2.0 g and 2.5 g of KNO_3 successively (the total amount of KNO_3 in the test tube is 6.5 g, 8.5 g and 11.0 g successively).

(7) According to the obtained data, take the temperature as the abscissa and the solubility as the ordinate to draw the dissolution curve. The relation between solubility and temperature can be clearly achieved from the figure.

Data Recording and Processing

The data recording and processing for determining the solubility of KNO_3 by the rising temperature method are shown in Table 3.12.

Table 3.12 Data recording and processing for determining the solubility of KNO_3 by the rising temperature method

Number		1	2	3	4	5
Mass of KNO_3/g						
Mass of H_2O/g						
Temperature of the crystal appearing in the solution/℃	t_1					
	t_2					
	Average value					
Solubility of KNO_3 in water/(g/100 g H_2O)						

Notes

(1) The test tube should be dry and clean.

(2) Use a paper slot to feed the solid into the bottom of the test tube.

(3) Use a thermometer to stir gently, carefully and slowly.

(4) When the room temperature is not low enough, the test tube can be soaked in cold water to be cooled. During the cooling process, use a glass rod to stir the solution gently and rub the tube wall to prevent the solution from being oversaturated.

(5) When recording the thermometer reading, grasp the moment at which the crystal just appears to avoid increasing the error.

Questions

(1) Do the amount of KNO_3 and the volume of water need to be accurate when measuring the solubility? What kind of glassware is suitable for the measuring device?

(2) What is the effect of water evaporation on the experiment when measuring the solubility? What measures should be taken?

(3) Is the stir required in the process of dissolution and crystallization?

(4) What steps should be taken to purify crude KNO_3, such as dissolution, evaporation and crystallization?

Exp. 14   Oxidation-reduction Reaction and Equilibrium

Objectives

(1) To master the direction, products and rate induced by the electrode potential, the acidity and alkalinity of the medium and the concentration of the compounds of redox couple.

(2) To learn to assemble the primary cell and understand the electromotive force of the chemical cell.

Principles

There are two categories of chemical reactions: one is that no electron transfer exists among reactants in the process of reaction, such as the neutralization reaction, precipitation reaction and coordination reaction; the other is that electron transfer exists among reactants in the process of reaction, which is the redox reaction.

The nature of redox reaction is electron transfer, namely the gain and loss or the deviation of electrons. The oxidation number will decrease after oxidants gain electrons, while the oxidation number will increase after reductants lose electrons. The relative value of electrode potential, which is composed of the oxidation state and reduction state of a substance, can be used to measure the capacity for the electronic gain and loss or the redox capacity of the substance. The smaller the electrode potential value (φ) of the redox couple is, the stronger the reducibility of the reduction state is; the bigger the electrode potential value (φ) of the redox couple is, the stronger the oxidizability of the oxidation state is. Therefore, the value of electrode potential can help us judge the direction of a redox reaction.

The concentration change of the oxidation state or the reduction state will affect the electrode potential, which will further affect the rate and product of the redox reaction. In particular, the presence of a precipitant or complexing agent, which can greatly reduce the concentration of an ion in the solution, can even change the direction of the reaction. The pH of the medium will affect the electrode potential when H^+ or OH^- is involved in the electrode reaction, which will further affect the direction and product of the redox reaction.

Experimental Items

Apparatuses and materials: test tube, centrifuge tube, voltmeter (or pH meter), watch glass, U-tube, electrode (zinc foil, copper foil), paper clip, red litmus paper (or phenolphthalein test paper), wire, sand paper, filter paper.

Reagents: HAc (6 mol/L), H_2SO_4 (1 mol/L), NaOH (6 mol/L), $NH_3·H_2O$ (concentrated), $ZnSO_4$ (1 mol/L), $CuSO_4$ (0.01 mol/L, 1 mol/L), KI (0.1 mol/L), KBr (0.1 mol/L), $KMnO_4$ (0.01 mol/L), Na_2SO_3 (0.1 mol/L), $FeCl_3$ (0.1 mol/L), $Fe_2(SO_4)_3$ (0.1 mol/L), $FeSO_4$ (0.1 mol/L), H_2O_2 (3%), KIO_3 (0.1 mol/L), bromine water, iodine water (0.1 mol/L), chlorine water (saturated), KCl (saturated), CCl_4, phenolphthalein indicator, starch solution (0.4%), agar, ammonium fluoride.

Procedures

1. Assembly of the Primary Cell and Effects of Concentration on the Electrode Potential

1) Assembly of the Primary Cell

Add about 20 mL of 1 mol/L $ZnSO_4$ solution into a small beaker and insert a zinc foil (sanding). Add about 20 mL of 1 mol/L CuSO₄ solution into another small beaker and insert a copper foil (sanding). Connect two beakers with a salt bridge to form a primary cell. Then connect the zinc and copper foils to the negative and positive electrodes of a voltmeter (or pH meter) respectively with a wire, and measure the potential difference between the two electrodes (Fig. 3.5).

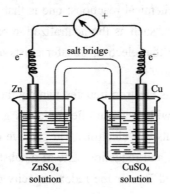

Fig. 3.5　Cu-Zn primary cell

2) Effects of Concentration on Electrode Potential

Add concentrated aqueous ammonia to the ZnSO₄ solution in the above-mentioned primary cell until the resulting precipitate is dissolved, and a colorless solution is produced:

$$Zn^{2+} + 4NH_3 \rightleftharpoons [Zn(NH_3)_4]^{2+}$$

Measure the potential difference and observe the change.

Add concentrated aqueous ammonia to the CuSO₄ solution until the resulting precipitate is dissolved, and a dark blue solution is produced:

$$Cu^{2+} + 4NH_3 \rightleftharpoons [Cu(NH_3)_4]^{2+}$$

Measure the potential difference and observe the change. Use Nernst equation to explain the experimental phenomena.

3) Design of the Concentration Cell

Design and measure the electromotive force of the following concentration cells, and compare the experimental value with the calculated value:

(−) Cu | CuSO₄ (0.01 mol/L) ‖ CuSO₄ (1 mol/L) | Cu(+)

2. Redox Reaction and Electrode Potential

(1) Add 0.5 mL of 0.1 mol/L KI solution into the test tube, and then add 2 drops of 0.1 mol/L FeCl₃ solution. After shaking well, add 0.5 mL of CCl₄ and oscillate the solution fully, then observe the color change of the CCl₄ layer.

(2) Replace the KI solution with the KBr solution to repeat the same experiment, and observe the phenomenon.

(3) Add about 0.5 mL of 0.1 mol/L FeSO₄ solution into two test tubes respectively, one

contains 3 drops of iodine water and the other contains 3 drops of bromine water. After shaking well, add 0.5 mL of CCl_4, oscillate the solutions fully, and then observe the color change of the CCl_4 layer.

According to the above experimental results, compare the electrode potentials of Br_2/Br^-, I_2/I^- and Fe^{3+}/Fe^{2+} qualitatively.

3. Effects of Acidity and Concentration on the Redox Reaction

1) Effects of Acidity

Add 0.5 mL of 1 mol/L H_2SO_4 solution, 0.5 mL of distilled water and 0.5 mL of 6 mol/L NaOH solution into three test tubes which all contain 0.5 mL of 0.1 mol/L Na_2SO_3 solution. After mixing them well, add 2 drops of 0.01 mol/L $KMnO_4$ solution into the three test tubes, observe the change and write down the reaction equations.

Add 10 drops of 0.1 mol/L KI solution and 2 drops of 0.1 mol/L KIO_3 solution into a test tube, then add a few drops of starch solution. Observe the color change after mixing the solution. Then add 2~3 drops of 1 mol/L H_2SO_4 solution to acidify the mixed solution and observe the change. Finally, add 2~3 drops of 6 mol/L NaOH solution to make the mixture alkaline and observe what changes. Write down the reaction equations.

According to the above experimental results, explain the effects of acidity on the redox reaction.

2) Effects of Concentration

Add 0.5 mL of 0.1 mol/L KI solution into the test tube which contains 0.5 mL of H_2O, 0.5 mL of CCl_4 and 0.5 mL of 0.1 mol/L $Fe_2(SO_4)_3$ solutions, oscillate the test tube and observe the color of the CCl_4 layer.

Add 0.5 mL of 0.1 mol/L KI solution to the test tube which contains 0.5 mL of CCl_4, 0.5 mL of 1 mol/L $FeSO_4$ and 0.5 mL of 0.1 mol/L $Fe_2(SO_4)_3$ solutions, oscillate the test tube, observe the color of the CCl_4 layer and compare it with the color in the previous experiment.

Add a little NH_4F solid into the test tube in step 1), oscillate it and observe the color of the CCl_4 layer.

According to the above experimental results, explain the effect of concentration on the redox reaction.

3) Effects of Acidity on the Redox Reaction Rate

Add 0.5 mL of 1 mol/L H_2SO_4 solution and 0.5 mL of 6 mol/L HAc solution into two test tubes containing 0.5 mL of 0.1 mol/L KBr solution respectively. Then add two drops of 0.01 mol/L $KMnO_4$ solution, and observe the fading speed of the fuchsia. Write down the reaction equations.

4. Oxidation-reduction Quality of the Substance Whose Oxidation Number is in the Middle

(1) Add 0.5 mL of 0.1 mol/L KI solution and 2 drops of 1 mol/L H_2SO_4 solution into a test tube, then add 2 drops of 3% H_2O_2 and observe the color change of the solution.

(2) Add 2 drops of 0.01 mol/L $KMnO_4$ solution into a test tube, then add 3 drops of 1 mol/L

H_2SO_4 solution, and after shaking well, add 2 drops of 3 % H_2O_2 and observe the color change of the solution.

Notes

(1) The use of voltmeters.

(2) The use of salt bridges.

Questions

(1) Does acidity affect the electrode potentials of Cl_2/Cl^-, Br_2/Br^-, I_2/I^-, Fe^{3+}/Fe^{2+}, Cu^{2+}/Cu and Zn^{2+}/Zn, and why?

(2) In which medium is the oxidability of $KMnO_4$ solution strongest, and why?

(3) Why is H_2O_2 both oxidative and reductive? Try to explain from the electrode potential.

Chapter 4　Elemental Inorganic Chemistry Experiments

Exp. 15　s-Block Elements (Alkali Metals and Alkaline Earth Metals)

Objective

(1) To compare the activity of alkali metals with that of alkaline earth metals.

(2) To compare the solubility of alkaline earth metal hydroxides with that of alkaline earth metal salts.

(3) To compare the similarity between lithium salts and magnesium salts.

(4) To learn operations of the flame reaction and be familiar with safety precautions for using potassium and sodium.

Experimental Items

Apparatuses and materials: evaporating dish, test tube, centrifuge, beaker, tweezers, sand paper, nickel wire, filter paper, spot plate, cobalt glass.

Reagents: potassium, sodium, magnesium, calcium, Na_2CO_3 (0.1 mol/L, 1 mol/L), LiCl (1 mol/L), NaCl (1 mol/L), NaF (1 mol/L), Na_2HPO_4 (1 mol/L), KCl (1 mol/L), $CaCl_2$ (1 mol/L), $SrCl_2$ (1 mol/L), $BaCl_2$ (1 mol/L), K_2CrO_4 (1 mol/L), $MgCl_2$ (0.5 mol/L, 1 mol/L), Na_2SO_4 (1 mol/L), $NaHCO_3$ (1 mol/L), $K[Sb(OH)_6]$ (saturated), $NaHC_4H_4O_6$ (saturated), $KMnO_4$ (0.01 mol/L), NH_4Cl (saturated), $(NH_4)_2C_2O_4$ (saturated), $(NH_4)_2CO_3$ (0.5 mol/L), Na_3PO_4 (0.5 mol/L), $(NH_4)_2SO_4$ (saturated), NaOH (2 mol/L, freshly prepared), $NH_3 \cdot H_2O$ (1 mol/L, 2 mol/L, freshly prepared), H_2SO_4 (1 mol/L), HCl (2 mol/L, 6 mol/L), HAc (2 mol/L), HNO_3 (concentrated).

Procedures

1. Activity of Alkali Metals and Alkaline Earth Metals

(1) Take a small piece of sodium, use the filter paper to absorb the kerosene on its surface, and then put the sodium at an evaporating dish to heat immediately. When the sodium begins burning, stop heating. Afterwards, observe the phenomenon and write down the reaction equation. As the product cooled, slightly crush it with a glass rod and transfer it into a test tube, and then add a small amount of water to dissolve it. Then cool it and observe if there is gas being released. Later, acidize the solution with 1 mol/L H_2SO_4, and then add a drop of 0.01 mol/L $KMnO_4$ solution. Finally, observe the phenomenon and write down the reaction equation.

(2) Take a small strip of magnesium, remove the oxide on its surface with sand paper and ignite the magnesium strip. Then observe the phenomenon and write down the reaction equation.

(3) Reaction of metals with water.

Fetch a small piece (the size of mung beans) of sodium and potassium respectively, use the filter paper to absorb the kerosene on their surfaces, and then put them respectively into two beakers containing water. For safety, cover the beakers with inverted funnels immediately. Afterwards, observe the phenomena, test the acidity and basicity of the solution and write down the reaction equations.

Take two small strips of magnesium and remove the oxide on their surfaces with sand paper. Afterwards, put them into two test tubes respectively, one contains cold water and the other contains hot water. Then contrast one phenomenon to the other and write down the reaction equations.

Put a small piece of calcium into a test tube and add a small amount of water. Afterwards, observe the phenomenon and test the acidity and basicity of the solution, and then write down the reaction equation.

2. Solubility of Alkaline Earth Metal Hydroxides

Take 1 mol/L $MgCl_2$, $CaCl_2$, $BaCl_2$ as well as the freshly prepared 2 mol/L NaOH solution and 2 mol/L $NH_3 \cdot H_2O$ as reagents to design the series experiments and explain the order according to the value of solubility.

3. Indissolvable Salts of Alkali Metals and Alkaline Earth Metals

1) Sparingly Soluble Salts of Alkali Metals

Test the reaction of a small amount of 1 mol/L LiCl solution with 1 mol/L NaF solution and Na_2HPO_4 solution respectively, then observe the phenomena (If necessary, heat the test tube gently to observe) and write down the reaction equations.

Add a small amount of $KSb(OH)_6$ saturated solution into a test tube containing a small amount of 1 mol/L NaCl solution, and then let it stand for several minutes. If there is no crystal being extracted, rub the inner wall of the test tube with a glass rod. Then observe the precipitation of crystalline $NaSb(OH)_6$.

Add 1 mL of saturated $NaHC_4H_4O_6$ solution into a test tube containing a small amount of 1 mol/L KCl solution, and observe the precipitation of $KHC_4H_4C_6$ crystal which is a kind of indissolvable salt.

2) Indissolvable Salts of Alkaline Earth Metals

(1) Carbonates: add 1 mol/L Na_2CO_3 solution to a small amount of 1 mol/L $MgCl_2$ solution, $CaCl_2$ solution and $BaCl_2$ solution respectively, and then centrifuge the precipitates. Afterwards, test the reaction of the precipitates with 2 mol/L HAc and HCl respectively, and observe whether the precipitates are dissolved.

Take another three tubes and add a small amount of 1 mol/L $MgCl_2$ solution, $CaCl_2$ solution and $BaCl_2$ solution respectively. Afterwards, add 1~2 drops of NH_4Cl saturated solution, 2 drops of 1 mol/L $NH_3 \cdot H_2O$ and 2 drops of 0.5 mol/L $(NH_4)_2CO_3$ to each test tube, and observe whether there is a precipitate produced. Write down the reaction equations and explain the experimental phenomena.

(2) Oxalates: add $(NH_4)_2C_2O_4$ saturated solution to 1 mol/L $MgCl_2$ solution, $CaCl_2$ solution and $BaCl_2$ solution, respectively. Afterwards, centrifuge the produced precipitates, and test the reaction of the precipitates with HAc and HCl respectively, then observe the phenomena and write down the reaction equations.

(3) Chromates: add 1 mol/L K_2CrO_4 solution to 1 mol/L $MgCl_2$ solution, $CaCl_2$ solution and $BaCl_2$ solution respectively, and observe whether there is a precipitate produced. Centrifuge the precipitates, test the reaction of the precipitates with 2 mol/L HAc and HCl respectively, then observe the phenomena and write down the reaction equations.

(4) Sulfates: add 1 mol/L Na_2SO_4 solution to 1 mol/L $MgCl_2$ solution, $CaCl_2$ solution and $BaCl_2$ solution respectively, and then observe whether there is a precipitate produced. Centrifuge the precipitates, test the solubility of precipitates in $(NH_4)_2SO_4$ saturated solution and concentrated HNO_3 respectively, then explain the phenomena and write down the reaction equations.

4. Similarity of Lithium Salts and Magnesium Salts

(1) Add 1 mol/L NaF solution to 1 mol/L LiCl solution and $MgCl_2$ solution, respectively, then observe the phenomena and write down the reaction equations.

(2) Test the reaction of 1 mol/L LiCl solution with 0.1 mol/L Na_2CO_3 solution and the reaction of 0.5 mol/L $MgCl_2$ solution with 1 mol/L $NaHCO_3$ solution. Observe the phenomena, and then write down the reaction equations.

(3) Add 0.5 mol/L Na_3PO_4 solution to 1 mol/L LiCl solution and 0.5 mol/L $MgCl_2$ solution, respectively. Observe the phenomena, and then write down the reaction equations.

Summarize the similarity of lithium salts and magnesium salts from the above experiments and explain it.

5. Flame Reaction

Take a glass rod with nickel wire, and bend the tip of the nickel wire into a small ring. Afterwards, dip 6 mol/L HCl solution with the glass rod which is then burnt in an oxidizing flame to almost colorless, and repeat this process two to three times for cleaning the nickel wire. After that, dip 1 mol/L LiCl solution, NaCl solution, KCl solution, $CaCl_2$ solution, $BaCl_2$ solution and $SrCl_2$ solution with the clean nickel wire which is then burnt in an oxidizing flame, respectively. Observe the flame color and record the results. Note that the flame color of the potassium ion should be observed by a piece of cobalt glass.

Questions

(1) If there is a fire accident caused by magnesium in the lab, can we use water or a carbon dioxide fire extinguisher to put the fire out, and why?

(2) How to separate Ca^{2+} and Mg^{2+}? Why are $Mg(OH)_2$ and $MgCO_3$ soluble in the NH_4Cl saturated solution?

Exp. 16　p-Block Nonmetallic Elements (I) (Halogen, O, S)

Objectives

(1) To master the oxidizability of halogens and reducibility of halogen ions.

(2) To master the oxidizability of hypohalites and halates.

(3) To learn the disproportionation reaction of halogens.

(4) To learn the properties of some metal halides.

(5) To master the reducibility of hydrogen sulfide and thiosulfates and the strong oxidizability of persulfates.

Experimental Items

Apparatuses and materials: test tube, potassium iodide-starch test paper, pH test paper.

Reagents: $KClO_3(s)$, $Na_2S_2O_3(s)$, $K_2S_2O_8(s)$, KBr (0.1 mol/L), NaF (0.1 mol/L), NaCl (0.1 mol/L), KI (0.01 mol/L, 0.1 mol/L), K_2CrO_4 (0.1 mol/L), $K_2Cr_2O_7$ (0.1 mol/L), $KMnO_4$ (0.1 mol/L), H_2O_2 (3%), $MnSO_4$ (0.002 mol/L, 0.1 mol/L), $AgNO_3$ (0.1 mol/L), chlorine water, iodine water, NaOH (2 mol/L, 40%), KOH (2 mol/L), H_2SO_4 (1 mol/L, 3 mol/L, 6 mol/L), HCl (2 mol/L), H_2S aqueous solution (saturated), HNO_3 (2 mol/L), $NH_3 \cdot H_2O$ (2 mol/L), carbon tetrachloride, diethyl ether, ethanol, fuchsin solution, starch solution.

Procedures

1. Oxidization Sequence of Chloride Water for a Mixture of Bromine Ions and Iodine Ions

Add 10 drops of 0.1 mol/L KBr solution, 2 drops of 0.01 mol/L KI solution and a dropper of CCl_4 in a test tube. After that, add chlorine water dropwise, then observe the color change of the CCl_4 layer and write down the reaction equations.

2. Oxidizability of Oxysalts of Chlorine

(1) Transfer 2 mol/L KOH to 4 mL of chlorine water with a dropper until the solution is weakly alkaline (tested by pH test paper), and divide the solution into four test tubes. Add 2 mol/L HCl solution to the first portion and choose suitable test paper to test the gaseous product. In the other three portions, fuchsine solution, a mixture of 3～4 drops of 0.1 mol/L KI solution and 1 drop of starch solution, and 0.1 mol/L $MnSO_4$ solution are added dropwise, respectively. Write down the reaction equations.

(2) Fetch a $KClO_3$ crystal of mung bean size, dissolve it with 1～2 mL of water, add a dropper of CCl_4 and 3～4 drops of 0.1 mol/L KI solution, then shake the test tube to observe the changes of organic phase and aqueous phase. Afterwards, add 2～3 drops of 6 mol/L H_2SO_4 solution to acidify the solution, what's happened to the solution? Write down the reaction equations.

3. Comparison of the Solubility of Halides

Drop 0.1 mol/L $AgNO_3$ solution into four test tubes containing 10 drops of 0.1 mol/L NaF solution, NaCl solution, KBr solution and KI solution, respectively, and obtain silver halide precipitates. After centrifugal separation, test the reaction of the precipitates with 2 mol/L HNO_3 solution, 2 mol/L $NH_3 \cdot H_2O$ solution and 0.5 mol/L $Na_2S_2O_3$ solution, respectively, and observe whether the precipitates are dissolved. Write down the reaction equations, explain the difference of solubility between fluoride and other halides, and summarize the transformation rules.

4. Light Sensitivity of Silver Halides

Evenly coat a piece of filter paper with the produced AgCl precipitate, and then put a key on the filter paper. After it being in light for about 10 min, fetch the key and the outline of the key can be seen clearly.

5. Identification and Properties of Hydrogen Peroxide

1) Identification of Hydrogen Peroxide

Add 10 drops of 3% H_2O_2 solution, 10 drops of diethyl ether, 3~4 drops of 1 mol/L H_2SO_4 solution (for acidification) and 2~3 drops of 0.1 mol/L K_2CrO_4 solution into a test tube successively, oscillate the test tube to observe the color changes of the water layer and diethyl ether layer, and then write down the reaction equations.

2) Acidity

Add 10 drops of 40% NaOH solution, 10 drops of 3% H_2O_2 solution and 10 drops of ethanol into a test tube, oscillate the test tube and observe the phenomenon, and write down the reaction equation.

3) Effects of the Acidity and Basicity of the Medium on the Oxidation-reduction Quality of H_2O_2

Add 2~3 drops of 2 mol/L NaOH solution to 10 drops of 3% H_2O_2 solution, and then add 5~6 drops of 0.1 mol/L $MnSO_4$ solution, observe the phenomenon and write down the reaction equations. After it has been rested for a while, decant the supernatant, add 2~3 drops of 3 mol/L H_2SO_4 solution, and then add 3% H_2O_2 solution dropwise. How is the phenomenon different? Write down the reaction equations and explain the change.

6. Properties of Thiosulfate

Dissolve a $Na_2S_2O_3 \cdot 5H_2O$ crystal of soybean size in about 3 mL of water, divide the obtained solution into four test tubes and carry out the following experiments:

(1) Add 2 mol/L HCl solution to the first test tube, observe the phenomenon and write down the reaction equation.

(2) Add iodine water to the second test tube, observe the phenomenon, and write down the reaction equation.

(3) Add chlorine water to the third test tube, try to identify the existence of SO_4^{2-} after the reaction and write down the reaction equation.

(4) Add $Na_2S_2O_3$ solution in the fourth test tube dropwise into a test tube containing 4 drops of 0.1 mol/L $AgNO_3$ solution, observe the phenomenon carefully, and write down the reaction equation.

7. Oxidizability of Potassium Persulfate

Add about 5 mL of 1 mol/L H_2SO_4 solution and $K_2S_2O_8$ solid of soybean size into the test tube containing 2 drops of 0.002 mol/L $MnSO_4$ solution, mix them well, and divide the solution into two test tubes. Afterwards, add 2 drops of 0.1 mol/L $AgNO_3$ solution into one of them, and then heat the two test tubes in a water bath together. Observe the color changes of the solutions, compare the phenomena and find the reasons, and write down the reaction equations.

8. The Reducibility of Hydrogen Sulfide

(1) Add 2 drops of 1 mol/L H_2SO_4 solution to the test tube containing 1 drop of 0.1 mol/L $KMnO_4$ solution. After acidification, add H_2S saturated solution, then observe the phenomenon and write down the reaction equations.

(2) Add 2 drops of 1 mol/L H_2SO_4 solution to the test tube containing 1 drop of 0.1 mol/L $K_2Cr_2O_7$ solution. After acidification, add H_2S saturated solution, then observe the phenomenon and write down the reaction equations.

Questions

(1) Why do laboratories often use solid peroxydisulfuric acid salts, instead of the pre-prepared solution?

(2) Design some experiments to indicate the oxidizability of NaClO and $KClO_3$ is strong or weak.

(3) According to the experimental results, compare ① the oxidizability of $S_2O_8^{2-}$ and MnO_4^-; ② the reducibility of $S_2O_3^{2-}$ and I^-.

Exp. 17 p-Block Nonmetallic Elements (II) (N, P, C, Si, B)

Objectives

(1) To master the main properties of ammonia, ammonium salts, nitric acid and nitrates.

(2) To master the main properties of phosphates.

(3) To master the properties of the nitrous acid and its salts.

(4) To master the adsorption effect of activated carbon and the mutual transformation conditions of carbon dioxide, carbonates and bicarbonates in an aqueous solution.

(5) To master the properties of silicates and borates.

Experimental Items

Apparatuses and materials: crucible, watch glass, test tube, beaker, thermometer, pH test paper.

Reagents: H_3BO_3 (s), $CaCl_2 \cdot 6H_2O$ (s), $CuSO_4 \cdot 5H_2O$ (s), $Co(NO_3)_2 \cdot 6H_2O$ (s), $NiSO_4 \cdot 7H_2O$ (s), $MnSO_4$ (s), $ZnSO_4 \cdot 7H_2O$ (s), $FeCl_3 \cdot 6H_2O$ (s), activated carbon, indigo solution, ethanol, $NaNO_2$ (0.1 mol/L, saturated), $KMnO_4$ (0.01 mol/L), Na_3PO_4 (0.1 mol/L), Na_2HPO_4 (0.1 mol/L), NaH_2PO_4 (0.1 mol/L), $Na_4P_2O_7$ (0.1 mol/L), $NaPO_3$ (0.1 mol/L), $CaCl_2$ (0.1 mol/L), $NaHCO_3$ (0.5 mol/L), $Pb(NO_3)_2$ (0.001 mol/L, 0.1 mol/L), $FeCl_3$ (0.2 mol/L), Na_2SiO_3 (20%), K_2CrO_4 (0.1 mol/L), Na_2CO_3 (1 mol/L), $MgCl_2$ (0.1 mol/L), $CuSO_4$ (0.1 mol/L), $AgNO_3$ (0.1 mol/L), KI (0.1 mol/L), $NaOH$ (6 mol/L), H_2SO_4 (1 mol/L, 3 mol/L, concentrated), $NH_3 \cdot H_2O$ (2 mol/L, concentrated), HCl (2 mol/L, concentrated), glycerin.

Procedures

1. Ammonia and Its Addition Reaction

Add 4~5 drops of concentrated aqueous ammonia into a crucible, cover the crucible with a beaker whose inner wall is moistened with concentrated hydrochloric acid, then observe the phenomenon and write down the reaction equation.

2. Detection of Ammonium Salts (air chamber method)

Add 4~5 drops of ammonium salt solution into the center of a watch glass and attach a small piece of wet pH test paper to the center of the other. Afterwards, add 6 mol/L NaOH solution to the ammonium salt solution until the solution is alkaline, and cover the first watch glass containing ammonium salt solution with the second watch glass to form an air chamber. Then heat the air chamber in a water bath and observe the color change of the pH test paper.

3. Formation and Decomposition of Nitrous Acid

Add 1 mL of $NaNO_2$ saturated solution and 1 mL of 3 mol/L H_2SO_4 solution into two test tubes respectively, cool the solutions with ice water for 2 min, then mix them well and observe the phenomenon. After being put aside for some time, what's happened to the solution? Write down the reaction equation.

4. Oxidation-reduction Quality of Nitrous Acid

1) Oxidizability of Nitrous Acid

Take 4 drops of 0.1 mol/L KI solution and acidify it with 2 drops of 1 mol/L H_2SO_4 solution. After acidification, add 0.1 mol/L $NaNO_2$ solution and observe the phenomenon and the product color. Then heat the test tube to lukewarm, what's happened to the solution? Write down the reaction equations.

2) Reducibility of Nitrous Acid

Take 1 drop of 0.01 mol/L $KMnO_4$ solution and acidify it with 2 drops of 1 mol/L H_2SO_4 solution. After acidification, add 0.1 mol/L $NaNO_2$ solution, then observe the phenomenon and write down the reaction equations.

5. Properties and Solubility of Phosphates

1) Properties of Phosphates

Test the pH values of orthophosphate, pyrophosphate and metaphosphate solutions with pH test paper.

Test the pH values of Na_3PO_4, Na_2HPO_4 and NaH_2PO_4 solutions with the same concentration of 0.1 mol/L with pH test paper. Afterwards, add 0.5 mL of 0.1 mol/L Na_3PO_4, Na_2HPO_4 and NaH_2PO_4 solutions into three test tubes, respectively, and then add moderate 0.1 mol/L $AgNO_3$ solution to each solution and observe whether there is a precipitate produced. What's happened to the pH values of the solutions? Please give your explanations.

2) Solubility of Phosphates

Add 0.1 mol/L $CaCl_2$ solution to Na_3PO_4, Na_2HPO_4 and NaH_2PO_4 solutions with the same concentration of 0.1 mol/L, respectively, and observe whether there is a precipitate produced. Afterwards, add 2 mol/L $NH_3 \cdot H_2O$ solution and observe the change, and then add 2 mol/L HCl solution and observe the change. Try to give your explanations about the changes and write down the reaction equations.

6. Adsorption Effect of Activated Carbon

(1) The adsorption effect on colored substances in the solution: add a small spoonful of activated carbon to 2 mL of indigo solution and oscillate the test tube. Afterwards, filter the solution to remove the activated carbon and observe the color change of the solution.

(2) The adsorption effect on inorganic ions: add a few drops of 0.1 mol/L K_2CrO_4 solution to 0.001 mol/L $Pb(NO_3)_2$ solution, observe the generation of yellow $PbCrO_4$ precipitate. Then add about 2 mL of 0.001 mol/L $Pb(NO_3)_2$ solution and a small spoon of activated carbon to another test tube, oscillate the test tube, and filter the solution to remove the activated carbon. After filtration, add a few drops of 0.1 mol/L K_2CrO_4 solution to the clear solution, and then observe and explain the phenomenon.

7. Reactions of Some Metal Ions with Carbonate

Add 1 mol/L Na_2CO_3 solution to 0.2 mol/L $FeCl_3$ solution and 0.1 mol/L $MgCl_2$ solution, $Pb(NO_3)_2$ solution, $CuSO_4$ solution, respectively, and then observe the phenomena. Afterwards, add 0.5 mol/L $NaHCO_3$ solution into another four test tubes containing these four solutions, respectively, observe the phenomena, determine the reactants by preliminary calculation, and write down the reaction equations.

8. Properties of Boron

(1) Dissolve a small amount of boric acid crystal (the size of mung bean) in about 2 mL of water (it can be heated gently for better dissolution), and measure the pH after the solution cools to room temperature. Afterwards, add 4~5 drops of glycerin to boric acid solution and measure the pH again. Please write down the reaction equation and give your explanations.

(2) Identification reaction of boric acid: place a small amount of boric acid crystal (the size of mung bean) in an evaporating dish, add 0.5 mL of ethanol and a few drops of concentrated H_2SO_4, mix them well and ignite them, and then observe the color of the flame and write down the reaction equation.

9. Formation of Indissolvable Silicates — "Garden in Water"

Add about 30 mL of Na_2SiO_3 solution (20%) into a beaker of 50 mL. Then place a grain of solid $CaCl_2$, $CuSO_4$, $Co(NO_3)_2$, $NiSO_4$, $MnSO_4$, $ZnSO_4$ and $FeCl_3$ respectively into the beaker, and keep some space between the solids and remember their locations. Place the beaker aside for one hour and then observe the phenomenon.

Questions

(1) In chemical reactions, why are nitric acid and hydrochloric acid not to be chosen for acidification generally?

(2) Boric acid is a kind of weak acid, but why does the acidity of boric acid solution increase after adding glycerol?

(3) Why can glassware with a frosted joint be used to store acid liquor rather than alkali liquor in the laboratory? Why should the containers containing sodium silicate be washed immediately after the experiment?

(4) How to distinguish among sodium carbonate, sodium silicate and sodium borate?

(5) Can the carbon dioxide extinguisher be used to extinguish the flame of magnesium metal, and why?

Exp. 18 p-Block Metallic Elements (Al, Sn, Pb, Sb, Bi)

Objectives

(1) To master the chemical properties of aluminum.

(2) To master the acidity and basicity of the hydroxides of Sn, Pb, Sb and Bi and the oxidation-reduction quality of different oxidation states of them.

Experimental Items

Apparatuses and materials: centrifuge.

Reagents: Al (s), PbO_2 (s), $K[Sb(OH)_6]$ (s), $NaBiO_3$ (s), $(NH_4)_2SO_4$ (saturated), $Al_2(SO_4)_3$ (1 mol/L), $FeCl_3$ (0.1 mol/L), $NaNO_3$ (0.5 mol/L), $SnCl_2$ (0.1 mol/L), $HgCl_2$ (0.1 mol/L), $Bi(NO_3)_3$ (0.1 mol/L), $MnSO_4$ (0.1 mol/L), $KMnO_4$ (0.1 mol/L), $SbCl_3$ (0.1 mol/L), $BiCl_3$ (0.1 mol/L), $AsCl_3$ (0.1 mol/L), KI (0.1 mol/L), K_2CrO_4 (0.1 mol/L), $Pb(NO_3)_2$ (0.1 mol/L), $NaHCO_3$ (saturated),

NaAc (saturated), $BaCl_2$ (0.1 mol/L), NaOH (2 mol/L, 6 mol/L, 40%), H_2SO_4 (1 mol/L, 2 mol/L, 3 mol/L), HCl (2 mol/L, concentrated), HNO_3 (6 mol/L), HAc (6 mol/L), CCl_4.

Procedures

1. Properties of Elementary Aluminum

After using sand paper to remove the surface oxide film of an aluminum sheet, test the reactions of it with:①hot water, ②cold water, ③2 mol/L HCl solution, ④2 mol/L NaOH solution, ⑤0.5 mol/L $NaNO_3$ solution in the adequate 40% NaOH solution, respectively, and confirm the generation of NH_3 in the product of reaction ⑤. Write down the reaction equations and have a brief summary about the properties of aluminum.

2. Reaction of Aluminum Salt into Alum

Add 1 mL of $(NH_4)_2SO_4$ saturated solution to 1 mL of 1 mol/L $Al_2(SO_4)_3$ solution, let it stand for a while, and observe the phenomenon. If the solution is clear, you can slightly rub the wall of the test tube. Observe the phenomenon and write down the reaction equation.

3. Reducibility of Sn (II)

(1) Add 0.1 mol/L $SnCl_2$ solution to 0.1 mol/L $FeCl_3$ solution, observe the phenomenon and write down the reaction equation. Try to add 1 drop of KSCN solution to test whether Fe^{3+} exists in the solution or not.

(2) Add 0.1 mol/L $SnCl_2$ solution to 0.1 mol/L $HgCl_2$ solution, observe the phenomenon, and write down the reaction equation.

(3) Add two drops of 0.1 mol/L $Bi(NO_3)_3$ solution to 0.1 mol/L $SnCl_2$ solution, observe the phenomenon, and write down the reaction equation.

Compare the reducibility of Sn (II) with that of Fe (II) and Hg (I) respectively through the above experiments.

4. Oxidizability of Pb (IV)

(1) Put a little PbO_2(s) into a test tube, add concentrated HCl solution dropwise, observe the phenomenon, and write down the reaction equation.

(2) Add 3 mol/L H_2SO_4 solution into a test tube containing a little PbO_2(s). After acidification, add one drop of 0.1 mol/L $MnSO_4$ solution, and then heat it in a water bath. Observe the phenomenon and write down the reaction equations.

5. Reducibility of Sb (III) and Bi (III)

(1) Add 2~3 drops of 0.1 mol/L $KMnO_4$ to 5 mL of 40% KOH solution and obtain K_2MnO_4 solution. Divide the solution into two parts, and add 0.1 mol/L $SbCl_3$ solution and $BiCl_3$ solution to the two parts respectively then observe the phenomenon and write down the reaction equation.

(2) After preparing $[Ag(NH_3)_2]^+$ solution in two test tubes, add a small amount of Na_3AsO_3 solution (home-made), Na_3SbO_3 solution (home-made) and 0.1 mol/L $Bi(NO_3)_3$ solution respectively. Heat the test tube gently, then observe the phenomenon and write down the reaction equation.

(3) Add 0.1 mol/L $AsCl_3$ and $SbCl_3$ solution into two test tubes respectively, and then add $NaHCO_3$ saturated solution to them until the solutions are weakly acidic. Add iodine water dropwise, then observe the phenomenon and write down the reaction equation.

(4) Take a small amount of 0.1 mol/L $Bi(NO_3)_3$ solution, add 6 mol/L NaOH solution until there is a white precipitate produced, then add chlorine water (or bromine water), observe the phenomenon and write down the reaction equation.

Explain the reductibility of Sb (III) and Bi (III) through the above experiments.

6. Oxidizability of Sb (V) and Bi (V)

(1) Add a small amount of water into two test tubes, one contains $K[Sb(OH)_6]$ (s) and the other contains $NaBiO_3$ (s), and then add dilute acid to acidify the solution (what kind of acid?). After acidification, add a small amount of 0.1 mol/L KI solution and carbon tetrachloride, observe the phenomena, and write down the reaction equations.

(2) Add two drops of 0.1 mol/L $MnSO_4$ solution into two test tubes, and then add 2 mol/L H_2SO_4 solution to acidify the solutions. After acidification, add a small amount of $K[Sb(OH)_6]$(s) and $NaBiO_3$(s) respectively in the two test tubes, observe the phenomena, and write down the reaction equations.

Explain the oxidizability of Sb (V) and Bi (V) through the above experiments.

7. Insoluble Substances

1) Halides

Add several drops of 0.1 mol/L $Pb(NO_3)_2$ solution to a small amount of water, and then add a few drops of 2 mol/L HCl solution and observe the phenomenon. Afterwards, heat the solution to see what happens, and then cool it and observe the new phenomenon. Try to give your explanations.

Add concentrated hydrochloric acid to 0.1 mol/L $Pb(NO_3)_2$ solution and observe the phenomenon. Afterwards, take a small amount of white precipitation, continue to add concentrated hydrochloric acid and observe the phenomenon, then dilute the solution with some water and observe the change. Write down the reaction equations.

Take a few drops of 0.1 mol/L $Pb(NO_3)_2$ solution, dilute the solution with a small amount of water, and then add 1~2 drops of 0.1 mol/L KI solution and observe the phenomenon. Afterwards, test the dissolution capability of the precipitate in hot water.

2) Oxyacid Salts of Lead

(1) Chromates: add several drops of 0.1 mol/L K_2CrO_4 solution to a small amount of 0.1 mol/L $Pb(NO_3)_2$ solution, and observe the formation of the precipitate $PbCrO_4$. Afterwards, test the dissolution capability of the precipitate in 6 mol/L HNO_3 solution, 6 mol/L NaOH solution, 6 mol/L HAc solution

and NaAc saturated solution, and then write down the reaction equations. Then use 0.1 mol/L $BaCl_2$ solution instead of $Pb(NO_3)_2$ solution to repeat the above experiment, observe the similarities and differences between two experimental phenomena, and write down the reaction equations.

(2) Sulfates: observe the color of the precipitate generated by the reaction of 0.1 mol/L $Pb(NO_3)_2$ with 1 mol/L H_2SO_4 solution. Afterwards, test the reaction of the precipitate in 2 mol/L NaOH solution and NaAc saturated solution, and then write down the reaction equations. Then use 0.1 mol/L $BaCl_2$ solution instead of $Pb(NO_3)_2$ solutions to repeat the above experiment, observe the phenomena and write down the reaction equations.

Questions

(1) Explain why the changes of the oxidation-reduction quality of Sn and Pb are different.

(2) Design experimental schemes to separate and identify the mixed solution of $SbCl_3$ and $Bi(NO_3)_3$.

Exp. 19　Properties of Element Compounds in the d-Block (I)

Objectives

(1) To master the acidity and basicity of certain hydroxides of elements in the d-block.

(2) To master the oxidation-reduction quality of the variable valence states of certain compounds of elements in the d-block.

(3) To understand the hydrolysis of certain metal ions of elements in the d-block.

Experimental Items

Reagents and materials: zinc granule (or zinc powder), $NaBiO_3$ (s), Na_2SO_3 (s), MnO_2 (s), HCl (2 mol/L, 6 mol/L, concentrated), H_2SO_4 (2 mol/L), NaOH (2 mol/L, 6 mol/L, 40%), $TiOSO_4$ (0.1 mol/L), $Cr_2(SO_4)_3$ (0.1 mol/L), $MnSO_4$ (0.1 mol/L), $FeCl_3$ (0.1 mol/L, 1.0 mol/L), $(NH_4)_2Fe(SO_4)_2$ (0.1 mol/L), $CoCl_2$ (0.1 mol/L), $NiSO_4$ (0.1 mol/L), $KMnO_4$ (0.01 mol/L), $(NH_4)_2MoO_4$ (saturated), Na_2WO_4 (saturated), NH_4VO_3 (saturated), H_2O_2 (3%), potassium iodide-starch test paper, bromine water.

Procedures

1. Acidity and Basicity of Hydroxides

Add moderate 2 mol/L NaOH solution to a small amount of 0.1 mol/L $TiOSO_4$ solution, $Cr_2(SO_4)_3$ solution, $MnSO_4$ solution, $(NH_4)_2Fe(SO_4)_2$ solution, $FeCl_3$ solution, $CoCl_2$ solution and $NiSO_4$ solution, respectively, and observe the formation of the precipitates. Divide the above precipitates into two parts, and add excessive 2 mol/L NaOH solution and HCl solution respectively to see whether the precipitates are dissolved. Then heat the solution in which the precipitate is dissolved in dilute alkali to boiling, observe the phenomena and write down the reaction equations.

2. Oxidation-reduction Quality of Certain Compounds

1) Reducibility of Iron (II), Cobalt (II) and Nickel (II)

Add a few drops of bromine water to 0.1 mol/L $(NH_4)_2Fe(SO_4)_2$ solution, $CoCl_2$ solution and $NiSO_4$ solution, respectively, then observe the phenomena and write down the reaction equations.

Add 6 mol/L NaOH solution to 0.1 mol/L $(NH_4)_2Fe(SO_4)_2$ solution, $CoCl_2$ solution and $NiSO_4$ solution, respectively, and observe the phenomena. After a while, observe the changes of the precipitates. Divide the precipitate produced by Co (II) and Ni (II) into two parts, add 3% H_2O_2 solution to one part and bromine water to the other part, what has changed? Write down the reaction equations.

Compare the reducibility of Fe (II), Co (II) and Ni (II) according to the experimental results.

2) Oxidability of Iron (III), Cobalt (III) and Nickel (III)

Prepare the precipitates of $Fe(OH)_3$, $CoO(OH)$ and $NiO(OH)$, add concentrated hydrochloric acid to them respectively, observe the phenomena, then test the produced gas by potassium iodide-starch test paper and write down the reaction equations.

Compare the oxidability of Fe (III), Co (III) and Ni (III) according to the experimental results.

3) Oxidation-reduction Quality of Manganese Compounds

(1) Reducibility of manganese (II): test the reaction of 0.1 mol/L $MnSO_4$ solution with air and bromine in the alkaline medium and with solid $NaBiO_3$ in the acidic medium, respectively, then observe the phenomena and write down the reaction equations.

(2) Oxidation-reduction quality of manganese (IV): add concentrated hydrochloric acid to a little solid MnO_2, heat the solution gently and test if there is chlorine gas produced, and write down the reaction equation.

Take a small amount of solid MnO_2, add a few drops of 40% NaOH solution and a small amount of 0.01 mol/L $KMnO_4$ solution, heat the mixture gently for a moment, then observe the phenomenon and write down the reaction equations.

(3) Oxidability of manganese (VII): test the reaction of Na_2SO_3 solution with $KMnO_4$ in the acid, neutral and alkaline medium respectively and write down the reaction equations.

4) Oxidation-reduction Quality of Chromium, Molybdenum and Tungsten

(1) Oxidation-reduction quality of different oxidation states of chromium: design series of test tube experiments by 0.1 mol/L $Cr_2(SO_4)_3$ solution, 3% H_2O_2 solution, 2 mol/L NaOH solution, 2 mol/L H_2SO_4 solution and other reagents to show the oxidation-reduction quality of different oxidation states of chromium in different media and the transformation conditions among them. Write down the reaction equations.

(2) Oxidability of molybdenum (VI) and tungsten (VI): take a small amount of $(NH_4)_2MoO_4$ saturated solution and add 6 mol/L hydrochloric acid solution for acidification, then add a grain of zinc granule (or zinc powder) to the solution, shake the solution and observe its color change. Put it aside for a while (a few drops of hydrochloric acid can be added in the further reaction process), then observe the color change and write down the reaction equations.

Use Na_2WO_4 saturated solution instead of the $(NH_4)_2MoO_4$ saturated solution to repeat the above experiment, observe the phenomena, and write down the reaction equations.

5) Oxidation-reduction Quality of Titanium and Vanadium

(1) Oxidation-reduction quality of titanium (IV) and titanium (III): add a grain of zinc granule (or zinc powder) to 0.1 mol/L $TiOSO_4$ solution and observe the phenomenon. After a period of time, divide the solution into two test tubes, test the reaction of the solutions with the air and a small amount of $CuCl_2$ solution, respectively, then observe the phenomena and write down the reaction equations.

(2) Colors and oxidation-reduction quality of the hydrated ions of the common oxidation states of vanadium: take NH_4VO_3 saturated solution, add 6 mol/L hydrochloric acid for acidification, and then add a small amount of zinc powder. Let the solution stand for a while and observe its color change. Then test the reaction of the solution with different amount of $KMnO_4$ solution to oxidize V^{2+} to V^{3+}, VO^{2+} and VO_2^+, observe their colors in the solution, and write down the reaction equations.

3. Hydrolysis of Metal Ions

1) Hydrolysis of Iron (III) Salts

After heating a small amount of distilled water to boiling, add a few drops of 1.0 mol/L $FeCl_3$ solution, heat the solution and keep it boiling for a while, then observe the phenomenon and write down the reaction equation.

2) Hydrolysis of Chromium (III) Salts

Add Na_2CO_3 solution to 0.1 mol/L $Cr_2(SO_4)_3$ solution, observe the phenomenon, then write down the reaction equation and explain the experimental results.

3) Hydrolysis of Titanium (IV) Salts

Take 1~2 drops of 0.1 mol/L $TiOSO_4$ solution, add an appropriate amount of distilled water, heat the solution to boiling, then observe the phenomenon and write down the reaction equation.

4. A Small Design Task

Design experimental schemes to separate out and detect Cr^{3+}, Fe^{3+} and Mn^{2+} from the mixed solution.

Notes

(1) Preparation of $TiOSO_4$ solution: add 30 mL of 6 mol/L H_2SO_4 to 2 mL of $TiCl_4$ solution, and then dilute the solution with water to 200 mL.

(2) Preparation of $Fe(OH)_2$: take 2 mL of distilled water in a test tube, add 2 drops of 2 mol/L H_2SO_4, boil the solution to get rid of the air, and then dissolve a little $(NH_4)_2Fe(SO_4)_2$ crystal in the solution. Take 2 mol/L NaOH solution in another test tube and boil it carefully. After the solution is cooled down, absorb the NaOH solution with a dropper which is then inserted into $(NH_4)_2Fe(SO_4)_2$ solution, then slowly release the NaOH solution (avoid the air

from entering the solution in the whole operation process), and at this time, the formation of an almost white $Fe(OH)_2$ precipitate can be observed.

Questions

(1) How to separate out Fe^{3+}, Al^{3+} and Cr^{3+} from the mixed solution?

(2) How to realize the mutual transformation of valence states between Cr^{3+} and CrO_4^{2-}, MnO_2 and Mn^{2+}, MnO_2 and MnO_4^{2-}, MnO_2 and MnO_4^{-}, MnO_4^{2-} and MnO_4^{-}, MnO_4^{-} and Mn^{2+}, etc.?

(3) How many common oxidation states do titanium and vanadium have? Indicate their states and colors in aqueous solution.

Exp. 20 Properties of Element Compounds in the d-Block (II)

Objectives

(1) To observe and master the colors of certain hydrated ions in the d-block.

(2) To understand the complexes of certain metal ions of d-block elements and the effect of forming complexes on their properties.

(3) To understand the application of certain complexes of d-block elements in identifying metal ions.

Experimental Items

Reagents: NaF (s), $Na_2C_2O_4$ (s), EDTA (s), HCl (2 mol/L), H_2SO_4 (6 mol/L), NaOH (2 mol/L, 6 mol/L), $NH_3 \cdot H_2O$ (2 mol/L, 6 mol/L), $Cr(NO_3)_3$ (1 mol/L), $TiOSO_4$ (0.1 mol/L), NH_4VO_3 (0.1 mol/L, saturated), $Cr_2(SO_4)_3$ (0.1 mol/L), $MnSO_4$ (0.1 mol/L), $FeCl_3$ (0.1 mol/L), $(NH_4)_2Fe(SO_4)_2$ (0.1 mol/L), $CoCl_2$ (0.1 mol/L), $NiSO_4$ (0.1 mol/L), KI (0.1 mol/L), $AgNO_3$ (0.1 mol/L), KSCN (saturated), H_2O_2 (3%), ethylenediamine (1%), dimethylglyoxime (1%), ether, pentanol, acetone, carbon tetrachloride.

Procedures

1. Observe and be Fmiliar with the Colors of the Following Hydrated Ions

1) Hydrate Cations
$[Ti(H_2O)_6]^{3+}$, $[Cr(H_2O)_6]^{3+}$, $[Mn(H_2O)_6]^{2+}$, $[Fe(H_2O)_6]^{2+}$, $[Co(H_2O)_6]^{2+}$ and $[Ni(H_2O)_6]^{2+}$.

2) Hydrate Anions
CrO_4^{2-}, $Cr_2O_7^{2-}$, MnO_4^{2-}, MnO_4^{-}, MoO_4^{2-}, WO_4^{2-}, VO_3^{-}.

Write down the results in a table format.

2. Color Changes of Certain Metal Ions

1) Hydrate Isomerism of Cr^{3+}

Take a small amount of 1 mol/L $Cr(NO_3)_3$ solution to heat and observe its color change before and after heating.

$$[Cr(H_2O)_6](NO_3)_3 \underset{\text{cooling}}{\overset{\text{heating}}{\rightleftharpoons}} [Cr(H_2O)_5(NO_3)](NO_3)_2 + H_2O$$

2) Observe the Color of Co(II) Complexes with Different Ligands

Add 0.1 mol/L $CoCl_2$ solution to the KSCN saturated solution until the solution appears blue purple, and then divide the solution into three test tubes, and two of them are add with distilled water and acetone respectively. Compare the color differences of solutions in three test tubes and give your explanations.

$$[Co(NCS)_4]^{2-} + 6H_2O \overset{\text{acetone}}{\rightleftharpoons} [Co(H_2O)_6]^{2+} + 4NCS^-$$

3. Some Metal Ion Complexes

1) Ammine

Add 6 mol/L $NH_3 \cdot H_2O$ solution to 0.1 mol/L $Cr_2(SO_4)_3$, $MnSO_4$, $FeCl_3$, $(NH_4)_2Fe(SO_4)_2$, $CoCl_2$ and $NiSO_4$ solutions respectively, observe the phenomena, write down the reaction equations and summarize the ability of the above metal ions to form ammine.

2) Effect of the Complex Formation on the Oxidation-reduction Quality

Add 0.1 mol/L $FeCl_3$ solution to the mixed solution of 0.1 mol/L KI and CCl_4 and observe the phenomenon. If adding a small amount of solid NaF before adding $FeCl_3$ solution, what's the difference? Give your explanations and write down the reaction equations.

Compare the reaction of 0.1 mol/L $(NH_4)_2Fe(SO_4)_2$ solution with 0.1 mol/L $AgNO_3$ solution in the presence and absence of EDTA at room temperature, and give your explanations.

3) Relationship of the Stability of the Complex with the Ligand

Add a small amount of solid $Na_2C_2O_4$ to 0.1 mol/L $Cr_2(SO_4)_3$ solution, oscillate the mixed solution and observe its color change, and then add 2 mol/L NaOH solution dropwise to see if there is a precipitate generated. Give your explanations and write down the reaction equations.

Add a small amount of KSCN saturated solution to 0.1 mol/L $FeCl_3$ solution and observe the phenomenon. Then add a small amount of solid $Na_2C_2O_4$ and observe the color change of the solution. Give your explanations and write down the reaction equations.

Add excess 2 mol/L $NH_3 \cdot H_2O$ solution to 0.1 mol/L $NiSO_4$ solution and observe the phenomenon. Then drop 1% ethylenediamine solution and observe the phenomenon.

4. Application of Complexes: Identification of Metal ions

1) Identification of Iron

Identify the iron (II), iron (III) according to learned knowledge.

2) Identification of Cobalt (II)

After the addition of pentanol (or propanol) in the 0.1 mol/L $CoCl_2$ solution, add KSCN saturated solution dropwise, and then observe the phenomena and write down the reaction equations.

3) Identification of Nickel (II)

Add 2 mol/L $NH_3 \cdot H_2O$ solution to 0.1 mol/L $NiSO_4$ solution until the solution is weakly alkaline. Then add a drop of 1% diacetyldioxime solution and observe the phenomenon. The equation is as follow:

$$Ni^{2+} + 2 \begin{matrix} CH_3-C=NOH \\ | \\ CH_3-C=NOH \end{matrix} + 2NH_3 = Ni\begin{pmatrix} CH_3-C=NOH \\ | \\ CH_3-C=NO \end{pmatrix}_2 \downarrow + 2NH_4^+$$

4) Identification of Chromium (III)

Add excess 6 mol/L NaOH solution to 0.1 mol/L $Cr_2(SO_4)_3$ solution, and then add 3% H_2O_2 solution dropwise and observe the phenomenon. Afterwards, acidify the solution with dilute H_2SO_4, add a small amount of ether (or propanol) and continue to add 3% H_2O_2 dropwise. Observe the phenomenon and write down the reaction equations.

5) Identification of Titanium (IV)

Add 3% H_2O_2 solution to a small amount of 0.1 mol/L $TiOSO_4$ solution and observe the phenomenon. Then add a little of 6 mol/L $NH_3 \cdot H_2O$ solution and observe the new phenomenon. The reaction equations are as follows:

$$TiO^{2+} + H_2O_2 = [TiO(H_2O_2)]^{2+} \text{ (orange red)}$$

$$[TiO(H_2O_2)]^{2+} + NH_3 \cdot H_2O = H_2Ti(O_2)O_2 \downarrow \text{ (yellow)} + NH_4^+ + H^+$$

6) Identification of Vanadium (V)

Take a small amount of NH_4VO_3 saturated solution and acidify it with hydrochloric acid. After acidification, add a few drops of 3% H_2O_2 solution and observe the phenomenon. The reaction equation is as follows:

$$NH_4VO_3 + H_2O_2 + 4HCl = [V(O_2)]Cl_3 + NH_4Cl + 3H_2O$$

5. A Small Design Task

Given that there is a mixed solution containing Fe^{3+}, Co^{3+} and Ni^{2+} ions, please design a scheme to detect them.

Questions

(1) Why do hydrated ions of d-block elements have colors?

(2) When using KI to quantitatively measure Cu^{2+}, the presence of impurity Fe^{3+} can cause interference. How to eliminate the interference?

(3) According to the electrode potential of the redox couple, Fe^{3+} is difficult to reduce Ag^+ to elementary silver at normal temperature, how to recycle silver in silver salt solution by Fe^{3+} according to the properties of complexes?

Exp. 21　ds-Block Metals (Cu, Ag, Zn, Cd, Hg)

Objectives

(1) To understand the acidity and basicity of copper, silver, zinc, cadmium, mercury oxides or hydroxides and the solubility of their sulfides.

(2) To master the properties and the mutual transformation of important compounds of Cu (Ⅰ) and Cu (Ⅱ).

(3) To test and be familiar with the coordination ability of copper, silver, zinc, cadmium and mercury ions, and the transformation between Hg^{2+} and Hg_2^{2+}.

Experimental Items

Apparatuses and materials: test tube, beaker, centrifuge, centrifuge tube, pH test paper, glass rod.

Reagents: KI, copper scraps, HCl (2 mol/L, concentrated), H_2SO_4 (2 mol/L), HNO_3 (2 mol/L, concentrated), NaOH (2 mol/L, 6 mol/L, 40%), $NH_3·H_2O$ (2 mol/L, concentrated), $CuSO_4$ (0.2 mol/L), $ZnSO_4$ (0.2 mol/L), $CdSO_4$ (0.2 mol/L), $CuCl_2$ (0.5 mol/L), $Hg(NO_3)_2$ (0.2 mol/L), $SnCl_2$ (0.2 mol/L), $AgNO_3$ (0.1 mol/L), Na_2S (1 mol/L), KI (0.2 mol/L), KSCN (0.1 mol/L), $Na_2S_2O_3$ (0.5 mol/L), NaCl (0.2 mol/L), metallic mercury, glucose (10%).

Procedures

1. Formation and Properties of Copper, Silver, Zinc, Cadmium and Mercury Hydroxides or Oxides

1) Formation and Properties of Copper, Zinc and Cadmium Hydroxides

Add the freshly prepared 2 mol/L NaOH solution to three centrifuge tubes containing 0.5 mL of 0.2 mol/L $CuSO_4$ solution, $ZnSO_4$ solution and $CdSO_4$ solution, respectively, and then observe the colors of the solutions and the formation of the precipitates.

Centrifuge the precipitates, wash them and then divide each precipitate into two portions: add 2 mol/L H_2SO_4 solution to one portion and continue to add 6 mol/L NaOH solution to the other portion. Observe the phenomena and write down the reaction equations.

2) Formation and Properties of Silver and Mercury Oxides

(1) Formation and properties of silver oxide: take 0.5 mL of 0.1 mol/L $AgNO_3$ solution, add 2 mol/L NaOH solution which is freshly prepared and observe the color and state of Ag_2O (why not AgOH). Then centrifuge and wash the precipitate, and divide the precipitate into two portions: add 2 mol/L HNO_3 solution to one portion and 2 mol/L aqueous ammonia to the other. Observe the phenomena and write down the reaction equations.

(2) Formation and properties of mercury oxide: take 0.5 mL of 0.2 mol/L $Hg(NO_3)_2$ solution, add 2 mol/L NaOH solution which is freshly prepared and observe the color of the solution and the state of the precipitate. Then centrifuge and wash the precipitate, and divide the precipitate into two portions: add 2 mol/L HNO_3 solution to one portion and 40% NaOH solution to the other. Observe the phenomena and write down the reaction equations.

2. Formation and Properties of Copper, Silver, Zinc, Cadmium and Mercury Sulfides

Add 1 mol/L Na_2S solution to five centrifuge tubes containing 0.5 mL of 0.2 mol/L $CuSO_4$ solution, 0.1 mol/L $AgNO_3$ solution, 0.2 mol/L $ZnSO_4$ solution, 0.2 mol/L $CdSO_4$ solution and

0.2 mol/L Hg(NO$_3$)$_2$ solution, respectively. Observe the formation and color of the precipitates.

Centrifuge and wash the precipitates, and then divide each precipitate into four portions: add 2 mol/L HCl solution, concentrated HCl solution, concentrated HNO$_3$ solution and aqua regia (self-made) to four portions and heat them in a water bath, respectively. Observe the dissolution of the precipitates.

Fill in Table 4.1 according to the experimental phenomena and relevant data, draw conclusions on the dissolution of copper, silver, zinc, cadmium and mercury sulfides, and write down the reaction equations.

Table 4.1 The physical properties and relevant data of the as-prepared sulfides

Sulfide \ Property	Color	Solubility				K_{sp}
		2 mol/L HCl solution	Concentrated HCl solution	Concentrated HNO$_3$ solution	Aqua regia	
CuS						
Ag$_2$S						
ZnS						
CdS						
HgS						

3. Complexes of Copper, Silver, Zinc and Mercury

1) Formation of Ammonia Compounds

Add 2 mol/L NH$_3$·H$_2$O dropwise to four test tubes containing 0.5 mL of 0.2 mol/L CuSO$_4$ solution, 0.2 mol/L ZnSO$_4$ solution, 0.2 mol/L Hg(NO$_3$)$_2$ solution and 0.1 mol/L AgNO$_3$ solution, respectively. Observe the formation of the precipitates. Continue to add 2 mol/L NH$_3$·H$_2$O (to be excessive), and observe the phenomena. Write down the reaction equations.

Compare the differences among the reactions of Cu^{2+}, Ag$^+$, Zn^{2+} and Hg^{2+} ions with ammonia water.

2) Formation and Application of Mercury Complexes

Add 0.2 mol/L of KI solution dropwise to 0.5 mL of 0.2 mol/L Hg(NO$_3$)$_2$ solution, and observe the formation and color of the precipitate. Then add a small amount of potassium iodide solid (not too much, stop adding when the precipitate is just dissolved) to the precipitate, and what is the color of the solution? Write down the reaction equation.

Add a few drops of 40% NaOH solution to the obtained solution, and then make the solution react with aqueous ammonia and observe the color of the precipitate.

Add 0.1 mol/L KSCN solution dropwise to 5 drops of 0.2 mol/L Hg(NO$_3$)$_2$ solution, and a white Hg(SCN)$_2$ precipitate is produced initially; continue to add KSCN solution, the precipitate is gradually dissolved and the colorless complex ion [Hg(SCN)$_4$]$^{2-}$ is formed; then add a few drops of

0.2 mol/L $ZnSO_4$ solution and observe the formation of white $Zn[Hg(SCN)_4]$ precipitate (the reaction can be used to qualitatively test Zn^{2+}), and if necessary, rub the wall of the test tube with a glass rod.

4. Redox Properties of Copper, Silver and Mercury

1) Formation and Properties of Cuprous Oxide

Take 0.5 mL of 0.2 mol/L $CuSO_4$ solution, add excess 6 mol/L NaOH solution dropwise to dissolve the initially formed blue precipitate into a dark blue solution. Then add 1 mL of 10% glucose solution, mix them well and heat the mixture gently. At this point, there is a yellow precipitate formed and then it turns red. Write down the reaction equations.

Centrifuge and wash the precipitate, and then divide the precipitate into two portions: add 1 mL of 2 mol/L H_2SO_4 solution to one portion, let it stand for a while and observe the change of the precipitate, and then heat the solution to boiling and observe the phenomenon; add 1 mL of concentrated $NH_3 \cdot H_2O$ to the other portion, let it stand for a while after oscillation, observe the color of the solution, and explain why the solution turns dark blue after a while.

2) Formation and Properties of Cuprous Oxide

Take 10 mL of 0.5 mol/L $CuCl_2$ solution, add 3 mL of concentrated HCl solution and a small amount of crushed scrap copper, heat the solution to boiling until it appears dark brown (the green color disappears completely). Afterwards, continue to heat the solution until it is nearly colorless, and then take a few drops of the above solution to 10 mL of distilled water. If there is a white precipitate produced, quickly pour all the solution into 100 mL of distilled water and wash the white precipitate until there is no blue.

Take a small amount of precipitate and divide it into two parts: make one part react with 3 mL of concentrated $NH_3 \cdot H_2O$ and observe the change; make the other part react with 3 mL of concentrated HCl solution and observe the change. Write down the reaction equations.

3) Formation and Properties of Cuprous Iodide

Add 0.2 mol/L KI solution into a test tube containing 0.5 mL of 0.2 mol/L $CuSO_4$ solution, and at the same time, oscillate the test tube and the solution turns brownish yellow (CuI is a kind of white precipitate, I_2 is soluble in KI and the solution is yellow). Then add an appropriate amount of 0.5 mol/L $Na_2S_2O_3$ solution dropwise to remove the iodine formed in the reaction. Observe the color and state of the product and write down the reaction equation.

4) Mutual Transformation of Mercury (II) and Mercury (I)

(1) Oxidability of Hg^{2+}: add 0.2 mol/L $SnCl_2$ solution (from moderate to excess) dropwise to 5 drops of 0.2 mol/L $Hg(NO_3)_2$ solution, observe the phenomenon, and write down the reaction equation.

(2) Transformation of Hg^{2+} to Hg_2^{2+} and disproportionation decomposition of Hg_2^{2+}: add 1 drop of metallic mercury to 0.5 mL of 0.2 mol/L $Hg(NO_3)_2$ solution, and then fully oscillate the mixture. Use a dropper to transfer the supernatant into two test tubes (the remaining mercury is to

be recovered), add 0.2 mol/L NaCl solution into one test tube and 2 mol/L $NH_3 \cdot H_2O$ into the other. Observe the phenomena and write down the reaction equations.

Questions

(1) If adding concentrated $NH_3 \cdot H_2O$ or concentrated HCl solution to the white cuprous chloride precipitate, what color will the solution be and why does the solution turn blue after being put aside for a while?

(2) What kind of substance is the dark brown solution in the above experiment? What happens when the nearly colorless solution is poured into distilled water?

(3) $Na_2S_2O_3$ solution is added to react with the iodine produced in the solution, and it is convenient to observe the white cuprous iodide precipitate; however, if $Na_2S_2O_3$ is excessive, there is no white precipitate. Why is that?

(4) What are the cautions when using mercury? Why should mercury be sealed up with water for safekeeping?

(5) According to the equilibrium principle, predict what the formed precipitate is after the introduction of H_2S gas into $Hg_2(NO_3)_2$ solution and explain it.

(6) In the preparation of cuprous chloride, is it possible to have a reaction of copper chloride with copper scraps under the weak acid conditions resulted by hydrochloric acid? Why is that? Can this reaction be carried out if the concentrated sodium chloride solution is used instead of hydrochloric acid, and why?

(7) Guess the activity order of sodium, potassium, calcium, magnesium, aluminum, tin, lead, copper, silver, zinc, cadmium and mercury based on their standard electrode potentials.

(8) What reaction will happen and what phenomenon can we see when SO_2 is introduced to a mixture of $CuSO_4$ saturated solution and NaCl saturated solution? Try to explain it and write down the corresponding reaction equation.

(9) What reagent is used to dissolve the following precipitates?
copper hydroxide, copper sulfide, copper bromide, silver iodide.

(10) There are three unlabeled bottles containing mercuric nitrate solution, mercurous nitrate solution and silver nitrate solution, respectively. Try to use at least two methods to identify them.

(11) Try to use an experiment to prove that brass is composed of copper and zinc (other components are not to be considered).

Exp. 22   Separation and Identification of Unknown Cations

Objectives

(1) To be familiar with the properties of common cations.

(2) To master the principle and method of separating and identifying the common cations.

(3) To master the basic operations such as transferring chemical reagents, heating in a water bath, centrifugal separation and precipitation washing.

Principles

The separation and identification of ions is based on the different reactions of each ion with the reagent, which are often accompanied by special phenomena such as the formation or dissolution of a precipitate, the emergence of a special color and the generation of the gas. The similarity and difference among effects of each ion on the reagent constitutes the basis of ion separation and detection method. In other words, the basic properties of ions are the basis of separation and detection.

The separation and detection of ions can only be performed under certain conditions. The certain conditions mainly refer to the acidity of the solution, the concentration of the reactants, the temperature of the reaction, the existence of substances that promote or hinder the reaction, etc. In order for the reaction to precede in the desired direction, appropriate reaction conditions must be selected. Therefore, in addition to being familiar with the properties of ions, we should also learn how to use the equilibrium laws of ions (acid-base equilibrium, solubility equilibrium of precipitation, redox equilibrium and coordination dissociation equilibrium) to control reaction conditions, which will be of great help in further understanding about how to select separation and detection conditions of ions.

The reactions of common cations with common reagents:

(1) Reaction with HCl

$$\left.\begin{array}{l}Ag^+\\Hg^{2+}\\Pb^{2+}\end{array}\right\} \xrightarrow{HCl} \left\{\begin{array}{l}AgCl \downarrow \text{ white, soluble in ammonia water}\\Hg_2Cl_2 \downarrow \text{ white, soluble in concentrated } HNO_3 \text{ and } H_2SO_4\\PbCl_2 \downarrow \text{ white, soluble in hot water, } NH_4Ac \text{ and NaOH}\end{array}\right.$$

(2) Reaction with H_2SO_4

$$\left.\begin{array}{l}Ba^{2+}\\Sr^{2+}\\Ca^{2+}\\Pb^{2+}\\Ag^+\end{array}\right\} \xrightarrow{H_2SO_4} \left\{\begin{array}{l}BaSO_4 \downarrow \text{ white, hardly soluble in acid}\\SrSO_4 \downarrow \text{ white, soluble in boiling acid}\\CaSO_4 \downarrow \text{ white, highly soluble, produced only when } Ca^{2+} \text{ concentration is very large}\\PbSO_4 \downarrow \text{ white, soluble in } NH_4Ac, NaOH \text{ and concentrated } H_2SO_4 \text{ but insoluble in dilute } H_2SO_4\\Ag_2SO_4 \downarrow \text{ white, produced in the concentrated solution and soluble in hot water}\end{array}\right.$$

(3) Reaction with NaOH

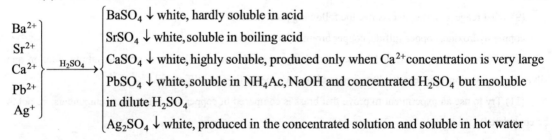

(4) Reaction with NH_3

$$\left.\begin{array}{l}Ag^+\\Cu^{2+}\\Cd^{2+}\\Zn^{2+}\end{array}\right\} \xrightarrow{\text{excessive } NH_3} \left\{\begin{array}{l}[Ag(NH_3)_2]^+\\ [Cu(NH_3)_4]^{2+} \text{ deep blue}\\ [Cd(NH_3)_4]^{2+}\\ [Zn(NH_3)_4]^{2+}\end{array}\right.$$

(5) Reaction with $(NH_4)_2CO_3$

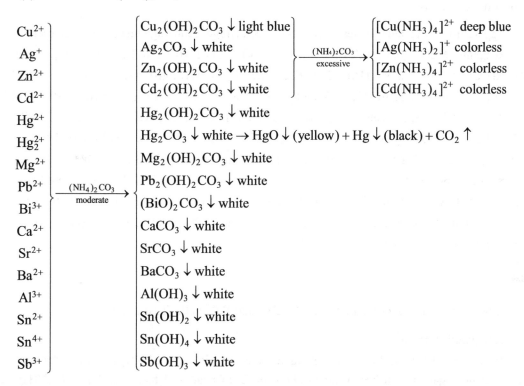

(6) Reaction with H_2S or $(NH_4)_2S$

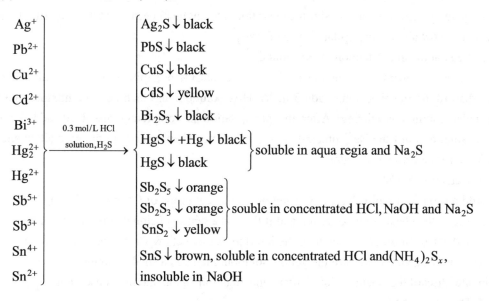

Experimental Items

Apparatuses and materials: test tube, beaker, centrifuge, centrifuge tube, alcohol burner.

Reagents: $MgCl_2$ (0.5 mol/L), $CaCl_2$ (0.5 mol/L), $BaCl_2$ (0.5 mol/L), $(NH_4)_2CO_3$ (1 mol/L), NH_3-NH_4Cl buffer solution, HAc (2 mol/L, 3 mol/L, 6mol/L), K_2CrO_4 (0.5 mol/L), $NH_3 \cdot H_2O$ (2mol/L, 6 mol/L), $(NH_4)_2C_2O_4$ (0.5 mol/L), $(NH_4)_2SO_4$ (1 mol/L), $(NH_4)_2HPO_4$ (0.5 mol/L), NaOH (6 mol/L), magnesium reagent, aluminum reagent, $AgNO_3$ (0.1 mol/L), $Ba(NO_3)_2$ (0.5 mol/L), $Cd(NO_3)_2$ (0.5 mol/L), $Al(NO_3)_3$ (0.5 mol/L), $NaNO_3$ (0.5 mol/L), HCl (6 mol/L), HNO_3 (2 mol/L, 6 mol/L), NaAc (2 mol/L), Na_2S (0.5 mol/L), H_2SO_4 (3 mol/L, 6 mol/L), Na_2CO_3 (saturated), $KSbC_4H_4O_6$ (saturated), $Fe(NO_3)_3$ (0.5 mol/L), H_2O_2 (6%), KSCN (1 mol/L), $FeCl_3$ (0.5 mol/L), $CoCl_2$ (0.5 mol/L), $Ni(NO_3)_2$ (0.5 mol/L), $MnCl_2$ (0.5 mol/L), $CrCl_3$ (0.5 mol/L), $ZnCl_2$ (0.5 mol/L), NH_4F (0.5 mol/L), NH_4SCN (saturated), $Pb(Ac)_2$ (0.5 mol/L), $(NH_4)_2Hg(SCN)_4$ (saturated), $NaBiO_3$ (s), pentanol, dimethylglyoxime (1%).

Procedures

1. Separation and Identification of Mg^{2+}, Ca^{2+} and Ba^{2+} in the Test Liquid

1) Separation of Ca^{2+}, Ba^{2+} and Mg^{2+}

Take 2 mL of test liquid into a centrifugal tube and add 1 mL of NH_3-NH_4Cl buffer solution, put the centrifugal tube in hot water at about 60℃ to heat, and then add 1 mol/L $(NH_4)_2CO_3$ solution until the precipitate is complete under stirring. After heating it for another few minutes, centrifuge the precipitate and transfer the clear liquid to another centrifuge tube. The filtrate is used for step 4) below and the precipitate is used for step 2).

2) Separation and Detection of Ba^{2+} and Ca^{2+}

Wash and centrifuge the obtained $CaCO_3$ and $BaCO_3$ precipitates with a small amount of water, discard the flushing liquid, add 3 mol/L HAc solution, and then heat the mixture in a water bath under continuous stirring. After the precipitation is completely dissolved, add 0.5 mol/L K_2CrO_4 solution until the Ba^{2+} precipitation is complete. Then centrifuge the precipitate and keep the filtrate for detecting Ca^{2+} later.

3) Detection of Ca^{2+}

Add a drop of 6 mol/L $NH_3 \cdot H_2O$ solution and a few drops of 0.5 mol/L $(NH_4)_2C_2O_4$ solution to the filtrate, then heat the mixture up, and if there is a white precipitate produced, the existence of Ca^{2+} is indicated. In order to eliminate the interference of yellow CrO_4^{2-} in observing the color of CaC_2O_4, centrifuge it and discard the yellow solution, then add a small amount of water to wash the precipitate, discard the washing liquid after centrifugation again, and then observe it.

4) Detection of Mg^{2+}

(1) Removal of residual Ba^{2+} and Ca^{2+}: add a drop of 0.5 mol/L $(NH_4)_2C_2O_4$ solution and 1 mol/L $(NH_4)_2SO_4$ solution to the filtrate obtained in step 1), respectively, heat the solution for a few minutes, and if the solution is cloudy, discard the precipitation after centrifuge separation and use the clear liquid to detect Mg^{2+}.

(2) Detection of Mg^{2+}: take 1 mL of clear liquid in a test tube, add 0.5 mL of 6 mol/L $NH_3 \cdot H_2O$ solution and 0.5 mL of 0.5 mol/L $(NH_4)_2HPO_4$ solution, then rub the wall of the test tube with a glass rod, and the formation of a white precipitate indicates the existence of Mg^{2+}.

Take 2 drops of clear liquid of 1), drop it on a spot plate, and then add 2 drops of 6 mol/L NaOH solution and 1 drop of magnesium reagent, and the formation of a blue precipitate indicates the existence of Mg^{2+}.

2. Separation and Identification of Ag^+, Cd^{2+}, Al^{3+}, Ba^{2+} and Na^+

The mixed ions are prepared by the corresponding nitrate solution. Add 2 drops of Ag^+ test solution and 5 drops of Cd^{2+}, Al^{3+}, Ba^{2+} and Na^+ test solutions respectively into a centrifuge test tube, mix them evenly, and then follow the following steps for separation and identification.

1) Separation and Identification of Ag^+

Add a drop of 6 mol/L of hydrochloric acid to the mixture and stir. When there is a precipitate formed, add a drop of 6 mol/L hydrochloric acid until the precipitation is complete. Stir the mixture for a moment, centrifuge the precipitate, then transfer the clear liquid to another centrifuge test tube and operate as described in step 2) below. Wash the precipitate with a drop of 6 mol/L hydrochloric acid solution and 10 drops of distilled water, separate it by centrifugation, and then add the washing liquid to the clear liquid above. Then add 2~3 drops of 6 mol/L ammonia water to the precipitate, stir it for dissolution, and add 1~2 drops of 6 mol/L HNO_3 solution to the clear liquid for acidification. If a white precipitate is produced, the existence of Ag^+ is indicated.

2) Separation and Identification of Al^{3+}

Add 6 mol/L ammonia water to the clear liquid of step 1) until it is alkaline and stir it for a moment. Centrifuge the precipitate, then transfer the clear liquid to another centrifuge tube and operate as described in step 3) below. Add 2 drops of 2 mol/L HAc solution and 2 mol/L NaAc solution in the precipitate, respectively, and add 2 drops of aluminum reagent, stir and heat it gently, and if a red precipitate is produced, the existence of Al^{3+} is indicated.

3) Separation and Identification of Ba^{2+}

Add 6 mol/L H_2SO_4 solution to the clear liquid of step 2) to produce a white precipitate, and then add two more drops and stir it for a moment. Centrifuge the precipitate, then transfer the clear liquid to another test tube and operate as described in step 4) below. Wash the precipitate with 10 drops of hot distilled water, separate it by centrifugation and then add the washing liquid to the clear liquid above. Add 3~4 drops of Na_2CO_3 saturated solution to the precipitate, stir it for a moment, then add 3 drops of 2 mol/L HAc solution and 2 mol/L NaAc solution, continue to stir it for a moment, and then add 2~4 drops of 0.5 mol/L K_2CrO_4 solution, and if a yellow precipitate is produced, the existence of Ba^{2+} is indicated.

4) Separation and Identification of Cd^{2+} and Na^+

Add a small amount of the clear liquid of step 3) into a test tube and then add 2~3 drops of 0.5 mol/L Na_2S solution. If a bright yellow precipitate is produced, the existence of Cd^{2+} is indicated. Add a small amount of the clear liquid of step 3) into another test tube and then add a few drops of tartaric acid antimony potassium saturated solution. If a white crystalline precipitate is produced, the existence of Na^+ is indicated.

3. Separation and Identification of Fe^{3+}, Co^{2+}, Ni^{2+}, Mn^{2+}, Cr^{3+} and Zn^{2+}

Take 3 mL of the above mixed reagent and add it to a centrifuge test tube, and then follow the following steps for separation and identification.

1) Separation of Fe^{3+}, Co^{2+}, Ni^{2+}, Mn^{2+}, Cr^{3+} and Zn^{2+}

Add H_2O_2 and sufficient NaOH to the solution, and the ions in this group can be divided into two groups. The precipitates are FeO(OH), CoO(OH), $Ni(OH)_2$ and $MnO(OH)_2$, while Cr^{3+} and Zn^{2+} become CrO_4^{2-} and $Zn(OH)_4^{2-}$ which remain in the solution. This grouping method is often referred to as "alkaline hydrogen peroxide method".

Add 5~6 drops of 6 mol/L NaOH solution to the solution until it is strong alkaline (pH > 12) and keep on adding 2~3 drops of NaOH solution. Afterwards, add 6% H_2O_2 solution dropwise while stirring the solution with a glass rod, and stop adding H_2O_2 when the precipitate turns brown and black. Continue to stir the mixture for 2~3 min, then heat it in a water bath to coagulate the gelatinous precipitate and make excess H_2O_2 decompose, and stop heating when no more bubbles are produced. After centrifuge separation, transfer the clear liquid to another centrifuge test tube, which is marked as clear liquid 1 and reserved for the following steps 7) and 8). In addition, use hot water to wash the precipitate once and centrifuge it, then discard the washing liquid.

2) Dissolution of the Precipitate

Add a few drops of 3 mol/L H_2SO_4 solution and 2 drops of 6% H_2O_2 solution to the precipitate obtained in the first step. After stirring, heat the centrifuge tube in a water bath to dissolve all the precipitates and make the excess H_2O_2 decompose. Then detect Fe^{3+}, Co^{2+}, Ni^{2+} and Mn^{2+} after the solution cools to the room temperature.

3) Detection of Fe^{3+}

Take 1 drop of the solution of step 2) into the notch of a spot plate. Afterwards, add 1 drop of 1 mol/L KSCN solution and the solution appears blood red. When adding 0.5 mol/L NH_4F solution, the blood red fades away, which indicates the existence of Fe^{3+}.

4) Detection of Co^{2+}

Add 2 drops of the solution of step 2) and a small amount of 0.5 mol/L NH_4F solution into a test tube, then add a small amount of pentanol, and finally add NH_4SCN saturated solution. If the pentanol layer appears blue (or blue-green), the existence of Co^{2+} is indicated. Use F^- and Fe^{3+} to form a colorless complex ion to mask Fe^{3+} so as to eliminate the interference of Fe^{3+} in identifying Co^{2+} with SCN^-.

5) Detection of Ni^{2+}

Add 2 drops of the solution of step 2) to a centrifuge tube and add a few drops of 2 mol/L $NH_3 \cdot H_2O$ solution until the solution is weakly basic (what is the precipitate at this point, and what is the disadvantage of adding too much ammonia water?). After centrifugal separation, add 1~2 drops of 1% dimethylglyoxime to the supernatant, and the formation of a peach red precipitate indicates the existence of Ni^{2+}.

6) Detection of Mn^{2+}

Take 1 drop of the solution of step 2), add a small amount of 2 mol/L HNO_3 solution and $NaBiO_3$ solid, stir it and let it stand, and the solution turns magenta, which indicates the existence of Mn^{2+}. If there is excess H_2O_2 in the solution of step 2), excess H_2O_2 will react with $NaBiO_3$ (redox reaction) and consume a small amount of $NaBiO_3$.

7) Detection of Cr^{3+}

Use 6 mol/L HAc solution to acidify the remaining clear liquid 1 from step 1), then reserve half of the obtained solution for detecting Zn^{2+} and add a few drops of 0.5 mol/L $Pb(Ac)_2$ solution to the rest. The formation of a yellow precipitate indicates that there is CrO_4^{2-}, that is, there is Cr^{3+} in the original solution.

If there is excess H_2O_2 in clear liquid 1, it will react with CrO_4^{2-} in the acidic medium so that the amount of CrO_4^{2-} is reduced and the detection sensitivity is decreased.

8) Detection of Zn^{2+}

Add the equal volume of $(NH_4)_2Hg(SCN)_4$ saturated solution in the solution left in step 7), shake the test tube, and the formation of a white $ZnHg(SCN)_4$ precipitate indicates that there is Zn^{2+}. If the phenomenon is not obvious, a glass rod can be used to rub the test tube wall so as to destroy the supersaturated solution.

Write down the reaction equations for each step.

Questions

(1) In the mixture of Ca^{2+} and Ba^{2+}, why can we detect Ba^{2+} with K_2CrO_4 in the existence of Ca^{2+}?

(2) If the reaction of a test solution which might contain Mg^{2+}, Ca^{2+} and Ba^{2+} with NaOH solution without CO_3^{2-} does not produce any precipitate, can the existence of Mg^{2+} be denied? If there is a white precipitate, can the existence of Mg^{2+} be confirmed, and why?

(3) When separating Fe^{3+}, Co^{2+}, Ni^{2+}, Mn^{2+}, Cr^{3+} and Zn^{2+}, why excessive NaOH is added, and why H_2O_2 is added at the same time? What are the disadvantages if the base is added too much, or if the excess H_2O_2 does not decompose completely?

(4) In order to dissolve the precipitation of FeO(OH), CoO(OH), $Ni(OH)_2$ and $MnO(OH)_2$, why H_2O_2 is added besides H_2SO_4?

(5) When detecting CrO_4^{2-} and Zn^{2+}, why should HAc first be used to acidify the solution?

Exp. 23 Separation and Identification of Unknown Anions

Objectives

(1) To be familiar with how to initially check the properties of common anions.

(2) To design the methods of separating and identifying nonmetallic anions in the mixed solution.

Principles

The elements that form anions are mostly non-metallic, such as S^{2-}, Cl^-, NO_3^- and SO_4^{2-}. Some metallic elements can also exist as complex anions, such as VO_3^-, CrO_4^{2-}, $Al(OH)_4^-$, but they are generally identified in the cationic analysis. There are eleven common important anions, they are Cl^-, Br^-, I^-, S^{2-}, SO_3^{2-}, $S_2O_3^{2-}$, SO_4^{2-}, NO_3^-, NO_2^-, PO_4^{3-} and CO_3^{2-}. Some general methods of separating and identifying them are introduced here.

Many anions exist or coexist only in alkaline solutions, and once the solution is acidified, they will decompose or react with each other. Under acidic conditions, NO_2^-, SO_3^{2-}, $S_2O_3^{2-}$, S^{2-} and CO_3^{2-} are easy to decompose, and oxidizing ions NO_3^-, NO_2^- and SO_3^{2-} can have redox reactions with reductive ions I^-, SO_3^{2-}, $S_2O_3^{2-}$ and S_2^-. Some ions are easily oxidized by air, for example, NO_2^-, SO_3^{2-} and S^{2-} are easy to be oxidized into NO_3^-, SO_4^{2-} and S. Improper analysis can also cause errors.

In fact, many ions have fewer opportunities to coexist with each other due to less mutual interference between anions. Therefore, most anionic analyses commonly adopt the separate analysis method, such as S^{2-}, SO_3^{2-} and $S_2O_3^{2-}$; only a few ions with mutual interference adopt the system analysis method, such as Cl^-, Br^- and I^-. The identification reactions of common anions are summarized in Table 4.2.

Table 4.2 Main identification reactions of common anions

Ion	Reagent	Identification reaction	Medium condition	Main interfering ion
Cl^-	$AgNO_3$	$Cl^- + Ag^+ == AgCl\downarrow$ (white) AgCl is soluble in excess ammonia water or $(NH_4)_2CO_3$. When HNO_3 is added, the precipitate is produced again	acid medium	
Br^-	chlorine water CCl_4 (or benzene)	$2Br^- + Cl_2 == Br_2 + 2Cl^-$ The precipitate Br_2 is soluble in CCl_4 (or benzene), which is aurantius (or orange red)	neutral or acidic medium	
I^-	chlorine water CCl_4 (or benzene)	$2I^- + Cl_2 == I_2 + 2Cl^-$ The precipitate I_2 is soluble in CCl_4 (or benzene), which is purple red	neutral or acidic medium	
SO_4^{2-}	$BaCl_2$	$SO_4^{2-} + Ba^{2+} == BaSO_4\downarrow$ (white)	acid medium	

Continued

Ion	Reagent	Identification reaction	Medium condition	Main interfering ion
SO_3^{2-}	dilute HCl	$SO_3^{2-} + 2H^+ = SO_2\uparrow + H_2O$ Test of SO_2: (1) SO_2 can reduce dilute $KMnO_4$ and make it fade (2) SO_2 can reduce I_2 to I^-, which fade starch-I_2 solution (3) SO_2 can fade the magenta solution	acid medium	$S_2O_3^{2-}$ and S^{2-} will interfere with the identification
	$Na_2[Fe(CN)_5NO]$ $ZnSO_4$ $K_4[Fe(CN)_6]$	red precipitates	neutral medium	The reaction of S^{2-} with $Na_2[Fe(CN)_5NO]$ forms a fuchsia complex, which interferes with the identification of SO_3^{2-}
$S_2O_3^{2-}$	dilute HCl	$S_2O_3^{2-} + 2H^+ = SO_2\uparrow + S + H_2O$ The solution becomes turbid due to the sulfur precipitation in the reaction	acid medium	SO_3^{2-} and S^{2-} will interfere with the identification
	$AgNO_3$	$S_2O_3^{2-} + 2Ag^+ = Ag_2S_2O_3\downarrow$ (white) The precipitate $Ag_2S_2O_3$ is unstable, and its hydrolysis occurs immediately after the formation, which is often accompanied by significant color changes: from white to yellow to brown, and finally to black $Ag_2S_2O_3 + H_2O = Ag_2S\downarrow(black) + 2H^+ + SO_4^{2-}$	neutral medium	S^{2-} will interfere with the identification
S^{2-}	dilute HCl	$2H^+ + S^{2-} = H_2S$ Test of H_2S gas: (1) according to the special smell of H_2S gas (2) H_2S gas can darken the test paper dipped with $Pb(NO_3)_2$ or $Pb(Ac)_2$	acid medium	SO_3^{2-} and $S_2O_3^{2-}$ will interfere with the identification
	$Na_2[Fe(CN)_5NO]$	$S^{2-} + [Fe(CN)_5NO]^{2-} = [Fe(CN)_5NOS]^{4-}$ (purple red)	alkaline medium	
NO_2^-	p-aminobenzene sulfonic acid and α-naphthylamine	$NO_2^- + H_2N-\bigcirc\hspace{-0.5em}\bigcirc + H_2N-\bigcirc-SO_3H + H^+ =$ $H_2N-\bigcirc\hspace{-0.5em}\bigcirc-N=N-\bigcirc-SO_3H + 2H_2O$	neutral or acetic acid medium	Strong oxidants such as $KMnO_4$ will interfere with the identification
NO_3^-	$FeSO_4$, concentrated H_2SO_4	$NO_3^- + 3Fe^{2+} + 4H^+ = 3Fe^{3+} + NO + 2H_2O$ $Fe^{2+} + NO = [Fe(NO)]^{2+}$ (brown) A brown ring is formed at the place where the mixture is stratified with concentrated sulfuric acid	acid medium	NO_2^- has the same reaction that interferes with the identification
CO_3^{2-}	dilute HCl	$CO_3^{2-} + 2H^+ = CO_2\uparrow + H_2O$ $CO_2 + 2OH^- + Ba^{2+} = BaCO_3\downarrow$ (white) $+ H_2O$	acid medium	
PO_4^{3-}	$AgNO_3$	$PO_4^{3-} + 3Ag^+ = Ag_3PO_4\downarrow$	neutral or faintly acid medium	CrO_4^{2-}, AsO_4^{3-}, AsO_3^{3-}, I^- and $S_2O_3^{2-}$ can react with Ag^+ to produce colored precipitation, which interferes with the identification

Ion	Reagent	Identification reaction	Medium condition	Main interfering ion
PO_4^{3-}	$(NH_4)_2MoO_4$ HNO_3	$PO_4^{3-} + 3NH_4^+ + 12MoO_4^{2-} + 24H^+ \rightleftharpoons$ $(NH_4)_3PO_4 \cdot 12MoO_3 \cdot 6H_2O\downarrow + 6H_2O$	HNO_3 medium	(1) SO_3^{2-}, $S_2O_3^{2-}$, S^{2-}, I^-, Sn^{2+} and other reductive substances are easy to reduce $(NH_4)_2MoO_4$ to the compound of low-valence molybdenum molybdenum blue, thus the solution becomes dark blue, which seriously interferes with the detection of PO_4^{3-} (2) The reaction of SiO_3^{2-} and AsO_4^{3-} with ammonium molybdate reagent can also form a similar yellow precipitate, which interferes with the identification (3) When there is a large amount of Cl^-, it can react with Mo (Ⅵ) to form the coordination compound and lower the sensitivity of reaction
SiO_3^{2-}	saturated NH_4Cl	$SiO_3^{2-} + 2NH_4^+ + 2H_2O \xrightarrow{\Delta} H_2SiO_3\downarrow$ (white colloidal) $+ 2NH_3\uparrow + H_2O$	alkaline medium	
F^-	concentrated H_2SO_4	$CaF_2 + H_2SO_4 \xrightarrow{\Delta} 2HF\uparrow + CaSO_4$ The resulting HF reacts with the silicate or SiO_2 and produces the SiF_4 gas, which immediately decomposes in water and transforms into insoluble silicic acid precipitation to make the water cloudy $Na_2SiO_3 \cdot CaSiO_3 \cdot 4SiO_2$ (glass) $+ 28HF \rightleftharpoons 4SiF_4\uparrow + Na_2SiF_6 + CaSiF_6 + 14H_2O$ $SiF_4 + 4H_2O \rightleftharpoons H_4SiO_4\downarrow + 4HF$	acid medium	

The analysis of unknown samples is not to detect all anions one by one by the identification reactions of these 11 anions, but some preliminary experiments should be done in advance to eliminate the possibility of the existence of some ions and simplify the analysis steps. Preliminary experiments are basically as follows:

(1) Reaction with dilute sulphuric acid: add dilute sulfuric acid to the sample and heat it. If there are bubbles, the sample may contain CO_3^{2-}, S^{2-}, SO_3^{2-}, $S_2O_3^{2-}$ and NO_2^-. If the sample is a solution and the ion concentration is not high, the bubbles maybe not so obvious.

(2) Reaction with $BaCl_2$ solution: add $BaCl_2$ solution to the neutral or weakly alkaline test solution, and if there is a white precipitate generated, the test solution may contain SO_4^{2-}, CO_3^{2-},

SO_3^{2-}, PO_4^{3-} and $S_2O_3^{2-}$; otherwise, the test solution does not contain SO_4^{2-}, CO_3^{2-}, SO_3^{2-} and PO_4^{3-}. The existence of $S_2O_3^{2-}$ cannot be determined because the precipitation is formed only when the $S_2O_3^{2-}$ concentration is high (above 4.5 g/L).

(3) Reaction with $AgNO_3$ and HNO_3: if there is a precipitate, the test solution may contain S^{2-}, $S_2O_3^{2-}$, Cl^-, Br^- and I^-. Further judgment can be made by the color of the precipitate.

(4) Test of reductive anions: strong reductive anion such as S^{2-}, SO_3^{2-} and $S_2O_3^{2-}$ can be oxidized by iodine, so the existence of anions can be determined according to whether the test solution fades after the iodine-starch solution is added. Cl^-, Br^-, I^- and NO_2^- can also react with the strong oxidant $KMnO_4$ solution.

(5) Test of oxidative anions: KI solution and CCl_4 are added to the acidified test solution. If the CCl_4 layer appears purple after the oscillation, there are oxidized anions. In our discussion, only NO_2^- has this reaction.

Some appropriate separation steps are also required in the identification of some ion interferences with each other. For example, when Cl^-, Br^- and I^- coexist, the following methods can be used for separation and identification:

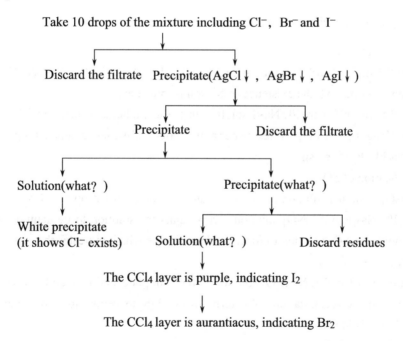

If S^{2-}, SO_3^{2-} and $S_2O_3^{2-}$ coexist in the solution, it is necessary to remove S^{2-} first because the existence of S^{2-} will prevent the identification of the other two ions. The removal method is as follows: $PbCO_3$ solid is added to the mixture, which will be converted into PbS precipitate with lower solubility, then centrifuge the precipitate and identify SO_3^{2-} and $S_2O_3^{2-}$ in the clear liquid respectively.

The existence of NO_2^- also interferes with the identification of NO_3^-. When using brown ring method to identify NO_3^-, it is necessary to remove NO_2^- first. The method is to add NH_4Cl

saturated solution to the mixture and then heat the mixture. The reaction is as follows:

$$NH_4^+ + NO_2^- = N_2\uparrow + 2H_2O$$

Experimental Items

Apparatuses and materials: test tube, centrifuge tube, centrifuge, spot plate, alcohol burner, $Pb(Ac)_2$ test paper.

Reagents: $PbCO_3$ (s), $FeSO_4$ (s), H_2SO_4 (2 mol/L, concentrated), HNO_3 (2 mol/L, 6 mol/L, concentrated), HCl (6 mol/L), HAc (6 mol/L), $NH_3 \cdot H_2O$ (6 mol/L), Na_2S (0.1 mol/L), Na_2SO_3 (0.1 mol/L), $Na_2S_2O_3$ (0.1 mol/L), Na_3PO_4 (0.1 mol/L), NaCl (0.1 mol/L), KBr (0.1 mol/L), KI (0.1 mol/L), KNO_3 (0.1 mol/L), $NaNO_2$ (0.1 mol/L), Na_2SO_4 (0.1 mol/L), $AgNO_3$ (0.1 mol/L), $BaCl_2$ (1 mol/L), $ZnSO_4$ (saturated), chlorine water, $(NH_4)_2MoO_4$ (saturated), $Na_2[Fe(CN)_5NO]$ (1%), $K_4[Fe(CN)_6]$ (0.1 mol/L), CCl_4, $(NH_4)_2CO_3$ (12%), *p*-aminobenzene sulfonic acid, *α*-naphthylamine.

Procedures

1. Identifying the Specific Anion

1) Identification of S^{2-}

(1) Drop 0.1 mol/L Na_2S solution on a spot plate, then drop 1% $Na_2[Fe(CN)_5NO]$ solution. If the solution turns purple red, the existence of S^{2-} ion is indicated.

(2) Add 0.5 mL of 0.1 mol/L Na_2S solution into a test tube, and then add 0.5 mL of 6 mol/L HCl solution. Hang the wet $Pb(Ac)_2$ test paper at the mouth of the test tube, heat it gently, and if the test paper turns black, S^{2-} exists.

2) Identification of SO_3^{2-}

Add $ZnSO_4$ saturated solution to the spot plate, and then add 1 drop of 0.1 mol/L $K_4[Fe(CN)_6]$ solution and 1% $Na_2[Fe(CN)_5NO]$ solution. Then adjust the solution to be neutral with $NH_3 \cdot H_2O$ and add 0.1 mol/L Na_2SO_3 solution to form a red precipitate which indicates the existence of SO_3^{2-}.

3) Identification of $S_2O_3^{2-}$

Add a drop of 0.1 mol/L $Na_2S_2O_3$ solution to the spot plate, and then add 2 drops of 0.1 mol/L $AgNO_3$ solution. If the precipitation color turns from white to yellow to brown and to black, the existence of $S_2O_3^{2-}$ is indicated.

4) Identification of SO_4^{2-}

Add 3~4 drops of 0.1 mol/L Na_2SO_4 solution in a centrifuge test tube, then add 1 drop of 1 mol/L $BaCl_2$ solution. After centrifugal separation, add several drops of 6 mol/L HCl solution to the precipitate, and if the precipitate is not dissolved, the existence of SO_4^{2-} is indicated.

5) Identification of PO_4^{3-}

Take a small amount of 0.1 mol/L Na_3PO_4 solution, add 10 drops of concentrated nitric acid and 20 drops of ammonium molybdate saturated solution, heat the mixed solution gently to 40~50 ℃, and the formation of a yellow precipitate can be observed.

6) Identification of Cl^-

Take 2 drops of 0.1 mol/L NaCl solution, add 1 drop of 2 mol/L HNO_3 solution and 2 drops of 0.1 mol/L $AgNO_3$ solution, and then observe the precipitation color. After centrifugal separation, discard the clear liquid, add a few drops of 6 mol/L ammonia water and the precipitate is dissolved. Then add 6 mol/L HNO_3 solution to acidify the solution and a white precipitate is formed. This method can be used to identify Cl^-.

7) Identification of Br^-

Take 2 drops of 0.1 mol/L KBr solution, add 1 drop of 2 mol/L H_2SO_4 solution and 5~6 drops of CCl_4, then add freshly prepared chlorine water dropwise while shaking, and if the CCl_4 layer appears brown to yellow, the existence of Br^- is indicated.

8) Identification of I^-

It's the same as the identification of Br^-. If the CCl_4 layer appears purple, the existence of I^- is indicated. If adding too much chlorine water, the purple fades away. This is because iodine generates IO_3^- and returns to the water phase.

9) Identification of NO_3^-: Brown Ring Experiment

Take 1 mL of 0.1 mol/L KNO_3, add 1~2 small grains of $FeSO_4$ crystal and oscillate it. After dissolution, add 5~10 drops of concentrated sulfuric acid along the wall of the test tube to observe whether there is a brown ring at the junction of the concentrated H_2SO_4 and the solution.

10) Identification of NO_2^-

Take 1 drop of 0.1 mol/L $NaNO_2$ solution, add 6 mol/L HAc solution to acidify the solution, and then add one drop of *p*-aminobenzene sulfonic acid and *α*-naphthylamine solution respectively. If the solution appears red immediately, the existence of NO_2^- is indicated.

The reaction is applicable to test a small number of NO_2^-. If the concentration of NO_2^- is too high, the solution color fades quickly and a yellow solution or brown precipitate is formed.

11) Identification of CO_3^{2-}

CO_3^{2-} is normally detected with a $Ba(OH)_2$ gas bottle, that is, CO_2 (gas) can make the aqueous solution of $Ba(OH)_2$ cloudy.

2. Separation and Identification of the Mixture Containing Known Anions

(1) Add 2 drops of 0.1 mol/L NaCl, KBr and KI solutions respectively into a test tube, and separate and identify Cl^-, Br^- and I^- according to the methods presented in the experimental procedure.

(2) Add 3 drops of 0.1 mol/L Na_2S, Na_2SO_3 and $Na_2S_2O_3$ solutions respectively into a test tube, separate out S^{2-} first, and then identify each ion according to the methods presented in the experimental procedure.

3. Analysis of the Mixture With Unknown Anions

Get an unknown solution containing three anions (it may contain NO_3^-, NO_2^-, S^{2-}, Cl^-, SO_4^{2-} and $S_2O_3^{2-}$). After the preliminary examination according to the methods given in the experimental procedure, design the experiment scheme of separation and identification by yourself and carry out the experiment.

According to the preliminary experimental results, please determine which anions may exist and explain it, write down the reaction equations and then draw a schematic of separation and identification and make a conclusion.

Questions

(1) Choose a reagent to distinguish the following five solutions: NaCl, $NaNO_2$, Na_2S, $Na_2S_2O_3$, Na_2HPO_4.

(2) Given that there is a basic colorless anionic mixture which becomes turbid after being acidified by H_2SO_4, which anions can be speculated in the mixture?

(3) When NO_3^- and NO_2^- coexist, how does the identification reaction proceed?

Chapter 5 Comprehensive and Design Experiments

Exp. 24 Preparation and Composition Measurement of a Cobalt (III) Complex

Objectives

(1) To master the commonly used methods of preparing metal complexes—substitution reactions and redox reactions in the aqueous solution.

(2) To infer the composition of complexes preliminarily.

(3) To learn the principle and method of using the conductivity meter to determine the composition of complexes.

Principles

1. Synthesis

The method which is in virtue of the substitution reaction of the aqueous solution to prepare metal complexes is actually using a kind of proper ligand to replace the coordinated water molecules of hydrated coordinated ions in the metal salt. Redox reaction is to prepare the metal complexes with a target valence state by oxidizing or reducing metal complexes with other oxidation states in the presence of ligands.

According to the standard electrode potentials, the divalent cobalt salt is more stable than the trivalent cobalt salt in an acidic solution; however, the trivalent cobalt is more stable than the divalent cobalt after the production of complexes. Therefore, cobalt (III) complexes are often prepared through oxidizing cobalt (II) complexes by air or hydrogen peroxide. Common cobalt (III) complexes are: $[Co(NH_3)_6]^{3+}$ (orange), $[Co(NH_3)_5H_2O]^{3+}$ (pink), $[Co(NH_3)_5Cl]^{2+}$ (fuchsia), $[Co(NH_3)_4CO_3]^+$ (fuchsia), $[Co(NH_3)_3(NO_2)_3]$ (yellow), $[Co(CN)_6]^{3-}$ (purple), $[Co(NO_2)_6]^{3+}$ (yellow), etc.

2. Component Analysis

If resorting to the chemical process to analyze and determine the composition of a complex, determine the outer ion of the complex sphere first, and then determine its inner component after destroying the complex ion. The stability of the complex ion is affected by many factors, such as heating and changing the alkalinity of the solution. In this experiment, we preliminarily adopt methods of qualitative, semi-quantitative and estimated analyses to deduce the chemical formula of the complex, and then use the conductivity meter to measure the conductivity of the complex solution with known concentration. Compared with the known conductivity of some electrolyte solutions, the quantity of ions existing in the chemical formula of the complex can be learned, and the formula can be further determined.

Free Co(II) can be used to generate the blue complex $[Co(NCS)_4]^{2-}$ when reacting with the potassium thiocyanate, which proves the existence of Co(II). $[Co(NCS)_4]^{2-}$ is usually prepared by the concentrated solution or powder of potassium thiocyanate due to its high solubility in water. At the same time, pentanol and diethyl ether are added in the preparation to make the complex stable.

Free NH_4^+ can be tested with the Nessler's reagent ($K_2[HgI_4]$ + KOH).

The conductivity of the electrolyte solution can be expressed by the conductance (G):

$$G = \gamma/K$$

In the formula, γ refers to the conductivity whose common unit is S/cm, and K refers to the conductivity cell constant whose unit is cm^{-1}. The conductivity cell constant K is calculated according to the above formula after measuring the conductance of an electrolyte solution with known conductivity. KCl solution is commonly used as the standard conductance solution.

Experimental Items

Apparatuses and materials: electronic platform scale, beaker, Erlenmeyer flask, measuring cylinder, funnel, ring stand, alcohol burner, test tube, dropper, test tube clamps, iron ring, thermometer, conductivity meter, pH test paper, filter paper, medicine spoon, asbestos net.

Reagents: ammonium chloride, cobalt chloride hexahydrate, potassium thiocyanate, concentrated ammonia water, concentrated nitric acid, concentrated hydrochloric acid, 6 mol/L HCl solution, 30% H_2O_2, 2 mol/L $AgNO_3$ solution, 0.5 mol/L $SnCl_2$ solution, Nessler's reagent, diethyl ether, pentanol.

Procedures

1. Preparation of Cobalt (III) Complexes

Add 1.0 g of ammonium chloride and 6 mL of concentrated ammonia water into a 100 mL Erlenmeyer flask and oscillate the Erlenmeyer flask to dissolve the solid. Divide 2.0 g of cobalt chloride hexahydrate powder into 4~5 portions and add them successively, and at the same time, shake the flask continuously to make the solution appear a kind of brown and thin pasty. Do not stop shaking the flask, add 3~5 mL of 30% H_2O_2 under the constant shaking, and then continue to shake for 10 min to completely dissolve the solid. When the solution stops bubbling, add 6 mL of concentrated hydrochloric acid slowly and shake, and then put the flask in a water bath at no more than 85 ℃ and shake for 10~15 min. After that, cool the reaction solution at room temperature, collect the precipitate by vacuum filtration, and wash it 2~3 times with 5 mL of cold water first and with 5 mL of 6 mol/L hydrochloric acid later. Finally, the solid product is dried at 105 ℃ and weighed.

2. Preliminary Inference of the Composition of Complexes

(1) Dissolve 0.2 g of the product by adding 30 mL of distilled water in a small beaker, then measure the pH value of the product solution to verify its acidity and basicity.

(2) Take 15 mL of the above solution, and slowly add 2 mol/L $AgNO_3$ solution until the precipitation is complete (how to judge the precipitation is complete?). After removing the precipitate by filtration, add 1~2 mL of concentrated nitric acid to the filtrate and stir, and then add $AgNO_3$ solution to observe whether there is a precipitate being formed. If so, compare its volume with that of the previous precipitate.

(3) Take 2 mL of the product solution in a test tube, add 5 drops of 0.5 mol/L $SnCl_2$ solution (why?) and oscillate. Afterwards, add a small amount of KSCN powder to the solution and oscillate. Then add 1 mL of pentanol and 1 mL of diethyl ether and oscillate. Observe the color of the organic layer (why?).

(4) Take 2 mL of the product solution in a test tube, and then add 2 drops of Nessler's reagent and observe the phenomenon.

(5) Heat the remaining product solution in (1) to make the solution completely brown and black (how to judge it?), cool it and test its acidity and basicity by pH test paper, and then filter the solution with double filter paper. Afterwards, take the obtained filtrate to repeat the experiment (3) and (4), and observe how the phenomena differ from the previous one. Infer the chemical formula of the complex preliminarily.

(6) According to the preliminarily inferred chemical formula, prepare 20~30 mL solution of the complex at a concentration of 0.01 mol/L, and use the conductivity meter to measure its conductivity. Then dilute the solution 10 times and measure its conductivity again. Compare with the data in Table 5.1 to determine the number of ions in the chemical formula.

Table 5.1 The conductivity of several electrolytes

Electrolyte	Type (ion number)	Conductivity / S	
		0.01 mol/L	0.001 mol/L
KCl	Type 1 - 1 (2)	1230	133
$BaCl_2$	Type 1 - 2 (3)	2150	250
$K_3[Fe(CN)_6]$	Type 1 - 3 (4)	3400	420

Questions

(1) What kind of reactions can occur when cobalt chloride hexahydrate is added to the mixture of ammonium chloride and concentrated ammonia water? What complex is produced? What roles does ammonium chloride play besides as a reactant?

(2) Why does cobalt chloride have to be added by several times?

(3) What is the effect of hydrogen peroxide in the experiments? Except for hydrogen peroxide, what other

substances can be used and how about the disadvantages of these substances? What is the role of concentrated hydrochloric acid in the experiments?

(4) Which steps are critical to obtain high productivity of products prepared in this experiment? Why is that?

Exp. 25 Preparation and Component Analysis of the Large Crystal of Cuprammonium Sulfate (II)

Objectives

(1) To master how to prepare cuprammonium sulfate (II) and how to determine its components.

(2) To master the basic operations of evaporation, crystallization, vacuum filtration, etc.

(3) To master the technique of determining ammonia by distillation method.

Principles

Cuprammonium sulfate ($[Cu(NH_3)_4]SO_4$) is commonly used as an insecticide and a mordant. It is also widely used in industry, such as being the main component of electroplate liquid during alkaline copper-electroplating. Cuprammonium sulfate is a kind of moderately stable blue crystal. It is liable to react with water and carbon dioxide in the air at room temperature and produces the basic salt of copper, which turns the crystal into green powder.

1. Preparation

$CuSO_4$ reacts with concentrated ammonia water to produce a copper dihydroxosulphate precipitate which is light blue. When ammonia is excessive, the precipitate eventually produces the cuprammonium sulfate solution which is dark blue.

$$2CuSO_4 + 2NH_3 \cdot H_2O \Longrightarrow Cu_2(OH)_2SO_4\downarrow \text{ (light blue)} + (NH_4)_2SO_4$$

$$Cu_2(OH)_2SO_4 + 8NH_3 \Longrightarrow 2[Cu(NH_3)_4]^{2+} + SO_4^{2-} + 2OH^-$$

Cuprammonium sulfate is easy to decompose and lose ammonia when heated, so conventional methods, such as evaporation and concentration, are inappropriate to prepare its crystal. The dark blue crystal of cuprammonium sulfate can be extracted by adding ethyl alcohol into its aqueous solution, due to the solubility of cuprammonium sulfate in ethyl alcohol is much less than that in water.

2. Component Analysis

1) Determination of NH_3 Content

$$Cu(NH_3)_4SO_4 + 2NaOH \Longrightarrow CuO\downarrow + 4NH_3\uparrow + Na_2SO_4 + H_2O$$

$$NH_3 + HCl \text{ (excessive)} \Longrightarrow NH_4Cl$$

$$HCl \text{ (surplus)} + NaOH \Longrightarrow NaCl + H_2O$$

2) Determination of SO_4^{2-} Content:
$$SO_4^{2-} + Ba^{2+} = BaSO_4\downarrow$$

Experimental Items

Apparatuses and materials: beaker, Erlenmeyer flask, watch glass, separating funnel, long neck funnel, alcohol burner, Büchner funnel, suction bottle, water pump, rubber stopper, wash bottle, medicine spoon.

Reagents: $CuSO_4 \cdot 5H_2O$ crystal, concentrated ammonia water, 95% ethanol, 0.5 mol/L HCl standard solution, 0.5 mol/L NaOH standard solution, 2 mol/L NaOH solution, 6 mol/L HCl solution, 0.1 mol/L $BaCl_2$ solution, 0.1% methyl red solution, 0.1 mol/L $AgNO_3$ solution.

Procedures

1. Preparation of Cuprammonium Sulfate (II)

Weigh 5.0 g of $CuSO_4 \cdot 5H_2O$ and dissolve it with 7 mL of water in a beaker, slowly add 10 mL of concentrated ammonia water, and stir the solution to make the precipitate fade away and the solution appear transparent blue. Then slowly add 18 mL of 95% ethanol along the wall of the beaker and cover with a watch glass to let the solution stand. Filter the solution under reduced pressure after there is a crystal being extracted, and wash the crystal several times by a mixture (5 mL) of 95% ethanol and concentrated ammonia water with a volume ratio of 1~2. Dry the crystal of $[Cu(NH_3)_4]SO_4 \cdot H_2O$ in the air naturally, then weigh it and save it for future use.

2. Component Analysis

1) Determination of NH_3 Content

Add 30 mL of HCl standard solution (about 0.5 mol/L) into an Erlenmeyer flask which is then placed in an ice-water bath for cooling. Next, add 0.30 g of sample into a 250 mL flask with a branch pipe and dissolve the sample with 80 mL of water. Afterwards, equip the Erlenmeyer flask with the separating funnel containing 10 mL of 2 mol/L NaOH solution with the help of the rubber stopper, make the lower end of the funnel insert into the sample solution 2~3 cm deep, and connect the beaker with the HCl standard solution in the ice-water bath by a gas-guided tube. Then heat the sample solution over high heat and open the funnel faucet to add NaOH solution. When the reaction solution is almost boiling, switch to the small fire to keep the solution slightly boiling. The ammonia in the sample can be removed by distillation for about an hour. After the distillation, take out the tube inserted into the HCl solution, wash its inside and outside with distilled water, and collect the flushing liquid in the absorption bottle containing ammonia. Finally, take out the absorption bottle from the ice-water bath, add 2 drops of 0.1% methyl red solution and use NaOH standard solution (0.5 mol/L) to titrate the remaining HCl solution.

The content of NH_3 is calculated according to the following formula:
$$w_{NH_3}/\% = (c_1V_1 - c_2V_2) \times 17.04/(m_s \times 1000) \times 100$$

In the formula, c_1 and V_1 refer to the concentration and volume of the HCl standard solution respectively; c_2 and V_2 refer to the concentration and volume of the NaOH standard solution respectively; m_s refers to the sample mass.

2) Determination of SO_4^{2-} Content

Weigh about 0.65 g (the sulphur content is about 90 mg) of sample and add it into a 400 mL beaker, add 25 mL of distilled water to dissolve it, and then dilute the solution with water to 200 mL.

(1) Preparation of precipitation: add 2 mL of 6 mol/L HCl solution to the above solution and cover with a watch glass, then heat it to near boiling. Add 30~35 mL of 0.1 mol/L $BaCl_2$ solution into a small beaker and heat it to near boiling. Afterwards, add the hot $BaCl_2$ solution to the sample solution dropwise by a dropper and stir continuously. When the addition of $BaCl_2$ solution is going to be finished, add 1~2 drops of $BaCl_2$ to the supernatant which is under the static state to judge the completion of precipitation based on whether there is white turbidity or not. Finally, cover the beaker with a watch glass, continue to heat the solution to age the precipitation for about 10 min, and then let it cool to the room temperature.

(2) Filtration and washing of precipitation: pour the supernatant into a long neck funnel by the tilt-pour process and collect the filtrate with a clean beaker (check whether there is a precipitate passing through the filter paper. If so, renew the filter paper). Wash the precipitate 3~4 times with a small amount of distilled water and transfer the precipitate carefully to the filter paper. Afterwards, wash the inwall of the beaker with blowing water by using a washing bottle and merge the washing liquid into the funnel. Then tear the corner of the filter paper to wipe the glass rod and the inwall of the beaker, and place the corner of filter paper into the funnel. Finally, use a small amount of distilled water to wash the precipitation on the filter paper until there is no Cl^- detected in the filtrate.

(3) Drying and burning of precipitation: remove the filter paper from the funnel to wrap the precipitation and put it in a crucible with constant weight. Gently heat the precipitation for drying and carbonization first, and then burn it with over heat for ashing of the filter paper. Afterwards, put the crucible into the Muffle furnace to burn for 30 min at 800~850℃. Later, take out the crucible, place it in the dryer after the heat diminishes, and weigh the precipitation after cooling. Then repeat the burning for 20 min and repeat the procedure of cooling, taking out and weighing until the precipitation achieves constant weight. According to the weight of precipitation ($BaSO_4$), the percentage of SO_4^{2-} in the sample is calculated.

Questions

(1) Why should the solution be heated to near boiling, but not complete boiling?

(2) Why should the burning temperature of $BaSO_4$ precipitation be controlled at 800~850℃?

Exp. 26 Preparation and Quality Analysis of Zinc Gluconate

Objectives

(1) To understand the biological significance of zinc and the preparation of zinc gluconate.

(2) To consolidate and comprehensively apply the operations of evaporation, concentration, filtration, recrystallization, titration, etc.

(3) To understand the quality analysis method of zinc gluconate.

Principles

Zinc, whose content in an adult body is about 2.0~2.5 g, is one of the essential trace elements in the human body and is the most abundant one in the brain. It is now known that zinc is present in more than 70 enzyme systems (over 200 enzymes) in the human body, such as respiratory enzymes, lactate dehydrogenase, carbonic acid dehydrogenase, superoxide dismutase, alkaline phosphatase, DNA and RNA polymerases. Zinc plays an important role in maintaining the normal physiological functions of the body. It is related to the synthesis of nucleic acid and protein, the metabolism of carbohydrates and vitamin A as well as the activity of pancreas, gonad and hypophysis, which can promote the normal development of skin, bone and sexual organs and maintain digestive and metabolic activities. Zinc deficiency will lead to slow growth, dermatitis, parageusia and heterosmia, vision deterioration, gonadal immature, gastrointestinal diseases, immune dysfunction, etc. While zinc supplementation can reduce the incidence of diabetes and heart disease, enhance the regeneration capacity of wound tissue, improve appetites and digestive functions, fight against the cold, strengthen the immune system, etc. But excessive zinc in the body can inhibit the utilization of iron, which is prone to cause refractory anemia and other diseases.

As a medicine adopted widely for zinc supplementation, zinc gluconate has advantages of quick effects, high absorptivity, small side effects, etc. The bioavailability of zinc gluconate is about 1.6 times that of zinc sulfate with an equal dose of zinc. Zinc gluconate can be produced by reactions between gluconic acid or calcium gluconate and zinc oxide or zinc salt. In this experiment, zinc gluconate is prepared by the direct reaction between equimolar calcium gluconate and zinc sulfate.

$$[CH_2OH(CHOH)_4COO]_2Ca + ZnSO_4 \rightleftharpoons [CH_2OH(CHOH)_4COO]_2Zn + CaSO_4\downarrow$$

As a by-product which has low solubility, the $CaSO_4$ precipitate can be removed by filtration. If concentrating the filtrate and adding ethanol, a kind of colorless or white zinc gluconate crystal which is tasteless, soluble in water and very difficult to dissolve in ethanol can be separated out.

The content of the target compound in the product is determined by coordination titration, zinc ion is titrated with EDTA standard solution in the presence of weak alkaline buffer of NH_3-NH_4Cl, and then the content of zinc gluconate in the product is calculated according to the consumption of the titrant EDTA.

Experimental Items

Apparatuses and materials: suction flask, Büchner funnel, water pump, acid burette, alcohol burner, transfer pipette, evaporating dish, volumetric flask, measuring cylinder, beaker, Erlenmeyer flask, thermometer, electronic platform scale, analytical balance.

Reagents: calcium gluconate, $ZnSO_4 \cdot 7H_2O$, 95% ethanol, 25% $BaCl_2$ solution, 2 mol/L HCl solution, potassium sulfate standard solution, 0.05 mol/L EDTA standard solution, NH_3-NH_4Cl buffer solution (pH=10.0), chromium black T indicator.

Procedures

1. Preparation of Zinc Gluconate

First of all, take 12 mL of distilled water in a 100 mL beaker, add 3.3 g of $ZnSO_4 \cdot 7H_2O$ and stir. After the solid is dissolved, place the beaker in a water bath with constant temperature and heat it to 85~90℃, and dissolve 4.5 g of calcium gluconate in 12 mL of distilled water to obtain a white paste. Afterwards, add the paste into the beaker slowly under continuous stirring (10~15 min) and keep the beaker at constant temperature for 20 min.

Later, while the solution is hot, filter the solution under diminished pressure with double filter paper, wash the beaker and precipitate with 2~3 mL of water, transfer the filtrate into an evaporating dish, and concentrate it in a boiling water bath to be viscous (until the volume is about 10 mL; if there is a precipitate in the concentrated solution, filter it out). Then transfer the filtrate into an ice-water bath after it cools to room temperature, add 20 mL of 95% ethanol slowly to the filtrate and stir continuously, and then a large number of white gelatinous zinc gluconate is gradually precipitated out. Let the supernate sit for a while before removing it by decantation. Once again add 20 mL of 95% ethanol to the colloidal precipitate (which can be added in 2 batches), keep stirring the precipitate with a glass rod in an ice-water bath to gradually transform it into a hard one, and then disperse it into white powder particles (as small, loose and non-stick as possible). After that, filter it and use 3~5 mL of ethanol to wash the product for 1 time. Finally, adsorb the remaining solvent on the surface of the product with filter papers, then weigh and keep the product for further use.

2. Inspection of Sulfate

Weigh 0.5 g of zinc gluconate, dissolve it in 20 mL of water, add 2 mL of 2 mol/L hydrochloric acid solution, and then transfer the solution into a 25 mL colorimetric tube to obtain the test solution. Afterwards, take another 2.5 mL of potassium sulfate standard solution, add 17 mL of water and 2 mL of 2 mol/L HCl solution, and then transfer the solution into a 25 mL colorimetric tube to obtain the control solution (0.05%). Then add 1 mL of 25% $BaCl_2$ solution to the test solution and the control solution, respectively, dilute them with water to 25 mL, shake the solution well and let it stand. After 10 min, observe and compare turbidity from the top of the colorimetric tube under the same black background.

3. Determination of Zinc Content

Weigh 0.7 g of the product accurately with an analytical balance, add 50 mL of water to dissolve the product completely, and then add 5 mL of NH_3-NH_4Cl buffer solution (pH = 10.0) and several drops of chromium black T indicator (or about 15 small grains of solid chromium black T). Then titrate the solution with 0.05 mol/L EDTA standard solution until the solution turns from purple to blue, and record the consumption of EDTA standard solution. After parallel determination of three portions, calculate the zinc content of the sample by the following formula:

$$w_{Zn}/\% = \frac{c_{EDTA} \times V_{EDTA} \times 65}{m \times 1000} \times 100$$

In the formula, c_{EDTA} and V_{EDTA} are the concentrations of the titrant and the used volume respectively; m is the mass of the sample (g).

Data Recording and Processing

(1) Inspection of sulfate: ①phenomenal description; ②inspection conclusion.

(2) Determination of zinc content in zinc gluconate (Table 5.2).

Table 5.2 Determination of zinc content in zinc gluconate

Experimental number	1	2	3
m_1 (weighing bottle + zinc gluconate) / g			
m_2 (weighing bottle + remaining zinc gluconate) / g			
m (zinc gluconate) / g			
$V_{initial}$ (EDTA) / mL			
V_{final} (EDTA) / mL			
ΔV (EDTA) / mL			
w_{Zn} /%			
w_{Zn} /% (mean)			

Notes

Preparation of potassium sulfate standard solution: weigh 0.181 g of potassium sulfate in a 1 L volumetric flask, add moderate water to dissolve the solid and dilute the solution to the scale, shake them well, and obtain the potassium sulfate standard solution (per millilitre corresponds to 100 μg of SO_4^{2-}).

Questions

(1) If gluconic acid is chosen as the raw material, which of the following four zinc compounds should be selected, and why?

①ZnO; ②$ZnCl_2$; ③$ZnCO_3$; ④Zn $(CH_3COO)_2$.

(2) If the results of determining zinc gluconate content do not meet the required standards, what are the possible reasons?

(3) What's the role of 95% ethanol which is added during the precipitation and crystallization of zinc gluconate?

(4) Why must the preparation of zinc gluconate be done in a hot-water bath?

Exp. 27 Preparation and Photocatalytic Performance of Titanium Dioxide Nanoparticles

Objectives

(1) To understand the principle of preparation and photocatalysis of TiO_2 nanoparticles.

(2) To master how to prepare TiO_2 nanophotocatalyst by controlled hydrolysis and how to degrade organic wastewater by photocatalytic performance.

(3) To learn to use the magnetic stirrer, Muffle furnace, centrifuge, ultrasonic cleaner, photochemical reaction device, UV-Vis spectrophotometer, etc.

Principles

1. Semiconductor Photocatalytic Technology and Titanium Dioxide Photocatalyst

Environmental pollution restricts the sustainable development of society and threatens the health of people. Organic pollutants, such as pharmaceutical intermediates and organic dyes, usually have stable chemical structures, high biological toxicity and easy migration. It is difficult to remove them effectively only by microbial decomposition, and it cannot fundamentally achieve the pollution-free disposal by physical methods such as adsorption and extraction. But the photocatalytic performance of semiconductor nanoparticles can degrade organics into small molecules and non-toxic material such as carbon dioxide and water. Therefore, the preparation of the semiconductor photocatalyst and its application in organic wastewater treatment in recent twenty years have received widespread attention and been thoroughly researched.

Titanium dioxide (TiO_2) is an important chemical product. Nano-sized TiO_2 has excellent properties such as large specific surface area, high photocatalytic activity, corrosion resistance to light and ignorable environmental toxicity. It is by far the most in-depth and the closest to commercial application of semiconductor photocatalysts. TiO_2 has three kinds of crystalline phase structures, namely anatase, rutile and brookite (metastable phase). Its application research mainly focuses on the anatase phase and rutile phase, both of which belong to tetragonal system. Generally, the rutile phase (the bandgap is about 3.0 eV) is considered to be the thermodynamic stable phase, while the anatase phase is the dynamic stable phase. Compared with the rutile phase, the anatase phase has a larger bandgap (3.2 eV) and stronger catalytic activity. In general, the amorphous hydrated titanium dioxide starts to transform into anatase-type TiO_2 at about 300 ℃ and rutile-type

TiO$_2$ at about 550℃; as the temperature rises, the proportion of rutile type is increased, and almost all the anatase-type TiO$_2$ transforms into the rutile-type TiO$_2$ when the temperature is higher than 700℃.

The principle of nano-sized TiO$_2$ photocatalysis is summarized as follows. There is a band gap between the valence band full of electrons and the empty conduction band in the semiconductor. When the photon energy of the irradiated light is greater than the energy of the band gap (anatase: $\lambda_{max} \leqslant 387$ nm; rutile phase: $\lambda_{max} \leqslant 410$ nm), electrons in the valence band jump to the conduction band after being excited, and the formed electron (e$^-$) and hole (h$^+$) will migrate to the surface of TiO$_2$ particles respectively. In the aqueous suspension of nano-sized TiO$_2$, h$^+$ is captured by H$_2$O and OH$^-$ in the solution to produce hydroxyl radicals (·OH) with high reaction activity and strong oxidability, and e$^-$ is captured by O$_2$ adsorbed on the surface of TiO$_2$ to produce superoxide anion radicals (·O$_2^-$). Through further reaction, oxidative free radicals such as ·OOH can be produced. These oxidative free radicals can oxidize and degrade organic molecules gradually, and finally mineralize the products into inorganic substances such as CO$_2$ and H$_2$O. Therefore, the essence of photocatalytic reaction is the process of converting the absorbed light energy into chemical energy and realizing the organic degradation in the nanometer semiconductor. The reactions of producing oxidative species in the photocatalytic process are as follows:

Formation of electron-hole pairs

$$TiO_2 + h\nu \longrightarrow TiO_2 + h^+ + e^-$$

Recombination of electron-hole pairs

$$h^+ + e^- \longrightarrow recombination + energy$$

Formation of hydroxyl free radicals

$$H_2O + h^+ \longrightarrow ·OH + H^+$$

$$OH^- + h^+ \longrightarrow ·OH$$

Mutual transformations of oxidative species

$$O_2 + e^- \longrightarrow ·O_2^-$$

$$H_2O + ·O_2^- \longrightarrow ·OOH + OH^-$$

$$2·OOH \longrightarrow H_2O_2 + O_2$$

$$·OOH + H_2O + e^- \longrightarrow H_2O_2 + OH^-$$

$$H_2O_2 + e^- \longrightarrow ·OH + OH^-$$

2. Preparation of Nano-Sized TiO$_2$ and Affecting Factors on Its Performance

The common preparation methods of nano-sized TiO$_2$ include solid phase method (e.g., crushing method), gas phase method (e.g., chemical vapor deposition method) and liquid phase method (such as sol-gel method and controlled hydrolysis method). The controlled hydrolysis method is to slowly hydrolyze phthalate ester or Ti(IV) salt in an aqueous or alcohol/water solution to obtain hydrated titanium dioxide, and calcine hydrated titanium dioxide at high temperature to produce TiO_2 nanoparticle with high crystallinity. This method has short reaction time, easy operation and economical process. The hydrolysis and condensation reactions of phthalate ester are as follows:

$$—Ti—OR + H_2O \longrightarrow —Ti—OH + ROH$$

$$—Ti—OH + RO—Ti— \longrightarrow —Ti—O—Ti— + ROH$$

However, the photocatalytic reaction of nano-sized TiO_2 needs to be realized by using ultraviolet light, and the proportion of UV-light in sunlight reaching the ground is very small (less than 5%). The use of artificial light increases the cost of photocatalytic reaction, which is not conducive to commercial application. Some strategies, such as doping of metal or non-metal elements, semiconductor recombination, dye-sensitization and noble metal recombination, can be adopted to prepare TiO_2 or its composite with a visible light response, which provides the possibility to realize degrade organic pollutants by using clean solar photocatalysis.

When doped with metal ions (such as Fe^{3+}) at moderate concentration, the density of charge defects inside the TiO_2 crystal increases. Meanwhile, the impurity level formed by Fe^{3+} is in the forbidden band of TiO_2, and it makes the absorption band of the catalyst redshift from the ultraviolet region to the visible region (over 400 nm) and broadens, which contributes to more effective use of visible light. Proper defect density can be obtained by adjusting the doping concentration, which will help to improve the photocatalytic activity of TiO_2. Factors affecting the efficiency of photocatalytic reaction include catalysts (size, crystal structure, morphology, specific surface area, quantity of hydroxyl groups on the catalyst surface, etc.), photoreaction conditions (wavelength and intensity of the radiation light) and reaction systems (initial pollutant concentration, pH value, additional catalyst or auxiliary, atmosphere, etc.).

Experimental Items

Apparatuses and materials: magnetic stirrer, air dry oven, Muffle furnace, high-speed centrifuge, ultrasonic cleaner, photochemical reactor, ultraviolet-visible spectrophotometer, circulating water pump, three-necked flask, porcelain crucible, measuring cylinder, beaker, constant voltage funnel, Büchner funnel, suction bottle.

Reagents: titanium butoxide, iron nitrate nonahydrate, anhydrous ethanol, glacial acetic acid, 0.1 mol/L nitric acid, rhodamine B.

Procedures

1. Preparation of TiO_2 Nanoparticles

Install a 250 mL three-necked flask containing 15 mL of distilled water on the magnetic stirrer.

Take 3 mL of titanium butoxide with a dry measuring cylinder and dissolve it in 30 mL of anhydrous ethanol (the beaker used here should also be dried). Drop titanium butoxide solution slowly into the three-necked flask with a funnel while stirring quickly (12~15 min). After further stirring for 10 min, add 0.5 mL of glacial acetic acid and continue to stir for 10 min. Then filter the obtained white precipitate by a vacuum filter, wash it once with 8 mL of anhydrous ethanol, air-dry it to make ethanol volatilize completely, and dry it in the air dry oven at 80℃. Finally, grind the obtained white solid into powder and calcinate it at 500℃ for 2 h (heating rate: 10℃/min) to obtain TiO_2 nanoparticles.

Preparation of iron-doped TiO_2 nanoparticles: 15 mL of ferric nitrate solution is used instead of 15 mL of distilled water, and the other processes are the same as above. The preparation method of ferric nitrate solution is as follows: dissolve a certain amount of iron nitrate nonahydrate with 2 mL of 0.1 mol/L nitric acid solution (the dosage can be referred to Table 5.3), drop the solution into a 250 mL three-necked flask containing 13 mL of distilled water, and stir them evenly.

Table 5.3 The dosage of iron nitrate nonahydrate for preparation of iron-doped TiO_2 nanoparticles

Doping percentage of Fe^{3+}/%	0.2	0.5	1.0	1.5
Mass of iron nitrate nonahydrate/mg	6	15	30	45

2. Photocatalytic Performance of TiO_2 Nanoparticles

Add 50 mg of self-made TiO_2 nanoparticles and 33 mL of water to a 100 mL beaker, then add 17 mL of rhodamine B solution (30 mg/L) into the beaker after ultrasonic dispersion for 10 min and stir the solution evenly so as to obtain a reaction mixture in which the mass concentration of TiO_2 nanoparticles and rhodamine B solution is 1.0 mg/mL and 10 mg/L respectively. Place the beaker on a magnetic stirrer shielded from light, stir the mixture for 20 min. Then take 5 mL of sample, turn on the power of the photochemical reactor to make the sample exposed to the UV or visible light, and turn off the photochemical reactor after 15 min, then repeat these operations 3 times (the total reaction time is 1 hour). Centrifuge the reaction liquids which are taken at different times, absorb 3 mL of supernatant, and test its absorbance A at 554 nm by UV-Vis spectrophotometer. In the concentration range of the experiment, the concentration of rhodamine B is proportional to the absorbance (following the Lambert-Beer's law). Calculate the decoloration rate of rhodamine B (D, the decreasing percentage of the absorbance of rhodamine B) to measure the degradation degree of rhodamine B. What's more, draw the decolorizing rate-time (D-t) curve to evaluate the performance of degradation of rhodamine B by TiO_2 nanosized photocatalyst.

$$D = (A_0 - A)/A_0 \times 100\% = (c_0 - c)/c_0 \times 100\%$$

In the formula, A_0 and A, c_0 and c represent the absorbances and corresponding concentrations of the solution at 0 and t min, respectively.

Then compare the photocatalytic performance of pure TiO_2 nanoparticles and TiO_2

nanoparticles with different concentrations of doped Fe under UV or visible light irradiation, respectively, and analyze the effect of Fe doping on the improvement of catalyst activity.

Notes

(1) After the use, the measuring cylinder and other apparatuses containing titanium butoxide solution should be washed with 5~8 mL of ethanol as soon as possible, and then washed with water.

(2) It is preferred to use double filter paper to prevent the filter paper from breaking during vacuum filtration.

(3) The use of photochemical reactor shall strictly observe the operation procedure of the instrument; the UV protection goggles, gloves and other protective equipment are needed during the operation; the photochemical reaction device should be turned off during the sampling.

(4) The XRD curve of pure TiO_2 and iron-doped TiO_2 prepared in this experiment can be identified by comparison with the standard card of anatase-phase TiO_2 (JCPDS 89-4921) and rutile-phase TiO_2 (JCPDS 89-4920).

(5) The maximum absorption of rhodamine B is at 554 nm.

Questions

(1) Why should the apparatus containing titanium butoxide or its solution be strictly dried?

(2) What is the purpose of dissolving titanium butoxide in anhydrous ethanol? What is the role of nitric acid in preparing iron-doped TiO_2 nanoparticles?

(3) How to design experiments to judge the performance of self-made TiO_2 nanocatalysts?

(Hint: ①P25 is a kind of commercial TiO_2 photocatalyst; it is commonly used as a reference to compare the performance of self-made photocatalysts; ②the organic dye rhodamine B may have different degrees of degradation in ultraviolet or visible light.)

References

Bian G Q, Ji S J. 2007. Comprehensive Chemical Experiment. Suzhou: Suzhou University Press
Tian Y M. 2008. New College Chemistry Experiment. Beijing: Science Press
Xu R R, Pang W Q. 2006. Inorganic Synthesis and Preparative Chemistry. Beijing: Higher Education Press
Zhang X. 2009. Inorganic Chemistry Experiment. Beijing: Metallurgical Industry Press

Exp. 28 Preparation of Zinc Vitriol

Objectives

(1) To learn how to extract and prepare $ZnSO_4·7H_2O$ from zinc ashes.

(2) To cultivate students' problem-solving ability by comprehensively applying basic inorganic chemical theories and basic experimental skills.

Principles

Zinc ash is an industrial residue in zinc smelting. Small and medium-sized enterprises generate hundreds of tons of industrial residues every year, which contain zinc (measured in ZnO) of 30%~40%, silicon (measured in SiO_2) of about 30%, and Fe, Al, Cd, Cu, Ni and other metallic elements. Traditional slag landfill not only pollutes the environment, but also causes waste of resources. In view of the important recycling value of Zn, Fe, Al, Cd, Cu, Ni and other elements in the zinc ash, it has important economic value to prepare $ZnSO_4 \cdot 7H_2O$ on the basis of component analysis of raw material by means of slag extraction, separation and purification of impurity elements and crystallization of products.

The preparation process of $ZnSO_4 \cdot 7H_2O$ from a kind of zinc ash with $Zn_5(OH)_8Cl_2$ as the main mineral component and high Fe and Si content is briefly described below: obtain a zinc-containing leaching solution from sulfuric acid leaching of zinc ash, then increase the pH value of the leaching solution by adding ammonia water to prompt the hydrolysis and polymerization of Al^{3+}, Fe^{3+} and Si(IV) soluble species, as well as the formation of the precipitate (Fe^{2+} is hydrolyzed after oxidation by hydrogen peroxide). After removing the above impurity sediment, use zinc powder to reduce the trace impurity Cd^{2+}, Cu^{2+} and Ni^{2+} in the filtrate and obtain the solution containing Zn^{2+}, NH_4^+, SO_4^{2-} and Cl^-. Precipitate Zn^{2+} with ammonium carbonate, wash the precipitate to remove impurity NH_4^+ and Cl^-, then dissolve the precipitate by dilute sulfuric acid and obtain zinc sulfate solution. In view of that the solubility of $ZnCl_2$ in ethanol is much higher than that of $ZnSO_4$, ethanol is added to the zinc sulfate solution to make $ZnSO_4 \cdot 7H_2O$ precipitate and further reduce the residual Cl^- content in the product.

Experimental Items

Apparatuses and materials: analytical balance, platform scale, burette, pipette, Erlenmeyer flask, volumetric flask, measuring cylinder, beaker, dropping bottle, iron support, alcohol burner, circulating water pump.

Reagents: zinc ash, 0.05 mol/L EDTA standard solution, ammonium carbonate, zinc powder, 30% hydrogen peroxide, anhydrous ethanol, 3 mol/L sulfuric acid, 6 mol/L ammonia water, chromium black T indicator, NH_3-NH_4Cl buffer solution (pH=10.0).

Procedures

1. Preparation of Zinc Vitriol

Weigh 10 g of Zn ashes, put them in a 250 mL beaker, and add 20 mL of distilled water and 10 mL of 3 mol/L H_2SO_4 solution. Afterwards, place the beaker in a water bath at the temperature of 60~70℃ while stirring continuously (a small amount of water can be added according to the reduction of liquid volume). Half an hour later, add an equal amount of distilled water and dilute sulfuric acid, then continue to heat and stir the mixture for half an hour so that the zinc in the sample is further leached.

Then stop heating, and after the solution cools down to room temperature, filter and wash the residue 1~2 times with 3~5 mL of water. Measure the pH value of the filtrate and adjust it to about 4.0 (using precise pH test paper) by adding 6 mol/L ammonia water. Afterwards, stir for 10 min, add H_2O_2 solution (30%, highly corrosive), and then add ammonia water to maintain the solution at the pH of about 4.0. Let the solution react for 20 min to make sure the ferrous ions in the solution can be completely oxidized and hydrolyzed to form $Fe(OH)_3$ precipitation, and then continue to add ammonia water into the beaker to pH = 5.2~5.4. After 15 min, filter the reaction mixture by a vacuum filter, add 0.1 g of zinc powder to the filtrate, stir for 5 min, and filter the excess zinc powder to get the filtrate.

Prepare 15~20 mL of ammonium carbonate saturated solution and add it slowly to the filtrate while stirring until no white precipitate is produced. Next, collect the precipitate by a vacuum filter and wash it repeatedly until there are no ammonia ions being detected in the new filtrate. Afterwards, put the cleaned precipitate into a beaker, add 3 mol/L H_2SO_4 solution slowly to dissolve the precipitate completely, and then add an appropriate amount of anhydrous ethanol (10~15 mL) and let the solution stand after stirring for 5 min. The crude product of $ZnSO_4 \cdot 7H_2O$ can be obtained by filtering the precipitate.

The purity of the product can be further improved by recrystallization: add an appropriate amount of distilled water to dissolve the crude product, then evaporate and concentrate the solution to crystallize the product; after cooling, filter the solution and take out the sample, air dry it at relatively low temperature, and finally weigh it and calculate the yield.

2. Product Analysis and Inspection

Weigh 0.7 g of sample (on an analytic balance) to an Erlenmeyer flask and add 50 mL of water. After the sample is completely dissolved, add 5 mL of NH_3-NH_4Cl buffer solution (pH = 10.0) and several drops of chromium black T indicator, titrate the solution with 0.05 mol/L EDTA standard solution until it turns from fuchsia to blue, and record the consumption of EDTA standard solution. After the parallel determination of three portions, calculate the average content of Zn in the sample and the purity of zinc vitriol.

Questions

(1) What is the amount of ammonia water added to adjust the pH of the solution, and what is going to be removed when the solution is at pH = 4.0 and 5.2~5.4 respectively?

(2) After the replacement reaction of Zn powder, why do we choose the method that we first add carbonate to produce Zn^{2+}, then dissolve the precipitate with sulfate acid and finally precipitate the product by using ethanol, instead of the method of crystallization after the direct evaporation and concentration of the filtrate?

Exp. 29 Preparation and Composition Analysis of Potassium Bis (oxalato) Copper (II) Dihydrate

Objectives

(1) To further grasp the basic operations of dissolution, precipitation, suction filtration, evaporation, concentration, etc.

(2) To learn the preparation and composition determination of potassium bis(oxalato) copper (II) dihydrate crystals.

Principles

There are many methods to prepare potassium bis(oxalato) copper(II) dihydrate. It can be prepared by directly mixing the copper sulfate and potassium oxalate. It can also be prepared by the reaction of copper hydroxide or copper oxide with potassium hydrogen oxalate. In this experiment, the crystal will be prepared by the reaction of copper oxide with potassium hydrogen oxalate. Under alkaline conditions, $CuSO_4$ can generate the precipitate $Cu(OH)_2$ which can convert to CuO by heating. The mixed solution of K_2CO_3 and KHC_2O_4 can be obtained after dissolving a certain amount of $H_2C_2O_4$ in water and adding K_2CO_3 solution, which can react with CuO to form $K_2[Cu(C_2O_4)_2]$. The blue $K_2[Cu(C_2O_4)_2] \cdot 2H_2O$ crystal can be obtained after processes of evaporation, concentration and cooling. The reactions involved are as follows:

$$CuSO_4 + 2NaOH = Cu(OH)_2 \downarrow + Na_2SO_4$$
$$Cu(OH)_2 = CuO + H_2O$$
$$2H_2C_2O_4 + K_2CO_3 = 2KHC_2O_4 + CO_2 + H_2O$$
$$2KHC_2O_4 + CuO = K_2[Cu(C_2O_4)_2] + H_2O$$

Weigh a certain amount of sample, dissolve it in ammonia water and keep constant volume. Afterwards, divide the solution into two portions: neutralize the first portion by H_2SO_4, and titrate $C_2O_4^{2-}$ of the original sample in sulfuric acid solution with $KMnO_4$; add PAR indicator to the second portion in HCl solution, heat the solution to nearly boiling under the condition of pH = 6.5~7.5, and while it is hot, titrate it with EDTA solution to green as the end point to determine Cu^{2+} of the crystal. According to the consumption and concentration of $KMnO_4$ and EDTA, the content of $C_2O_4^{2-}$ and Cu^{2+} is calculated, and the component ratio of $C_2O_4^{2-}$ and Cu^{2+} is determined, and then the empirical formula of the product is calculated.

The PAR [4-(2-pyridyl azo) resorcinol] indicator belongs to pyridyl azo compounds, and its structural formula is as follows:

Compared with PAN, PAR has more hydrophilic group which makes itself and its chelates show good water-solubility, and it has much more obvious end point of titration to Cu^{2+} when pH = 5~7.

Experimental Items

Apparatuses and materials: platform scale, balance, beaker, measuring cylinder, suction device, volumetric flask, evaporating dish, pipette, acid burette, Erlenmeyer flask.

Reagents: PAR indicator, $CuSO_4 \cdot 5H_2O$, $H_2C_2O_4 \cdot 2H_2O$, K_2CO_3, copper (primary standard substance), NaOH (2 mol/L), HCl (2 mol/L), H_2SO_4 (3 mol/L), ammonia water (1 : 1), $KMnO_4$ standard solution (0.01 mol/L), EDTA standard solution (0.02 mol/L).

Procedures

1. Synthesis of Potassium Bis(Oxalato) Copper(II) Dihydrate

1) Preparation of CuO

Weigh 2.0 g of $CuSO_4 \cdot 5H_2O$ and add it into a 100 mL beaker, add 40 mL of water to dissolve it, and then add 10 mL of 2 mol/L NaOH under stirring. Afterwards, heat the solution over a low flame until the precipitate turns black (to form CuO), and keep it boiling for about 20 min. Then filter the solution with double filter paper after it cools a little and wash the precipitate 2 times with a small amount of deionized water.

2) Preparation of KHC_2O_4

Weigh 3.0 g of $H_2C_2O_4 \cdot 2H_2O$ and add it into a 250 mL beaker, then add 40 mL of deionized water and heat the beaker gently to dissolve the solid (the temperature should not exceed 85 ℃ to avoid the decomposition of $H_2C_2O_4$). When the solution cools down slightly, add 2.2 g of anhydrous K_2CO_3 by several times, and the mixed solution of KHC_2O_4 and $K_2C_2O_4$ is obtained after the dissolution.

3) Preparation of Potassium Bis(Oxalato) Copper(II) Dihydrate

Add CuO together with the filter paper to a beaker containing the mixed solution of KHC_2O_4 and $K_2C_2O_4$, heat the beaker in a water bath to dissolve the precipitate as much as possible (about 30 min). Filter the solution while it is hot (re-filter it if the solid breaks through the filter paper), wash the precipitate twice with a small amount of boiling water, and transfer the filtrate into an evaporating dish. Afterwards, concentrate the filtrate to about 1/3 of the original volume by vapour-bath heating, and then put it aside for 10 min before cooling it completely with cold water. After most of the precipitate has been separated out, absorb the water of the crystal with the filter paper after vacuum filtration, weigh it and calculate the yield, and preserve the product for the composition analysis.

2. Composition Analysis of the Product

1) Preparing the Sample Solution

Weigh 0.95~1.05 g of the synthetic crystal sample accurately (accurate to 0.0001 g) and add it into a 100 mL beaker, and add 5 mL of $NH_3 \cdot H_2O$ to dissolve it. Afterwards, add 10 mL of water and stir the sample solution well. Then transfer it into a 250 mL volumetric flask and add water to the scale.

2) Determining the Content of $C_2O_4^{2-}$

Take 25.00 mL of sample solution and add it into a 250 mL Erlenmeyer flask, add 10 ml of 3 mol/L H_2SO_4 solution, heat the mixed solution in a water bath to 75~85℃, and then keep it in the water bath for 3~4 min. While it is hot, titrate it with 0.01 mol/L $KMnO_4$ standard solution to pale pink. If the color does not fade after 30 s, the end point has been reached. Record the consumption of $KMnO_4$ standard solution. Titrate it three times in parallel.

3) Determining the Content of Cu^{2+}

Take another 25.00 mL of sample solution, and add 1 mL of 2 mol/L HCl solution, 4 drops of PAR indicator and 10 mL of buffer solution (pH = 7) in sequence. Heat the mixed solution to near boiling. While it is hot, titrate it with 0.02 mol/L EDTA standard solution to yellow-green. If the color does not fade after 30 s, the end point has been reached. Record the consumption of EDTA standard solution. Titrate it three times in parallel.

Data Recording and Processing

(1) Product record.

Morphology and color:

Mass:

Theoretical yield:

Yield:

(2) Titration Data record.

(3) Calculating the composition of the synthetic product.

Calculate w of $C_2O_4^{2-}$:

$$w_{C_2O_4^{2-}} = c_{KMnO_4} \times V_{KMnO_4} \times 88.02 \times 250 \times 5 \times 100\% / (m_{sample} \times 1000 \times 25.00 \times 2)$$

Calculate w of Cu^{2+}:

$$w_{Cu^{2+}} = c_{EDTA} \times V_{EDTA} \times 63.55 \times 250 \times 100\% / (m_{sample} \times 1000 \times 25.00)$$

Definite the composition of the synthetic product according to the molar ratio of Cu^{2+} and $C_2O_4^{2-}$.

$$\text{The molar ratio} = w_{Cu^{2+}} \times 88.02 \times 100\% / (63.55 \times w_{C_2O_4^{2-}})$$

Notes

(1) The solubility of potassium oxalato copper in water is very small. It can be dissolved by adding an appropriate amount of ammonia water, because ammonia water can help Cu^{2+} form cupric ammonia ions, and the pH of the solution is about 10 when dissolution occurs. The buffer solution which is mixed of 2 mol/L NH_4Cl solution and 1 mol/L ammonia water can also be taken as the solvent.

(2) The indicator itself is yellow under the titration condition, while Cu^{2+} and EDTA are blue and the end point is yellow-green. In addition to Cu^{2+}, PAR can be used as an indicator for the elements such as bismuth, aluminum, zinc, cadmium, copper, erbium, thorium and thallium, and it changes from red to yellow as reaching the end point.

Questions

(1) Please design other schemes to prepare potassium bis(oxalato) copper(II) dihydrate crystal by copper sulfate.

(2) Why not use potassium hydroxide to react with oxalic acid to form potassium hydrogen oxalate in this experiment?

(3) What's the principle of determining $C_2O_4^{2-}$ and Cu^{2+} respectively? In addition to the method in this experiment, are there any other analytical methods?

(4) How does the color change before and after titration with PAR as the indicator?

(5) What's the effect of the pH value on the sample analysis process?

References

Sun S J. 2004. Continuous determination of free nitric acid and copper silver in silver electrolyte. Gansu Chemical Engineering, 2: 47-48

Wei S G, Meng R Z, Cheng X M, et al. 2003. Studies on determination of oxalate and copper in $K_2[Cu(C_2O_4)_2]$. Journal of Guangxi Normal University (Natural Science Edition), 21(4): 316-317

Yan X M. 2002. Preparation and determination of potassium diaguotrioxalatocopperate(II). Journal of Southwest University for Nationalities, 28(1): 80-83

Zhang H. 2008. Principle and application of coordination chemistry. Beijing: Chemical Industry Press

Exp. 30 Synthesis and Composition Analysis of Potassium Trioxalatoferrate (III)

Objectives

(1) To learn how to synthesize potassium trioxalatoferrate (III).

(2) To master the basic principle and method of determining the chemical formula of compounds.

(3) To consolidate the basic operations of inorganic synthesis, titrimetric analysis and gravimetric analysis.

Principles

Potassium trioxalatoferrate (III) is a kind of bright-green monoclinic crystal, which is soluble in water but insoluble in the organic solvent such as ethanol and acetone. When heated, it will lose crystal water at 110℃ and decompose at 230℃. It is a kind of photosensitive material that is easy to decompose when exposed to light.

In this experiment, FeC_2O_4 firstly is prepared by the reaction of $(NH_4)_2Fe(SO_4)_2$ with $H_2C_2O_4$:

$$(NH_4)_2Fe(SO_4)_2 + H_2C_2O_4 \rightleftharpoons FeC_2O_4\downarrow + (NH_4)_2SO_4 + H_2SO_4$$

In the presence of excess $K_2C_2O_4$, the product can be obtained by oxidizing FeC_2O_4 with H_2O_2:

$$6FeC_2O_4 + 3H_2O_2 + 6K_2C_2O_4 \rightleftharpoons 4K_3[Fe(C_2O_4)_3] + 2Fe(OH)_3\downarrow$$

In the presence of $H_2C_2O_4$ and $K_2C_2O_4$, the intermediate product $Fe(OH)_3$ in the reaction can also be transformed into the product:

$$2Fe(OH)_3 + 3H_2C_2O_4 + 3K_2C_2O_4 \rightleftharpoons 2K_3[Fe(C_2O_4)_3] + 6H_2O$$

Each component content of the complex can be determined by the following analytical methods, and then its chemical formula can be obtained by calculation.

(1) Determination of the crystal water content by gravimetric analysis.

Dehydrate some of the product at 110℃, then the crystal water content can be calculated according to the weight loss.

(2) Determination of $C_2O_4^{2-}$ content by potassium permanganate titration.

$C_2O_4^{2-}$ can be quantitatively oxidized by MnO_4^- in acid media:

$$5C_2O_4^{2-} + 2MnO_4^- + 16H^+ \rightleftharpoons 2Mn^{2+} + 10CO_2 + 8H_2O$$

Use $KMnO_4$ standard solution with known concentration to titrate $C_2O_4^{2-}$, and the $C_2O_4^{2-}$ content can be calculated according to the consumption of $KMnO_4$.

(3) Determination of iron content by potassium permanganate titration.

Reduce Fe^{3+} to Fe^{2+} by Zn powder, and then use $KMnO_4$ standard solution to titrate Fe^{2+}:

$$5Fe^{2+} + MnO_4^- + 8H^+ \rightleftharpoons 5Fe^{3+} + Mn^{2+} + 4H_2O$$

The Fe^{3+} content can be calculated according to the consumption of $KMnO_4$.

(4) Determination of potassium content.

The K^+ content can be obtained by subtracting the content of crystal water, $C_2O_4^{2-}$ and Fe^{3+} from the complex.

Experimental Items

Apparatuses and materials: analytical balance, drying oven.

Reagents: H_2SO_4 (6 mol/L), $H_2C_2O_4$ (saturated), $K_2C_2O_4$ (saturated), H_2O_2 (w is 0.05), C_2H_5OH (w is 0.95 and 0.5), $KMnO_4$ standard solution (0.02 mol/L), $(NH_4)_2Fe(SO_4)_2\cdot 6H_2O(s)$, Fe powder, acetone, ice.

Procedures

1. Preparation of Potassium Trioxalatoferrate(III)

Dissolve 5.0 g of $(NH_4)_2Fe(SO_4)_2\cdot 6H_2O(s)$ in 20 mL of water, add 5 drops of 6 mol/L H_2SO_4 solution for acidification, and then heat the solution. After the solid is dissolved, add 25 mL of $H_2C_2O_4$ saturated solution, and heat the solution to boiling. Then let it stand and discard the supernatant after the precipitation of yellow FeC_2O_4 is complete. Wash the precipitate 2~3 times by decantation, and the water consumption is about 15 mL each time.

Add 10 mL of $K_2C_2O_4$ saturated solution to the precipitate, and heat the solution to 40 ℃ in a water bath. Use a dropper to slowly add 12 mL of the H_2O_2 solution with a mass fraction of 0.05 while stirring and maintain the temperature at around 40 ℃, and a brown precipitate [$Fe(OH)_3$] is produced in the solution at this time. After adding H_2O_2, heat the solution to boiling, and add 8 mL of $H_2C_2O_4$ saturated solution in two batches (5 mL is added first, then 3 mL is added slowly). At this time, the system should become a transparent bright green solution (control the volume at about 30 mL). If the system is turbid, filter it when it is hot. Add 10 mL of ethanol with a mass fraction of 0.95 to the filtrate, and if the solution is turbid at this time, heat it gently to make it clear. Then place the solution in the dark place for cooling and crystallization. After that, filter the solution, wash the crystal with an ethanol solution whose mass fraction is 0.5, rinse the crystal twice with a small amount of acetone, and then drain the solvent and dry the crystal in the air. Finally, weigh it and calculate the yield. Note that the product should be kept away from light.

2. Composition Analysis

1) Determination of Crystal Water Content

The crystal water content in the product is determined by the self-designed analysis scheme.

2) Determination of $C_2O_4^{2-}$ Content

The $C_2O_4^{2-}$ content in the product is determined by the self-designed analysis scheme.

3) Determination of Iron Content

The iron content in the reserved solution is determined by the self-designed analysis scheme.

4) Determination of Potassium Content

The content of K^+ can be calculated by the total content minus those of H_2O, $C_2O_4^{2-}$ and Fe^{3+}, and then the chemical formula of the complex can be determined.

Notes

(1) All crystal water will be lost when the product is being dried at 110 ℃ for 1 h.

(2) When using potassium permanganate to titrate $C_2O_4^{2-}$, in order to speed up the reaction, the temperature should be raised up to 75~85 ℃, but not more than 85 ℃, otherwise $H_2C_2O_4$ is easy to decompose.

(3) After the titration, keep the titration solution for measuring the iron content.

(4) The reducing agent iron powder should be excess, and the reaction system needs to be heated in order to ensure that iron can completely reduce Fe^{3+} to Fe^{2+}. Besides reacting with Fe^{3+}, iron powder also reacts with H^+ in the solution. Therefore, the solution must maintain enough acidity to avoid the hydrolysis and separation of Fe^{3+} and Fe^{2+}.

(5) The excess iron powder should be filtered out before titration. During the filtration process, Fe^{2+} should be transferred to the filtrate quantitatively. Therefore, iron powder in the funnel should be washed after filtration. The cleaning solution and filtrate are merged for titration. In addition, dilute H_2SO_4 should be used instead of water for washing (why?).

Questions

(1) Why do we boil the solution after dropping H_2O_2 during synthesis?

(2) What is the role of adding ethanol with a mass fraction of 0.95 in the final step of the synthetic product? Can the product be obtained by evaporating the solution to dryness, and why?

(3) Why does the product need to be washed several times? What effects does insufficient washing have on the determination of its composition?

(4) The crystal water content of $K_3[Fe(C_2O_4)_3] \cdot 3H_2O$ can be determined by heating and dehydration. Can all substances containing crystal water be determined by this method, and why?

Exp. 31 Preparation and Content Analysis of Calcium Peroxide

Objectives

(1) To master the principle and method of how to prepare calcium peroxide.

(2) To mater the method of how to analyse the content of calcium peroxide.

(3) To consolidate the basic operations of inorganic preparation and chemical analysis.

Principles

Calcium peroxide is a kind of white or light-yellow crystalline powder. It is stable at room temperature. When heated to 300 ℃, it can decompose into calcium oxide and oxygen. It is insoluble in water, but it is soluble in dilute acid and form hydrogen peroxide. Besides, it is widely used as fungicide, preservative, antacid, oil bleacher, nontoxic disinfectant of seeds and grains, as well as additives in food, cosmetics, etc.

Calcium peroxide can be prepared by the reaction of calcium chloride with hydrogen peroxide and alkali, or calcium hydroxide and ammonium chloride with hydrogen peroxide. The precipitate formed in the aqueous solution is $CaO_2 \cdot 8H_2O$. After it is dehydrated and dried at about 150 ℃, the product can be obtained.

The content of calcium peroxide can be determined by the reaction of calcium peroxide with acid under acidic conditions to form hydrogen peroxide and the titration with $KMnO_4$ standard solution.

$$CaCl_2 + H_2O_2 + 2NH_3 \cdot H_2O + 6H_2O = CaO_2 \cdot 8H_2O + 2NH_4Cl$$
$$5CaO_2 + 2MnO_4^- + 16H^+ = 5Ca^{2+} + 2Mn^{2+} + 5O_2 \uparrow + 8H_2O$$

Experimental Items

Apparatuses and materials: analytical balance, acid burette.

Reagents: $CaCl_2 \cdot 2H_2O(s)$, H_2O_2 (w=0.30), $NH_3 \cdot H_2O$ (concentrated), HCl (2 mol/L), $MnSO_4$ (0.05 mol/L), $KMnO_4$ standard solution (0.02 mol/L), ice.

Procedures

1. Preparation of CaO_2

Dissolve 7.5 g of $CaCl_2 \cdot 2H_2O$ in 5 mL of water, add 25 mL of H_2O_2 whose w is 0.30, and then add the mixture of 5 mL of concentrated ammonia water and 20 mL of cold water while stirring. Afterwards, cool the solution in an ice-water bath for 0.5 h, filter it, and wash the solid crystal 2~3 times with a small amount of cold water. Then drain the water and put the crystal into an oven at 150℃ for 0.5~1 h. As the crystal cools, weigh it and calculate the yield.

2. Content Analysis of CaO_2

Weigh two portions of products for about 0.15 g each with an analytical balance, and put each of them into a 250 mL Erlenmeyer flask. Add 50 mL of distilled water and 15 mL of 2 mol/L HCl solution to dissolve the product, respectively, and then add 1 mL of 0.05 mol/L $MnSO_4$ solution. Afterwards, titrate the solution with 0.02 mol/L $KMnO_4$ standard solution until the solution appears reddish, and the end point is when the solution color doesn't fade away in 30 s. Calculate the mass fraction of CaO_2. If the relative average deviation of the measured value is greater than 0.2%, a further titration is required.

Questions

(1) What are the main impurities in the product? How to improve its yield and purity?

(2) $KMnO_4$ is one of the most commonly used oxidants in redox titration, which is usually used in the acidic solution (dilute H_2SO_4 is generally used). Why doesn't this experiment use dilute H_2SO_4? Does the substitution of dilute HCl for dilute H_2SO_4 affect the determination results? How to prove it?

Exp. 32   Preparation of Basic Copper Carbonate

Objectives

Study the reasonable ratio of reactants and determine the suitable temperature conditions of the preparation reaction through the exploration of the preparation conditions of basic copper carbonate and the analysis of the color and state of the product, thereby developing abilities of designing experiments independently.

Principles

Basic copper carbonate $[Cu_2(OH)_2CO_3]$ is dark green or pale bluish, which is the main component of natural malachite green. It will decompose when heated to 200℃ and it has poor solubility in water. The newly prepared sample is easy to decompose in boiling water.

Experimental Items

Students should make a list of apparatuses, reagents and other materials needed for the experiment by themselves, and then do the experiment with the consent of teachers.

Procedures

1. Preparation of the Reactant Solution

Prepare 0.5 mol/L $CuSO_4$ solution and 0.5 mol/L Na_2CO_3 solution with 100 mL each.

2. Exploration of the Condition of the Preparation Reaction

1) Exploration of the Ratio of $CuSO_4$ and Na_2CO_3 Solutions

Add 2.0 mL of 0.5 mol/L $CuSO_4$ solution into four test tubes respectively, and then add 1.6 mL, 2.0 mL, 2.4 mL and 2.8 mL of 0.5 mol/L Na_2CO_3 solution into another four test tubes respectively. Afterwards, put these eight test tubes in a water bath at 75 ℃. Several minutes later, pour $CuSO_4$ solutions successively into Na_2CO_3 solutions with different volume respectively, oscillate the test tubes and compare the speed of the generation, quantities and colors of precipitates in each test tube, then conclude the best ratio of the two reactants.

2) Exploration of Reaction Temperature

Add 2.0 mL of 0.5 mol/L $CuSO_4$ solution into three test tubes, respectively, then add 0.5 mol/L Na_2CO_3 solution with the best reactant ratio obtained by the above experiment into another three test tubes. Afterwards, take one from these two groups of test tubes respectively and place them at room temperature. Several minutes later, pour $CuSO_4$ solution into Na_2CO_3 solution, oscillate the test tube, and observe the phenomenon. Place the other test tubes in a water bath at 50 ℃ or 100 ℃, repeat the procedure, and determine the appropriate temperature of the preparation reaction by experimental results.

3. Preparation of Basic Copper Carbonate

Take 60 mL of 0.5 mol/L $CuSO_4$ solution and prepare basic copper carbonate according to the appropriate reactant ratio and temperature determined by the above experiments. After the precipitate is complete, filter it under diminished pressure and wash it several times with distilled water until the final filtrate does not contain SO_4^{2-}, then collect the solid powder and drain it with filter paper. Dry the product in an oven at 100 ℃, and then weigh it and calculate the yield after the product cools to room temperature.

Questions

(1) Which copper salts are suitable for preparing basic copper carbonate? Please write down the chemical reaction equation of the reaction of copper sulfate solution with sodium carbonate solution.

(2) Do the reaction conditions such as the temperature of reaction, the concentration of reactants and the ratio of reactants affect the reaction product?

(3) Why do the precipitates in each tube have different colors? In which color does the product contain the highest amount of basic copper carbonate?

(4) If Na_2CO_3 solution is poured into $CuSO_4$ solution, will the result be different?

(5) At what temperature will some brown (or brownish black) substance be produced in the reaction? What is the brown substance?

(6) Besides the reactant ratio and the reaction temperature, will the reactant type, the reaction time and other factors affect the quality of the product?

(7) Please design an experiment to determine the content of copper and carbonate in the product and analyze the quality of prepared basic copper carbonate.

Exp. 33 Determination of the Composition and Stability Constant of Titanium (IV) Peroxide Complex

Objectives

(1) To grasp the principle and method of spectrophotometry to determine the concentration of complexes.

(2) To be familiar with how to use the equimolar series method to determine the composition of complexes and calculate their stability constants.

(3) To learn the construction and usage of spectrophotometer.

Principles

The determination of the composition and stability constants of complexes plays an important role in understanding the properties of complexes and inferring their structures. At present various methods of determination have been established, such as photometric method, pH potentiometric titration, ion selective electrode method, polarographic method and extraction method.

Photometric method is one of the most commonly used methods of determination. It is simple and fast, especially suitable for low-concentration solutions, and the range of solvent selection is relatively large. The principle is based on the characteristic maximum absorption of the complexes at a certain wavelength, and the solution of the complexes has different absorption of light from the solution of the original ligands and metal ions. The relation between the absorbance value A and the solution composition at a certain wavelength is consistent with the absorption law under ideal conditions:

$$A = b \sum_{i=0}^{n} \varepsilon_i c_i$$

In the formula, b is the optical path length of the cuvette; and ε_i is the molar absorption coefficient of the i particle at the concentration of c_i. The molar absorption coefficient and concentration of the complexes with various coordination ratios that may form in the system are difficult to measure. Therefore, the equilibrium concentration of the complexes cannot be calculated directly from the absorbance value of the solution to calculate their stability constants. If the target complex is stable enough and the ligand concentration is high enough, the molar absorption coefficient of the target complex which is saturated can be measured. Complexes can exist in various forms in the solution, so there are also various methods of data processing.

The equimolar series method (also known as the Job's method) is used in this experiment to determine the composition and stability constant of the complex. It changes the molar ratio of M and L successively under the circumstance that the volume of the solution is fixed and the total number of moles of the metal ion M and the ligand L remains unchanged. As the molar ratio is different, the amount of the saturated target complex is different. Prepare a series of solutions with different molar ratios of M/L to form a group of test solutions and determine their absorbance values, and then draw the molar ratio-absorbance curves of the solutions. If the produced complex is stable, the curve has an obvious maximum value [as shown in Fig. 5.1(a)], then the composition of the complex can be determined. If the produced complex is not very stable, that is, it has a certain degree of dissociation, the maximum value of the curve is not obvious [as shown in Fig. 5.1(b)], and at this point, draw the tangent lines to the curve from two endpoints of abscissa, the maximum value of the curve can be determined by the intersection point of the tangent lines, and then the composition of the complex can be determined by its maximum value.

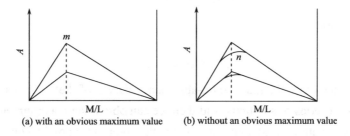

(a) with an obvious maximum value (b) without an obvious maximum value

Fig. 5.1 The functional relation between absorbance value A and the M/L molar ratio

1. Determination of the Complex Composition

The coordination reaction of the complex in the system is:

$$M + nL \rightleftharpoons ML_n \tag{5.1}$$

Suppose $[M] + [L] = c$. In the formula, $[M]$ and $[L]$ are the initial concentrations of the metal ion and ligand in the system respectively and c is the constant in the experiment.

The mole fraction of L is x. At equilibrium, the concentration of the metal ion is c_M, and then the concentration of the ligand c_L and the concentration of the complex y are:

$$c_M = c(1-x) - y \tag{5.2}$$

$$c_L = cx - ny \tag{5.3}$$

$$y = \beta c_M c_L^n \tag{5.4}$$

In the formula, β is the stable constant of the complex.

The differential formulas of the above are:

$$\frac{dc_M}{dx} = -c - \frac{dy}{dx} = 0 \tag{5.5}$$

$$\frac{dc_L}{dx} = c - n\frac{dy}{dx} \tag{5.6}$$

$$\frac{dy}{dx} = \beta c_L^n \frac{dc_M}{dx} + n\beta c_M c_L^{n-1} \frac{dc_L}{dx} \tag{5.7}$$

When the concentration of the complex ML_n is very high, $\frac{dy}{dx} = 0$. The formula (5.7) is:

$$c_L \frac{dc_M}{dx} + nc_M \frac{dc_L}{dx} = 0 \tag{5.8}$$

Plug the $\frac{dc_M}{dx}$ and $\frac{dc_L}{dx}$ in the formula (5.5) and (5.6) into the formula (5.8):

$$c_L = nc_M$$

Then plug c_L into the formula (5.3):

$$nc_M = cx - ny \tag{5.9}$$

Multiply (5.2) by n:

$$nc_M = nc - ncx - ny \tag{5.10}$$

(5.9) minus (5.10):

$$n = \frac{cx}{c(1-x)} = \frac{x}{1-x}$$

Therefore, we can draw the molar ratio-absorbance curves of the solutions by measuring the absorbance value of a series of solutions, and then we can obtain the maximum absorbance value x and calculate n (coordination ratio), thus determining the composition of the complex, ML_n.

If no other complexes are formed within the range of different concentrations, the position of maximum value remains unchanged. The lower the concentration is, the more obvious the complex dissociation is and the flatter the curve is. If the ligand and metal ion also absorb the light at the same wavelength, the absorbance value of the ligand and metal ion should be deducted from the total absorbance value. This method can not only determine the composition of the complex, but also calculate its stability constant.

2. Calculation of the Stability Constant of the Complex

The coordination and dissociation equilibrium of the complex in the system is as follows:

$$M + nL \rightleftharpoons ML_n$$

So the stability constant β of the complex ML_n is:

$$\beta = \frac{[ML_n]}{[c_M][c_L]^n}$$

If two solutions with different composition have the same absorbance value, the concentrations of the complexes in the two solutions must be the same (a and b or a' and b' as shown in Fig. 5.2).

Suppose the concentration of the complex is c_x, [M] and [L] are the initial concentrations of the metal ion and ligand respectively. Then

$$\beta = \frac{c_x}{(M_a - c_x)(L_a - nc_x)^n} = \frac{c_x}{(M_b - c_x)(L_b - nc_x)^n} \tag{5.11}$$

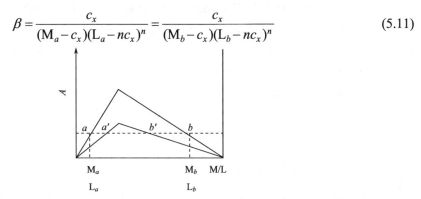

Fig. 5.2　The functional relation between absorbance value A and M/L molar ratio

Two points with the same absorption value on the same molar ratio-absorbance curve can be taken and substituted into equation (5.11) to calculate the concentration of the complex c_x and the stability constant of the complex β. If the composition of the complex is ML (i.e., $n = 1$), β cannot be calculated directly from two points on a molar ratio-absorbance curve. It should be calculated by finding two points of the same extinction value from different molar ratio-absorbance curves (a and a' or b and b' as shown in Fig. 5.2).

This method is simple, quick and reliable. However, when the stability constant of the complex is too large or too small, the system with too high coordination number cannot get correct results. The composition and stability constant of Ti (IV)-H_2O_2 complex in this experiment are determined by the equimolar series method.

Experimental Items

Apparatuses and materials: UV-Vis spectrophotometer, analytical balance, electric furnace, burette stand, Erlenmeyer flask, brown reagent bottle, beaker, volumetric flask, acid burette, graduated pipette, measuring cylinder.

Reagents: titanium potassium oxalate [$K_2TiO(C_2O_4)_2 \cdot 2H_2O$], hydrogen peroxide ($H_2O_2$, 30%), potassium permanganate ($KMnO_4$), 0.2 mol/L manganese sulfate solution, oxalic acid ($H_2C_2O_4 \cdot 2H_2O$), sulfuric acid.

Procedures

1. Preparation of Ti (IV) Solution

Accurately weigh 0.3542 g of titanium potassium oxalate, dissolve it with 20 mL of 1 : 1 (V/V)

H_2SO_4 by gentle heating. After cooling, transfer the solution into a 100 mL volumetric flask, dilute it to the scale with water and shake well, then the solution with a Ti (IV) concentration of 1.00×10^{-2} mol/L is obtained.

2. Preparation and Calibration of Hydrogen Peroxide Solution

(1) Pipette 1 mL of 30% H_2O_2 and dilute it with 2 mol/L H_2SO_4 solution to 100 mL. Take 10 mL of the above H_2O_2 solution in an Erlenmeyer flask, dilute it with distilled water to about 20 mL, add 2~3 drops of 0.2 mol/L manganese sulfate solution, and use the standard solution of potassium permanganate to calibrate it.

(2) Preparation of H_2O_2 solution with a concentration of 1.00×10^{-2} mol/L: absorb a certain amount of H_2O_2 solution with known concentration in a 100 mL volumetric flask, then dilute it to the scale with 2 mol/L H_2SO_4 solution and shake it well.

3. Preparation and Calibration of Potassium Permanganate Solution

(1) Weigh 1.70 g of $KMnO_4$ solid in a 250 mL beaker, dissolve it with 200 mL of boiling water, pour the supernatant liquid into a brown bottle, and then add 300 mL of distilled water and shake well. After resting for two days, siphon the supernatant liquid into a 600 mL beaker and discard the leftovers in the bottle. Then wash the brown bottle, pour the potassium permanganate solution into it and keep it in the dark place. The concentration is to be calibrated.

(2) Accurately weigh 0.15 g of $H_2C_2O_4 \cdot 2H_2O$, put it into two Erlenmeyer flasks, add 25 mL of distilled water and 5 mL of 2 mol/L H_2SO_4 solution to dissolve it, and slowly heat it until the solution is steaming hot (70~80℃). While it is hot, titrate it with potassium permanganate solution, and the titration speed may not be too fast. At the end of the titration, shake the solution well to prevent it from exceeding the end point. If the last half drop of potassium permanganate keeps reddish in half a minute after shaking the solution well, it indicates that the titration has reached the end point. Write down the consumption of potassium permanganate solution and repeat the calibration two to three times.

(3) Calculate the concentration of potassium permanganate solution.

4. Selection of the Determination Wavelength of Ti (IV)-H_2O_2 Complex

The sulfuric acid solution and hydrogen peroxide solution of Ti (IV) do not have absorption in the visible region, and the wavelength at which Ti (IV)-H_2O_2 has the maximum absorption can be selected as the determination wavelength. Pipette 2.5 mL of 1.00×10^{-2} mol/L Ti (IV) solution and 2.5 mL of 1.00×10^{-2} mol/L H_2O_2 solution into a 25 mL volumetric flask, dilute it to the scale with 2 mol/L H_2SO_4 solution, shake it well, and the solution to be tested is obtained. Take the 2 mol/L H_2SO_4 solution as the reference solution and determine the absorbance value of Ti (IV)-H_2O_2 solution by a 1 cm cuvette and within the wavelength range of 360~600 nm. Then draw an absorption curve with the determined absorbance value as the ordinate and the wavelength as the abscissa, and select the appropriate determined wavelength.

5. Determination of the Composition and Stability Constant of Ti(IV)-H_2O_2 Complex

Prepare a series of mixed solutions according to the equimolar series method: add 1.00×10^{-2} mol/L Ti(IV) solution and 1.00×10^{-2} mol/L H_2O_2 solution with different volume ratios (i.e., molar ratios) into different 25 mL volumetric flasks, dilute the solutions to the scale with 2 mol/L H_2SO_4 solution, and then determine their absorbance values.

Data Recording and Processing

1. Drawing the Absorption Curve of Ti(IV)-H_2O_2 Complex

The absorbance values of the complex at different wavelengths are measured and recorded in Table 5.4.

Table 5.4 The absorbance values of the complex at different wavelengths

Wavelength λ/nm	360	380	400	410	420	440	460	480	500	520	560	600
Absorbance A												

The absorption curve of Ti(IV)-H_2O_2 complex is made according to absorbance value A and wavelength λ, which is used to further determine the measured wavelength of the complex.

2. Determination of the Composition of Ti(IV)-H_2O_2 Complex

The absorbance values at different molar ratios are recorded in Table 5.5.

Table 5.5 The absorbance values of the complex at different molar ratios

Solution number	1	2	3	4	5	6	7	8	9	10
Volume of Ti(IV) solution/mL										
Volume of H_2O_2 solution/mL										
Absorbance A										

Draw molar ratio-absorbance curve of Ti(IV)-H_2O_2 complex with A as ordinate and Ti(IV)/H_2O_2 molar ratio as abscissa, and the composition of the complex is determined by the maximum position of the curve.

3. Calculation of Stability Constant β of Ti(IV)-H_2O_2 Complex

Find the solution composition corresponding to any two points with the same absorbance value on the molar ratio absorbance curve, calculate the concentration of the complex by the formula of (5.11), and then calculate the stability constant β of the complex.

Notes

(1) Hydrogen peroxide solution should be prepared temporarily.

(2) The solution with different molar ratios of M and L must be accurately prepared under the condition that the total concentration of M and L remains unchanged.

(3) It is recommended to prepare 8~10 solutions with 5.0 mL of total volume of 5.0 mL and 1 : 9~9 : 1 of molar ratio.

Questions

(1) Explain the scope of application and the cause of the error of the equimolar series method for determining the stability constant.

(2) When calculating the stability constant β of the complex ML_n, two points with equal absorbance value on the same molar ratio-absorbance value curve can be taken for $n \neq 1$ type complexes, while two points on different molar ratio-absorbance value curves must be taken for $n=1$ type complexes. Why is that?

References

Institute of Coordination Chemistry, Nanjing University. 1984. Coordination Chemistry (Inorganic Chemistry Series, Volume 12). Beijing: Science Press
Wang B K, Qian W Z, et al. 1984. Intermediate Inorganic Chemistry Experiment. Beijing: Higher Education Press
Yangzhou University, Xuzhou University, Yancheng Normal University, etc. 2010. New College Chemistry Experiment (II). Beijing: Chemical Industry Press

Exp. 34 Preparation and Content Analysis of Cobalt Ferrite $CoFe_2O_4$

Objectives

(1) To study the principle of preparing ferrite and metal oxide by coprecipitation.

(2) To design the preparation of the magnetic composite oxide $CoFe_2O_4$ by coprecipitation.

(3) To be familiar with: the characterization of the sample phase by XRD; the observation of appearance, grain size and size distribution of the sample by transmission electron microscopes and scanning electron microscopes; the principle and method of determining the magnetic parameters of the sample by vibrating sample magnetometers and testing its magnetic properties qualitatively by magnets.

Principles

Ferrite has been widely used in permanent magnets, magnetic recording, magnetic fluids, anti-electromagnetic interference and electronic equipment due to its high cubic magnetocrystalline anisotropy, high coercivity and moderate saturation magnetization. Ferrites are both a magnetic medium and a dielectric medium. It has multiple absorbing mechanisms and loss mechanisms, such as resonance absorption, scattering absorption, single domain absorption, magnetic loss and dielectric loss. It has now become one of the promising magnetic functional materials and electromagnetic wave absorption materials. Magnetic material is the pillar industry in Zhejiang Province, with a priority development status. Therefore, it is of great strategic significance to be familiar with and understand the structure, preparation and characterization method of magnetic oxides such as $CoFe_2O_4$.

Under certain conditions, the method which simultaneously precipitates two or more metal ions in one solution with a precipitant is called coprecipitation. Take the preparation of $CoFe_2O_4$ as an example, the preparation principle is as follows:

$$Co^{2+} + 2Fe^{3+} + 8OH^- \longrightarrow Co(OH)_2 + 2Fe(OH)_3$$

$$Co(OH)_2 + 2Fe(OH)_3 \xrightarrow{\text{thermal decomposition}} CoFe_2O_4 + 4H_2O$$

$CoFe_2O_4$ has a cubic spinel structure. It can be seen as Co^{2+} replacing Fe^{2+} of Fe_3O_4, in which two Fe^{3+} occupy the tetrahedron (A) and octahedron (B) of spinel cubic cells, respectively. Its single electron orbital spins in parallel and the spin magnetic moment cancels out with each other. Therefore, the magnetism of $CoFe_2O_4$ is derived from Co^{2+} and belongs to the ferromagnetic class.

Experimental Items

Apparatuses and materials: balance, magnetic stirrer, thermometer, beaker, electric furnace, oven, Muffle furnace, X-ray powder diffractometer, scanning electron microscope, transmission electron microscope, vibrating sample magnetometer and other instruments for characterization.

Reagents: cobalt nitrate hexahydrate, iron nitrate nonahydrate, sodium hydroxide, polyethylene glycol (or vinyl alcohol), etc.

Procedures

1. Preparation of $CoFe_2O_4$

According to the stoichiometric ratio of preparing 0.01 mol $CoFe_2O_4$, accurately weigh moderate metallic nitrates and add 40 mL of distilled water (add 2%~3% polyethylene glycol solution) while stirring continuously to make reactants completely dissolved. Transfer the solution into a three-necked flask, put the flask in a water bath at 60°C, and add 1 mol/L NaOH solution while stirring continuously to adjust the solution pH to around 9. Stop adding NaOH solution when there is a large amount of brown precipitate in the solution. Then raise the temperature to 80°C and stir for about 2 h. After the product cools down, transfer it into a beaker, let the solution stand for a while to make it layered, then use a permanent magnet to help the precipitate stay at the bottom of the beaker. Discard the supernatant liquid, use distilled water to wash the solid precipitate several times until the supernatant liquid is neutral, and then wash the precipitate 2~3 times with anhydrous ethanol and obtain the precursor of $CoFe_2O_4$. Put the precursor in an oven at 300°C for 2 h, and then the $CoFe_2O_4$ powder is obtained (if you want to get a product with a better crystal form and better magnetic properties, you can sinter the powder at 500°C for 2 h).

2. Microstructure and Magnetic Characterization

(1) Determine the content of iron (take salicylic acid as the indicator) and cobalt (take xylenol orange as the indicator, in the presence of methenamine, at pH = 5~6) by EDTA.

(2) Determine the crystal phase of $CoFe_2O_4$ by a X-ray powder diffractometer.

(3) Observe the morphology of $CoFe_2O_4$ by a scanning electron microscope.

(4) Observe the size and size distribution of $CoFe_2O_4$ grains by a transmission electron microscope.

(5) Determine the magnetic parameters of $CoFe_2O_4$ by the vibration sample magnetometer or test its magnetic property by the magnet.

3. Results and Discussion

(1) Calculate the yield of $CoFe_2O_4$:

$$\text{Yield} = m_{CoFe_2O_4 \text{(actual)}} / m_{CoFe_2O_4 \text{(theoretical)}} \times 100\%$$

(2) Calculate the content of $CoFe_2O_4$:

$$\text{The mass fraction of } CoFe_2O_4 = c_{EDTA} \times V_{EDTA} \times M_{CoFe_2O_4} / 2m \times 100\%$$

In the formula, m is the mass of the sample used for the titration.

(3) Analyze the reasons for the difference between the weight analysis results and the capacity analysis results.

Notes

(1) The key to the preparation of $CoFe_2O_4$ is the regulation of pH and reaction temperature.

(2) If there are impurity peaks in the XRD pattern of the prepared powder sample, it is indicated that the hydroxide in the sample has not been fully decomposed or $CoFe_2O_4$ has poor crystalline performance. The sample can be sintered at 400℃ in the Muffle furnace for 1 h.

(3) The characteristic peaks of XRD pattern are: 2θ=18.37°, 30.34°, 35.52°, 43.17°, 53.70°, 57.14° and 62.76° (JCPDS-ICDD 22-1086).

Questions

(1) Why do we prepare $CoFe_2O_4$ by the coprecipitation method?

(2) When $CoFe_2O_4$ is prepared, why do we adjust the pH of the solution to around 9?

Exp. 35　Preparation, Optical Absorption Properties and Stability of Gold (Silver) Colloids

Objectives

(1) To understand the preparation methods, properties and applications of noble metal

nanoparticles.

(2) To master the chemical reduction method and photo-chemistry method to prepare gold and silver colloidal solutions, learn to use photochemical reaction device, UV-Vis spectrophotometer, laser particle analyzer, etc., and to process the data representation of the product.

(3) To analyze how different stabilizers help the colloidal solution resist the coagulation effect of electrolytes.

Principles

1. Overview of the Properties of Noble Metal Nanoparticles

The noble metal nanoparticles and the corresponding metal bulk materials show very different properties in many aspects. It is well known that the bulk metal has a yellow metallic luster, while gold particles of 3~20 nm present red affected by the surface plasmon resonance (SPR)[1], and their colors vary with the size and morphology of the particle involving orange, wine red, purple, etc. The position of SPR absorption peak is also related to the particle shapes, pH values of solutions, types of protectants, etc. In terms of catalytic activity, nanoparticles with specific morphologies and structures, such as platinum, gold, rhodium and other nanoparticles, show much higher catalytic performance than micron or millimeter scale particles due to more highly active surfaces being exposed. Based on the high chemical stability, good biocompatibility, special surface chemical properties, optical and electrical properties, catalytic properties and other characteristics, gold colloids and silver colloids have important applications in optoelectronics, sensors, biomedicine and industrial catalysis[2].

2. Preparation of Noble Metal Nanoparticles

The commonly methods are to be used to prepare noble metal nanoparticles include chemical reduction method, photochemical method, radiation chemical method, acoustic chemical method, impregnation method, etc. The used precursor compounds of noble metals include silver nitrate, chloroauric acid, chloroplatinic acid, rhodium chloride, etc., which are reduced to a single atom by different methods in the solution or emulsion state. The particles with specific size and morphology can be obtained by controlling the material concentration and reaction rate to regulate the nucleation and growth process of particles. At the same time, the use of stabilizers can ensure a certain stability of noble metal nanoparticle dispersions. Large size ions, surfactants and polymers can be used as stabilizers of noble metal nanoparticles. Citrate, polyvinyl alcohol, hexadecyl trimethyl ammonium bromide and mercaptan are commonly used to prepare gold or silver colloids.

As for Au, the electrode potential of $[AuCl_4]^- + 3e^- \rightleftharpoons Au(s) + 4Cl^-$ is 0.93 V. The reaction can be completed within 1 min when using the stronger reductant sodium borohydride at normal temperature, while the $HAuCl_4$ solution is required to react for 15~30 min in the boiling state when using the weaker reductant sodium citrate (Frens method). Silver or gold colloids can also be prepared by using UV irradiation to reduce silver nitrate or chloroauric acid[3-5]. The use of ethanol, ethanediol and polyvinyl alcohol can accelerate the process of photochemical reduction reaction, which can affect the size of gold colloids. The mechanism of preparing gold nanoparticles by UV irradiation can be found in the reference[6]. The oxidability of Ag^+ is weaker than that of Au (the electrode potential of $Ag^+ + e^- \rightleftharpoons Ag$ is 0.799 V), so the above synthetic method can also be used to prepare the silver colloidal solution[7].

3. Stability of Colloidal Solution

The colloidal system formed by the dispersion of nanoparticles is an unstable thermodynamic system. On the one hand, particles tend to coalesce with each other to reduce their surface energy; on the other hand, the electrostatic repulsion between particles and the steric effect of surface modified molecules both contribute to inhibit the agglomeration of particles and keep the colloidal solution stable. But the stability can be damaged by changing the temperature of the system and the concentration of the electrolyte, which will cause the particles to gather and sink. At this time, both the position and intensity of the SPR absorption peak of noble metal nanoparticles will change. Therefore, the occurrence of colloid agglomeration can be learned by detecting the UV-Vis absorption spectra of the sol. In addition, the dynamic light scattering provided by the laser particle analyzer can be used to test the hydraulic particle size distribution diagram of noble metal nanoparticles in the dispersion medium, thus the average hydraulic particle size and multiple dispersion coefficient of the particles can be obtained. The UV-Vis absorption spectra of colloidal solution may not change significantly when the electrolyte causes the slight aggregation of particles. However, the dynamic light scattering can accurately detect the increase of the average hydraulic particle size and multiple dispersion coefficients, thus determining the stability of the colloidal solution.

Experimental Items

Apparatuses and materials: laser particle analyzer, UV-Vis spectrophotometer, UV chemical reaction device (with a 254 nm UV lamp), magnetic stirrer, air dry oven, quartz test tube, cuvette, beaker, test tube, pipette.

Reagents: silver nitrate solution ($AgNO_3$, 2.5 mol/L), chloroauric acid solution ($HAuCl_4$, 2.5 mmol/L), sodium citrate solution (25 mmol/L), carboxymethyl chitosan solution (CCT, 1%), sodium borohydride, polyethylene glycol 2000 (PEG 2000), hexadecyl trimethyl ammonium bromide (CTAB), ethanol, hydrochloric acid, nitric acid, sodium chloride (1 mol/L).

Procedures

1. Preparation of Gold (or Silver) Colloidal Solution

(1) Chemical reduction method: add 2 mL of 2.5 mmol/L $HAuCl_4$ (or $AgNO_3$) solution, 16 mL of water, 2 mL of sodium citrate solution and 1 grain of magnetic stir bar to a 50 mL beaker. Place the beaker on a magnetic stirrer and quickly add 0.6 mL of 1% $NaBH_4$ solution which is newly

prepared with ice water. After the color of the solution is stable (about 30 s), stop stirring and obtain the gold (or silver) colloidal solution.

(2) Photochemical method: after mixing 2mL of CCT solution, 16 mL of water and 2 mL of $HAuCl_4$ (or $AgNO_3$) solution well, put the solution into a quartz test tube and fix it in the photochemical reaction device. Afterwards, turn on the UV light for irradiation and prepare the stable gold (or silver) colloidal solution of CCT. Then take 3 mL of sample after 15 min, 30 min and 40 min respectively, observe the colors of samples and determine the absorption spectrum. Replace the CCT solution with 2 mL of 15 mg/mL PEG 2000 or CTAB solutions and repeat the above experiment, then the stable gold (or silver) colloidal solution of PEG 2000 or CTAB can be obtained.

2. Optical Absorption Properties and Stability Testing of Colloids

(1) Determine the UV-Vis absorption spectra of gold (or silver) colloids prepared in above different reaction conditions, take the water as reference, and the range of wavelength scanning is 300~800 nm.

(2) Transfer three gold (or silver) colloidal solutions (each one is 2 mL) prepared by different stabilizers to three clean test tubes respectively, and observe whether there is a Tyndall effect by a laser pen in a dark place. Afterwards, add 1 mol/L NaCl solution to the above colloidal solutions respectively, slightly oscillate to mix them evenly, and observe whether the color of the solution changes and the amount of NaCl solution used when the color changes. Compare the stability of the colloid in the presence of high concentration electrolyte, and scan the UV-Vis absorption spectra of the colloidal solution after the addition of NaCl.

(3) Select the sample with the least obvious color change in step (2), place 1.2 mL of original colloidal solution and 1.2 mL of colloidal solution with NaCl in two sample pools respectively. Then use the laser particle analyzer to test hydraulic particle size distribution diagram of the colloid, and obtain the hydraulic particle size distribution diagram, average hydraulic particle size and multiple dispersion coefficient.

Notes

(1) After the experiment, the reaction vessels and cuvette used for preparation and testing of colloidal solutions need to be soaked in aqua regia for 20 min to remove the colloidal particles, and the instrument should be flushed repeatedly with tap water and distilled water before drying. Wear goggles and anti-acid gloves and do experiments in the fume hood when using aqua regia. The aqua regia used to soak the instrument should be recycled to the designated container.

(2) The UV chemical reaction device used in this experiment is equipped with 8 W low-pressure mercury lamp whose wavelength is 253.7 nm. The use of photochemical reaction device should strictly follow its operation procedures to avoid the harm caused by ultraviolet light.

(3) Please take the chloroauric acid and silver nitrate solutions as required to avoid waste. After using, wrap the bottle with foil paper as soon as possible and put it back into the refrigerator to avoid photolysis caused by

long-time exposure.

Questions

(1) According to the data of the test results, use Excel or Origin to draw the UV-Vis absorption spectra of the colloidal solution to analyze and discuss the factors that affect the position of plasma resonance absorption peak on the surface of colloidal particles.

(2) Explain the effect differences of different stabilizers on colloidal solutions to resist the electrolyte coagulation.

(3) Analyze the test results of the laser particle analyzer, judge whether the samples are agglomerated and give your reasons with the results of the UV-Vis spectrophotometer.

References

[1] Daniel M C, Astruc D. 2004. Gold nanoparticles: assembly, supramolecular chemistry, quantum-size-related properties, and applications toward biology, catalysis, and nanotechnology. Chem Rev, 104: 293-346

[2] Dreaden E C, Alkilany A M, Huang X H, et al. 2012. The golden age: gold nanoparticles for biomedicine. Chem. Soc. Rev., 41(7): 2740-2779

[3] Wu H C, Dong S A, Dong Y N, et al. 2007. Photochemical synthesis and the seeding-mediated growth of gold nanoparticles under the sunlight radiation. Chemical Journal of Chinese Universities, 28(1): 10-15

[4] Chen Y M, Wang L Y, Li F H. 2010. Preparation and characterization of silver nanoparticles by using polyvinylpyrrolidone as stabilizer. Polymer Materials Science & Engineering, 26(10): 137-139

[5] Liu Q Y, Qin A M, Jiang Z L, et al. 2005. Photochemical preparation of Au nanoparticles with polyethyelene glycol and its resonance scattering spectral study. Spectroscopy and Spectral Analysis, 25(11): 1857-1860

[6] Eustis S, Hsu H Y, El-Sayed M A. 2005. Gold Nanoparticle formation from photochemical reduction of Au^{3+} by continuous excitation in colloidal solutions. A Proposed Molecular Mechanism. J Phys Chem B, 109(11): 4811-4815

[7] Yao S W, Cao Y R, Zhang W G. 2006. Preparation of silver nanoparticles of different shapes via photoreduction method. Chinese Journal of Applied Chemistry, 23(4): 438-440

Exp. 36 Interaction of Metal Ions and Serum Albumin

Objectives

(1) To master the usage of the UV-Vis spectrophotometer and fluorescence spectrophotometer.

(2) To calculate the relevant data of the fluorescence spectrophotometer.

(3) To detect the effects of common metal ions on serum albumin in organisms.

Principles

Metal ions play a key role in many life processes. The study of the interaction between metal ions and protein is an important part of life science and a frontier field of chemistry and life science research. Serum albumin, a kind of globulin which can combine with inorganic ions, organic compounds and small-molecule drugs, plays an important role in organisms[1]. Many studies about the interaction between serum albumin and metal ions (or drug molecules) from the 1950s (late

1980s in China) have been investigated to reveal the mystery of life processes, which depends on the physiological importance, easy separation and purification of serum albumin.

At present, the research methods used to study the interaction between metal ions and serum albumin include: ①ultraviolet-visible absorption spectrometry; ②fluorescence spectrometry; ③balanced dialysis method; ④capillary electrophoresis; ⑤electrophoresis, etc.

Ultraviolet (UV) spectra and fluorescence spectra are important methods to study the interaction between biomolecules and small molecules[2,3]. The protein spectral variation of small-molecule perturbation can be used to deduce not only the change of chromophore microenvironment and protein structure but also the interaction between small molecules and protein. However, the UV spectrum produced by electron transition is a wide band, which is difficult to distinguish the vibrational structure of tryptophan residues, tyrosine residues and phenylalanine residues in the protein. The characteristics of emission peaks, fluorescence polarization, energy transfer and fluorescence life in fluorescence spectra can provide useful information for the structure and microenvironment of fluorescent chromogenic groups in the protein.

The UV-Vis spectrometry and fluorescence spectrometry are described below.

1. UV-Vis Spectrometry

The serum albumin usually has three distinct ultraviolet absorption bands: ①the absorption below 210 nm is from the peptide bond and many conformational factors; ②the absorption of 210~250 nm is from multiple factors, such as the absorption of aromatic and other residues, the absorption of certain hydrogen bonds and the interaction with other conformations and spirals; ③the absorption near 250~290 nm is from residues of aromatic groups, among which tyrosine residues have strong absorption near 278 nm (Tyr, 260~290 nm), and tryptophan (Trp) residues have strong absorption near 290 nm, while phenylalanine residues (Phe, 250~260 nm) have weak absorption. External factors such as solvent polarity and pH will also affect the absorption spectrum.

When metal ions combine with protein, the strength or band position of the absorption spectrum of protein or metal ions will change: ①the variation of metal ion spectra induced by protein can be used to deduce the coordination environment of metal ions; ②the variation of protein spectra induced by metal ions can be used to deduce the change of the chromophore microenvironment and protein structure. The combination of metal ions and protein can be deduced by comparing and calculating the spectra. If the absorption peak of protein enhances, it can be considered that small molecules enter the hydrophobic cavity of the protein, resulting in the extension of the peptide chain and the decline of the hydrophobic environment. If there is only solvent effect, the blue shift of peak indicates that the chromophore buried in the non-polar region of the protein is exposed to a polar solvent, while the red shift indicates that the chromophore is flipped to a less polar region[4].

2. Fluorescence Spectrometry

Fluorescence spectrometry is an effective method to study the conformation of protein molecules, which has many advantages such as high sensitivity, high selectivity, low sample consumption and simple method. The residues in serum albumin, namely tryptophan, tyrosine and phenylalanine, can induce the formation of nature fluorescence. Generally speaking, the fluorescence intensity ratio of tryptophan, tyrosine and phenylalanine is 100 : 9 : 0.5. So the natural fluorescence of protein is mainly from tryptophan. The fluorescence of protein and few metal ions will change induced by the interaction between metal ions and serum albumin, which can be used for qualitative and quantitative researches. The fluorescence quenching method is often used to determine the binding constant and site number. The thermodynamic data can be used to calculate the interaction force between metal ions and serum albumin. The Forster resonance energy transfer can be used to determine the distance between metal ions and tryptophan residues.

1) Fluorescence Quenching

The fluorescence quenching is caused by the interaction between the quenching agent and the fluorescent substance in the solution, which will decreases the fluorescence efficiency or shortens the lifespan of excited state. Specific quenching is caused by the specific chemical action between the fluorescent material and the quenching material. The fluorescence quenching can be caused by many processes, such as excited state reaction, energy transfer, complex formation and collision quenching. The mechanism of fluorescence quenching mainly includes static quenching and dynamic quenching.

Dynamic quenching, also known as collision quenching, is caused by the diffusion between the fluorescence body and the quenching body. It is usually assumed that the quenching is dynamic quenching and K_q is calculated according to the Stern-Volmer equation [5]:

$$F_0/F = 1 + K_q \tau_0 [Q] = 1 + K_D [Q] \qquad (5.12)$$

In the formula, F_0 is the fluorescence intensity of the fluorescence body without the quenching; F is the fluorescence intensity after adding the quenching body; K_q is the bimolecular quenching constant; $[Q]$ is the concentration of the quenching body; τ_0 is the fluorescence lifespan of the fluorescence body without the quenching; K_D is the Stern-Volmer constant, and $K_D = K_q \tau_0$.

The quenching constant K_q is calculated by the fluorescence lifespan. The fluorescence lifespan of biomolecules is about 10 ns, and the maximum diffusion collision quenching constant of various quenching agents is about 2.0×10^{10} L/(mol·s) [6]. When K_q is greater than 2.0×10^{10} L/(mol·s), it is considered as static quenching.

2) Binding Constant K_A and Binding-site Number n of the Interaction between Metal Ions and Protein

It is assumed that the small molecule has n identical and independent binding sites in protein molecules. According to the literature [7], the formula is derived:

$$\lg(F_0 - F)/F = \lg K_A + n\lg[Q] \qquad (5.13)$$

Draw a straight line according to lg $(F_0 - F) / F$ and lg $[Q]$. The binding constant K_A and the binding-site number n can be calculated by the intercept and slope of the straight line.

3) Energy Transfer and Binding Distance between Protein (Donor) and Metal Ions (Receptor)

When the emission spectrum of protein molecules overlaps with the absorption spectrum of small molecules, the energy is transferred from tryptophan (donor) to small molecules (receptor) through the intermolecular dipole-dipole resonance coupling. According to Forster's resonance energy transfer theory[5], the distance r between small molecules and tryptophan residues in protein can be calculated, and the effect of other substances on the conformation of serum albumin can be analyzed by the change of r. The relationship among the energy transfer efficiency E, donor-receptor distance r and energy transfer distance R_0 is as follows:

$$E = 1 - F / F_0 = R_0^6 / (R_0^6 + r^6) \tag{5.14}$$

$$R_0^6 = 8.8 \times 10^{-25} K^2 N^{-4} \Phi J \tag{5.15}$$

In the formulas, R_0 is the critical distance when the transfer efficiency is 50%; K^2 is the dipolar space orientation factor; N is the refractive index of the medium; Φ is the fluorescence quantum yield of the receptor. The average value of K^2 is 2/3, N is 1.36 which is the average value of water (1.33) and organic matter (1.39), and Φ is 0.118[8]. J is an overlap integral between the fluorescence emission spectrum of the donor (protein) and the absorption spectrum of the receptor, which can be expressed as:

$$J = \frac{\sum F(\lambda) \cdot \varepsilon(\lambda) \cdot \lambda^4 \Delta\lambda}{\sum F(\lambda) \cdot \Delta\lambda} \tag{5.16}$$

In the formula, $F(\lambda)$ is the fluorescence intensity of the fluorescence donor at the wavelength λ; and $\varepsilon(\lambda)$ is the molar extinction coefficient of the receptor at wavelength λ. The overlap integral J is obtained according to the formula, and R_0, E and r are calculated combined with the fluorescence data.

4) Determination of Acting Force Types between Metal Ions and Protein

The acting forces between small molecules and protein include hydrogen bond, van der Waals force, electrostatic forces and hydrophobic forces. According to the relative value of ΔH and ΔS, the main acting force types between small molecules and protein can be determined. When $\Delta H > 0$ and $\Delta S > 0$, it is hydrophobic force; when $\Delta H < 0$ and $\Delta S > 0$, it is electrostatic force; when $\Delta H < 0$ and $\Delta S < 0$, it is hydrogen bond and van der Waals force[9]. When the temperature does not change much, ΔH can be regarded as a constant. According to the thermodynamic formula:

$$\ln(K_2 / K_1) = \Delta H (1/T_1 - 1/T_2) / R \tag{5.17}$$

$$\Delta G = -RT\ln K \tag{5.18}$$

$$\Delta S = (\Delta H - \Delta G) / T \tag{5.19}$$

ΔH and ΔS can be obtained according to the binding constant K_A at different temperatures, which can be used to determine the types of acting force between metal ions and protein.

5) Protein Conformation Induced by Metal Ions

The synchronized fluorescence spectrum of fixed wavelength is often used to determine the change of protein conformation. The fluorescence of tyrosine residues is shown when $\Delta\lambda$ is 15 nm, while the fluorescence of the tryptophan residues is shown when $\Delta\lambda$ is 60 nm[5]. The change of protein conformation can be determined by the change of emission wavelength due to the relation between the maximum emission wavelength of amino acid residues and the hydrophobicity of their environment. The synchronized fluorescence spectrum ($\Delta\lambda$ = 15 nm and $\Delta\lambda$ = 60 nm) is scanned by gradually increasing the concentration of metal ions without changing the protein concentration, which can be used to deduce that how metal ions affect protein conformation.

Experimental Items

Apparatuses and materials: UV-Vis spectrophotometer, fluorescence spectrophotometer, biological ice machine, refrigerator, automatic triple water distiller, drying oven, water bath kettle (6~8 holes), pH meter, test tube, trace syringe (50 μL), analytical balance, absorption cell, beaker, volumetric flask, pipette, lens paper.

Reagents: $Cu(Ac)_2 \cdot H_2O$, $FeCl_3$, NaCl, HCl (6 mol/L), NaOH (6 mol/L), bovine serum albumin (BSA), tris (hydroxymethyl) aminomethane (Tris-HCl).

Procedures

1. Preparation of Solution

(1) Prepare a buffer solution containing 50 mmol/L Tris-HCl and 100 mmol/L NaCl to maintain the pH (= 7.0) and the ionic strength. The buffer solution is used to prepare BSA reserve solution and stored at 4℃.

(2) Weigh Cu $(Ac)_2 \cdot H_2O$ and $FeCl_3$ accurately. First dissolve them in a beaker with a little water, then transfer the solution into a volumetric flask, dilute the solution to 2 mmol/L, and store it at room temperature.

2. Determination of Interaction between Metal Ions and Protein by UV-Vis and Fluorescence Spectra

Add BSA solution and metal ion (Cu^{2+} and Fe^{3+}) solutions in a series of 5 mL colorimetric tubes, dilute the mixed solution with Tris-HCl buffer solution to the scale: ①determine the UV-Vis spectra at 200~400 nm after incubation for 30 min in a thermostatic waterbath (310 K) with Tris-HCl buffer solution as reference (the final concentration of metal ions is 0, 2.0, 4.0, 8.0, 12.0, 16.0 and 20.0 ($\times 10^{-5}$ mol/L); ②determine the fluorescence spectra and synchronous fluorescence spectra at 298 K and 308 K, the fixed concentration of BSA is 1.87×10^{-7} mol/L, the final concentration of metal ions is 0, 2.0, 4.0, 8.0, 12.0, 16.0 and 20.0 ($\times 10^{-6}$ mol/L).

Notes

(1) Experimental requirements: ①groups of four students; ②simple analysis of the ultraviolet spectrum; ③related computation of the fluorescence spectrum.

(2) Try to keep the scanning conditions of instruments consistent.

Questions

(1) Why should the determination be carried out at 37℃?

(2) Why does the pH of the buffer solution remain at 7.0~7.4? What's the role of NaCl in the buffer solution?

(3) How to distinguish between dynamic quenching and static quenching?

References

[1] Robert E B, Steven L B. 1997. Molecular recognition of protein-ligand complexes: applied to drug design. Chem Rev, 97: 1359-1472

[2] Tian L F, Liu Z F, Hu X L, et al. 2012. Proteins fluorescence quenching by $[Hg(SCN)_4]^{2-}$ and its analytical application. Chemical Journal of Chinese Universities, 33(1): 59-65

[3] Wang Y J, Zhang Y Q, Li Y X, et al. 2013. Study on the interaction between rhizoxin and human serum albumin by spectrophotometry. Journal of Instrumental Analysis. 32(2): 239-243

[4] Hu Y J, Liu Y, Pi Z B, et al. 2005. Interaction of cromolyn sodium with human serum albumin: a fluorescence quenching study. Bioorg Med Chem, 13(24): 6609-6614

[5] Chen G Z, Huang X Z, Xu J G. 1990. Fluorescence Analysis. 2nd ed. Beijing: Science Press

[6] Lu J X, Zhang G Z, Zhao P, et al. 1997. Studies on interaction between adriamycin and serum albumin as well as effect of ions on the reaction. Acta Chimica Sinica, 55: 915-920

[7] Xie M X, Xu X Y, Wang Y D, et al. 2005. Spectroscopic investigation of the interaction between 2,3-dihydro-4′,5,7-trihydroxyflavone and human serum albumin. Acta Chimica Sinica, 63(22): 2055-2062

[8] Yang M M, Yang P, Zhang L W. 1994. The effect of caffeic acid on albumin by fluorescence method. Chinese Science Bulletin, 39 (1): 31-35

[9] Song Y M, Wu J X. 2006. Synthesis, characterization and interaction of rare earth complexes of rutin with HSA and BSA. Chinese Journal of Inorganic Chemistry, 22(12): 2165-2172

Chapter 6 Chemistry in Life

Exp. 37 Preparation and Application of $Na_2S_2O_3$

Objectives

To learn the preparation, properties and application technology of $Na_2S_2O_3$, and consolidate some operations such as filtration, evaporation and crystallization.

Principles

(1) $Na_2SO_3 + S + 5H_2O \xrightarrow{\Delta} Na_2S_2O_3 \cdot 5H_2O$

(2) $2Na_2S + Na_2CO_3 + 4SO_2 \xrightarrow{\Delta} 3Na_2S_2O_3 + CO_2\uparrow$

(3) Development reaction.

Experimental Items

Apparatuses and materials: filter paper, watch glass, beaker, evaporating dish, Büchner funnel, suction flask, vacuum pump, asbestos net, alcohol burner.

Reagents: $Na_2SO_3(s)$, $S(s)$, activated carbon, 2 mol/L H_2SO_4, 1 mol/L KBr, 0.1 mol/L Na_2S, iodine water, 0.5% starch solution.

Procedures

1. Preparation of $Na_2S_2O_3$

Take a 100 mL beaker, weigh 8.0 g of anhydrous sodium sulfite (molecular weight is 126) and 2 g of sulfur powder, wet them with a small amount of water (ethanol), then add 50 mL of water, heat the solution and let it boil for 30 min while stirring continuously, and meanwhile, timely replenish water. After the reaction is completed, add 1～2 g of activated carbon (decolorizer, for adsorbing excess sulfur powder) in the boiling solution, continue to boil for about 10 min under continuous stirring, and then filter it immediately and discard impurities. Afterwards, heat the filtrate in an evaporating dish (slight boiling), and stop the evaporation when crystallization just occurs, then cool it down to make sodium thiosulfate crystal precipitate. After the suction filtration, the white crystal of $Na_2S_2O_3 \cdot 5H_2O$ (molecular weight is 248.2) can be obtained. Dry the crystal in an oven at 40°C for 40～60 min, and then weigh it and calculate its yield (the theoretical yield is 15.75 g).

2. Application of $Na_2S_2O_3$

1) Application of $Na_2S_2O_3$ in the Photographic Development and Fixing

a. The Basic Principle of $Na_2S_2O_3$ in the Photographic Development and Fixing

In the process of photos development, the photographic paper (photosensitive material) only gets latent images through the sensitization of photographic plate. Only through developing by the developer (such as Haider and Metol) can the invisible latent images be developed as visible images. The main reaction is as follows:

$$HOC_6H_4OH + 2AgBr = OC_6H_4O + 2Ag + 2HBr$$

But when the photographic paper is in the emulsion layer, there exists a large amount of non-photosensitive silver bromide. On the one hand, the photographic cannot get a transparent image; on the other hand, silver bromide will continue to change when exposed to light during storage to make the image unstable. Therefore, the photographic paper must have the fixing process after development.

Sodium thiosulfate (common name is Hypo or sodium hyposulfite) can serve the function of fixation in that it can react with the silver bromide to produce a kind of coordination compound which is easily soluble in water. The fixing process can be represented by the following reactions:

$$AgBr + 2Na_2S_2O_3 = Na_3[Ag(S_2O_3)_2] + NaBr$$
$$AgNO_3 + KBr = AgBr\downarrow \text{ (pale yellow)} + KNO_3$$

Note: $AgNO_3$ may not be excessive.

$$2Ag^+ + S_2O_3^{2-} = Ag_2S_2O_3\downarrow$$

b. Picture Development

In the dark room, cover the photographic plate with the photographic paper directly in the light-sensitive box for photosensitive process, and the sensitive time is selected according to the situation of negatives. Then put the sensitized photographic paper in the developer for developing. When images have been basically clear, take out the photographic paper with tweezers and clean it in water, then put it into the fixing solution for 10~15 min. Again, take out the photographic paper, then put it in water and flush it with water. After that, dry and glaze the photographic paper by the glazing machine or attach it to the plate glass for air dry and glazing, and finally cut off the deckle edge.

2) Application of Sodium Thiosulfate in the Quantitative Analysis (Iodometric Method)

Sodium thiosulfate standard solution has a quantitative reaction with quantitative iodine, regarding the starch as an indicator. The titration reaches the end point when the blue color of the solution just disappears. The equation is:

$$I_2 + 2S_2O_3^{2-} = 2I^- + S_4O_6^{2-}$$

Calculate the amount of iodine based on the volume and concentration of consumed sodium thiosulfate standard solution.

Notes

$Na_2S_2O_3 \cdot 5H_2O$, a colorless transparent or white monoclinic crystal with density of 1.685 g/cm^3 (20℃), is easily soluble in water but hardly soluble in ethanol (its aqueous solution is alkaline). It is easily weathered in dry air at above 33℃, and it decomposes into sodium sulfate and sodium sulfide while being burnt. Moreover, it remains stable in alkaline or neutral solutions while it breaks down quickly in acid solutions. The two sulfur atoms of $S_2O_3^{2-}$ present different oxidation states: one is +6 and the other is –2, so it is a moderate-intensity reducing agent. In addition to the above applications, it can react with elementary Cl_2, which can be applied as the antichlor after the bleach of cottons or applied in tanning, electroplating, etc.

$$S_2O_3^{2-} + 2H^+ \Longrightarrow S\downarrow + SO_2\uparrow + H_2O$$

It can be heated to lukewarm and the sulfur is gradually separated out.

$$Na_2S_2O_3 + 4Cl_2 + 5H_2O \Longrightarrow 2H_2SO_4 + 2NaCl + 6HCl$$

Developer formula

D-72	Metol	Anhydrous sodium sulfite	Hydroquinone	Anhydrous sodium carbonate	Potassium bromide
	0.75 g	11.25 g	3 g	16.88 g	0.5 g

It is diluted with cold water to 250 mL.

Fixer formula

F-5	Hypo	Anhydrous sodium sulfite	Acetic acid (28%)	Boric acid	Potassium vanadium
	60 g	3.75 g	11.75 mL	1.88 g	3.75 g

It is diluted with cold water to 250 mL.

Exp. 38 Home-made Plant Acid-base Indicator

Objectives

(1) To enhance the knowledge of acid-base indicators and understand the principle of discoloration.

(2) To select the plants that you are familiar with, prepare the acid-base indicator and determine its color range.

Principles

An acid-base indicator refers to a kind of substance that appears as different colors in an acidic and alkaline solution. Pigments contained in flowers, fruits, stems and leaves of many plants show different colors in acidic or alkaline solutions, which can be used as acid-base indicators.

Experimental Items

Apparatuses and materials: test tube, measuring cylinder, glass rod, mortar, dropper, spot plate, funnel, gauze.

Reagents: petals (such as morning glories), leaves (such as purple cabbages), radishes (such as carrots, sweet pink-fleshed radishes), alcoholic solution (the volume ratio of ethanol and water is 1 : 1), dilute hydrochloric acid, dilute NaOH solution.

Procedures

(1) Take some petals, leaves, radishes, etc., mash them in a mortar, respectively, then add 5 mL of alcohol solution and stir them. Filter them with 4 layers of gauze respectively and obtain alcohol solutions of petaline pigment, leaf pigment, radish pigment, etc. Afterwards, add them into different test tubes.

(2) Drop some dilute hydrochloric acid, dilute NaOH solution and distilled water into the holes of a white spot plate respectively, and then add 3 drops of the alcohol solution of petaline pigment into each hole. Observe the phenomena.

(3) Replace the alcohol solution of petaline pigment with the alcohol solution of leaf pigment and the alcohol solution of radish pigment to do the above experiment, and observe the phenomena.

Data Recording and Processing

(1) Write down other acid-base indicators you know and their ranges of discoloration.

(2) Draw a coordinate graph (horizontal axis: the amount of acetic acid; vertical axis: changes of the solution pH) to describe the change of the solution pH after acetic acid is added slowly to sodium hydroxide solution.

(3) Bring the pH test paper back home, take a plant you are familiar with as an indicator and determine its sensitivity (i.e., its color range). Then compare it with the discoloration range of phenolphthalein and fill in the following table.

Indicator	Phenothalin	Self-made indicator
Range of discoloration	8~10	

Notes

Litmus, a common acid-base indicator, is a plant extract. Other common acid-base indicators include:

Thymol blue	Methyl orange	Methyl red	Bromothymol blue	Thymol blue	Phenothalin	Thymolphthalein
pH 1.2~2.8 from red to yellow	3.1~4.4 from red to yellow	4.4~6.2 from red to yellow	6.2~7.6 from yellow to blue	8.0~9.6 from yellow to blue	8.0~9.6 from colorless to red	9.4~10.6 from colorless to blue

References

Zheng C L. 2009. Chemistry Experiment Course and Teaching Theory. Beijing: Higher Education Press

Exp. 39 Qualitative Identification of Iodine in Kelp and Salt

Objectives

To understand the basic properties of the major oxidized compounds of iodine, and learn how to separate iodine from kelp and identify iodized salt.

Principles

1. Extraction and Identification of Iodine in Kelp

Iodine content is high in kelp of which organic matter can be decomposed by oxidation and burned to ashes by heating and burning (carbonization and ashing). After the nitration by nitric acid and further leaching with an aqueous solution, the solution containing iodate radicals can be obtained, elementary iodine can be obtained based on the following reaction, and the qualitative identification can be achieved by the property that starch solution appears blue when adding elementary iodine:

$$KIO_3 + 5KI + 3H_2SO_4 = 3I_2 + 3H_2O + 3K_2SO_4$$

2. Identification of Iodized Salt and Non-Iodized Salt

Most of iodized salt in the market contains iodine (potassium iodate). According to the provisions of the relevant state departments, the iodine content of iodized salt must be 20~50 mg per kilogram, and the minimum detectable amount of elementary iodine is 1×10^{-7} g/mL if using starch as the indicator. Iodine ions react with iodate radicals to form elementary iodine, which can be used to qualitatively identify whether the salt contains iodine or not in an easy way.

Experimental Items

Apparatuses and materials: test tube, beaker, mortar, suction flask, Büchner funnel, evaporating dish, crucible tongs, alcohol burner, watch glass, tweezers, filter paper, pH test paper.

Reagents: iodized salt, sodium chloride, dry kelp, 2 mol/L H_2SO_4 solution, 3% H_2O_2, 0.5 mol/L KIO_3 solution, starch solution, 0.1 mol/L KI, carbon tetrachloride.

Procedures

(1) Self-design an experimental schema to extract iodine from kelp and complete the qualitative identification.

(2) Design a simple experiment to identify iodized salt and non-iodized salt.

Alternative Experimental Schemes

1. Extraction and Identification of Iodine in Kelp

(1) Weigh 5~10 g of dry kelp, ignite or roast it for complete ashing.

(2) Pour kelp ash into a 50 mL beaker, and add 3~5 mL of distilled water to boil and soak it for 3~5 min. Afterwards, add 3~5 mL of distilled water and repeat the above leaching once, then filter the solution and collect the filtrate. The total volume should not exceed 10 mL (note: do not wash the precipitate, lest the total iodide concentration of the leachate is too low).

(3) Add 1~2 mL of leaching solution, 10 drops of 2 mol/L H_2SO_4 and 3% H_2O_2 solution successively into a test tube, oscillate and mix them well, then add a few drops of starch solution. If the color turns blue, the existence of iodine in kelp can be proved.

Control experiment: add 1 mL of 0.1 mol/L KI solution, 10 drops of 2 mol/L H_2SO_4 and 3% H_2O_2 solution into a test tube, oscillate and mix them well, and then add a few drops of starch solution to observe the phenomenon.

(4) Add 2 mol/L H_2SO_4 to the remaining filtrate and acidify it to a neutral pH value carefully. Afterwards, add 1 mL of 3% H_2O_2 solution and oscillate the solution. (note: kelp ash contains potassium carbonate. Acidification makes it neutral or weak acid, which is beneficial to the next step of oxidation of iodine precipitation. But excess sulfuric acid is easy to lead to loss of the hydrogen iodide which oxidizes and form iodine). Add 2~3 mL of carbon tetrachloride to the solution, and extract the iodine in the aqueous solution by oscillating the test tube. Then let it stand to make the solution layered and observe the color of carbon tetrachloride.

2. Identification of Iodized Salt and Non-Iodized Salt

Take 2 test tubes and add a small amount of iodized salt and non-iodized salt respectively. Then add 1 mL of 2 mol/L H_2SO_4 and 10 drops of 0.1 mol/L KI solution. After the solid is dissolved by stirring, add a few drops of starch solution and observe the phenomena. The solution turns blue contains iodized salt while the other whose color doesn't change contains non-iodized salt.

Control experiment: take 10 drops of 0.5 mol/L KIO_3 solution, followed by adding 10 drops of 2 mol/L H_2SO_4 and 0.1 mol/L KI solution, oscillate and mix them well, and then add a few drops of starch solution to observe the phenomenon.

Exp. 40　Interesting Chemical Experiment (I)

Objectives

(1) To make students realize that there is chemistry everywhere, stimulate students' interest in learning and train students' ability of using chemical knowledge to analyze and solve problems.

(2) To master the experiment principles and test methods about how to identify the fingerprinting, drunk driving, murals discoloration, invisible ink, ethanol and methanol.

Principles

1. Fingerprint Identification (Iodine Smoked Method)

When heated, iodine will sublimate into iodine vapor which can dissolve in grease and other secretions on your finger and form a brown fingerprint. Everyone's fingers always contain grease, mineral oil and water. When pressing a finger on the paper, the grease, mineral oil and sweat from the fingerprint will stay on, but the human eye cannot see them. When we put the paper with the fingerprint on the mouth of the test tube containing iodine and heat the test tube, the iodine starts to sublimate and turns into a red purple vapor (note that the iodine vapor is toxic and may not be inhaled). As the grease and mineral oil from the fingerprint on the paper are both organic solvents, the iodine vapor can be dissolved in such oil matter after rising to the mouth of the test tube, and the fingerprint can also be displayed.

2. Identification of Drunk Driving

Ethanol, the major component of wine, is reductive. While potassium dichromate ($K_2Cr_2O_7$, orange-red) has strong oxidation that oxidizes ethanol to acetic acid under acidic conditions, and at the same time, potassium dichromate itself is reduced to trivalent chromium (dark green). Therefore, it can be judged from the color of the solution whether the exhaled air or blood contains ethanol. This reaction is very sensitive, and ethanol can be detected as long as its content is greater than 0.2%.

$$2K_2Cr_2O_7 + 8H_2SO_4 + 3CH_3CH_2OH =\!=\!= 3CH_3COOH + 2Cr_2(SO_4)_3 + 11H_2O + 2K_2SO_4$$

3. Murals Discoloration

Gray-black (not white) substances on murals are lead sulfide (PbS). The white pigment is white lead [basic lead carbonate, $2PbCO_3 \cdot Pb(OH)_2$], which has a strong covering power and appears snow-white when painted on murals. However, the following reaction takes place due to the long-term effect of trace H_2S in air:

$$2PbCO_3 \cdot Pb(OH)_2 + 3H_2S =\!=\!= 3PbS\downarrow + 2CO_2\uparrow + 4H_2O$$

To restore the picture, simply dab the picture with a soft cloth dipped in some hydrogen peroxide, and then the following reaction can occur:

$$PbS + 4H_2O_2 =\!=\!= PbSO_4 + 4H_2O$$

4. Invisible Ink

(1) If using $CoCl_2$ solution to write on white paper, the words do not show up after drying. But if it is heated (you can bake it or use a hair dryer), the blue words can show up; when contacting

with steam (place it on the mouth of the cup filled with hot water and move around), the words will be invisible again. The following reactions occur:

$$CoCl_2 \cdot 6H_2O \text{ (pink)} \rightarrow CoCl_2 \cdot 2H_2O \text{ (fuchsia)} \rightarrow CoCl_2 \cdot H_2O \text{ (bluish violet)} \rightarrow CoCl_2 \text{ (blue)}$$

$CoCl_2 \cdot 6H_2O$ solution is light colored when it is at a low concentration and almost colorless when it is dry, but it will lose water when exposed to heat and become blue $CoCl_2$. When contacting with steam, the above-mentioned changes and reactions proceed and the solution becomes light-colored $CoCl_2 \cdot 6H_2O$ again.

(2) If using the alcoholic solution containing 1% phenolphthalein to write on white paper, the words do not show up after drying, but if the surface is covered with low-concentration alkali solution, the red handwriting will show up. If it is smeared with acetic acid, the handwriting will fade from red to colorless. It is because phenolphthalein is a kind of acid-base indicator that is colorless when pH < 8.2 and red when pH > 10.

(3) 1% starch solution or milk can also be used as invisible ink. They appear colorless while writing, appear blue when painted with iodine solution and fade after being baked or heated gently. This is because iodine appears blue when contacting with the starch, and elementary iodine is easy to sublimate and volatile when being heated.

5. Identification of Ethanol and Methanol

Methanol and ethanol are both colorless and transparent liquids that are volatile and have a special aroma, so they are difficult to distinguish in appearance. However, methanol and its metabolites called formaldehyde, formic acid are highly toxic to humans. Drinking methanol will cause blindness, and even death.

The method of identifying methanol and ethanol is to oxidize the alcohol to aldehydes first, and identify them according to the reaction characteristics of formaldehyde. Methanol and ethanol are easily oxidized by potassium permanganate under acidic conditions and become formaldehyde and acetaldehyde. Aldehydes can undergo an addition reaction with Schiff reagent to form a red purple adduct with blue shadow. After adding sulfuric acid, the color which formaldehyde shows does not disappear, while the color shown of other aldehydes such as acetaldehyde will fade. Therefore, this reaction can be used as a method of distinguishing formaldehyde from other aldehydes.

In summary, the way to identify ethanol and methanol is to oxidize the alcohol to aldehydes which then react with Schiff reagent (add sulfuric acid). The higher the methanol content is, the darker the color will be. The minimum limit of detection of methanol via this method is 0.2 g/L.

$$5CH_3OH + 2MnO_4^- + 6H^+ =\!=\!= 5HCHO + 2Mn^{2+} + 8H_2O$$
$$5CH_3CH_2OH + 2MnO_4^- + 6H^+ =\!=\!= 5CH_3CHO + 2Mn^{2+} + 8H_2O$$

Experimental Items

Apparatuses and materials: white paper, beaker, tweezers, watch glass, measuring cylinder, test tube, test tube rack, dropper, filter paper, glass rod, hair dryer, alcohol burner, etc.

Reagents: 0.1 mol/L $K_2Cr_2O_7$, H_2SO_4 (1 mol/L), basic lead carbonate solid [$2PbCO_3 \cdot Pb(OH)_2$], 5% TAA (thiacetamide), 3% H_2O_2, 0.2 mol/L $CoCl_2$, 1% phenolphthalein, 1% starch, 0.1 mol/L iodine water, 1 mol/L acetic acid, 1 mol/L NaOH, iodine (s), ethanol, methanol, ethanol mixed with methanol, $KMnO_4$, $H_2C_2O_4$ solution, Schiff reagent (fuchsin sulfonate, colorless).

Procedures

1. Fingerprint Identification (Iodine Smoked Method)

Take a small piece of white paper, and press with your finger to leave a fingerprint. Put 2~3 g of iodine in a small beaker and cover it with a watch glass to heat. After subliming iodine into iodine vapor, remove the watch glass, place the white paper on the mouth of the beaker with the fingerprint facing down, and slowly the white paper will show a brown fingerprint.

Due to the volatilization of iodine, the experiment should be carried out in a fume hood.

2. Identification of Drunk Driving

Add 1 drop of 0.1 mol/L $K_2Cr_2O_7$ solution, 2 drops of 1 mol/L H_2SO_4 solution and 1 drop of ethanol in a test tube, and then observe the color change of the solution.

3. Murals Discoloration

Place the filter paper coated with a layer of basic lead carbonate on the mouth of the test tube which contains thioacetamide solution (add dilute H_2SO_4 to acidize), and H_2S produces at this time, observe the changes of the filter paper (white → black). Then drop 3% H_2O_2 solution in the black place and observe the color change. This experiment should be carried out in a fume hood.

4. Invisible Ink

(1) If writing with 0.2 mol/L $CoCl_2$ solution on white paper, the handwriting appears blue after blow drying by a hair dryer. Then put the paper on the water bath to contact with water vapor and the blue handwriting fades away.

(2) If writing with 1% phenolphthalein on white paper and coating the handwriting with 1 mol/L NaOH solution, the handwriting appears red. Then use 1 mol/L acetic acid solution to paint on the red words, the handwriting turns colorless.

(3) Similarly, if writing with 1% starch on white paper and coating the handwriting with iodine solution, the handwriting appears blue. If heating the paper for a while, the blue handwriting fades away. This experiment should be carried out in a fume hood.

5. Identification of Ethanol and Methanol

Take three test tubes, add 0.5 mL of ethanol, methanol and ethanol mixed with methanol respectively, then add a small amount of dilute H_2SO_4 solution and 1~2 drops of $KMnO_4$ solution, and mix them well. After 10~20 s, add 1~2 drops of $H_2C_2O_4$ solution respectively and oscillate them until the solution fades. Then add 3~4 drops of Schiff reagent respectively, oscillate the solution, let it stand for 5~10 min and observe the phenomena. Methanol is reddish-purple (with blue shadow), ethanol is colorless, and ethanol mixed with methanol is blue.

Notes

Iodine vapor is highly toxic, so experiments should be carried out in a fume hood.

Exp. 41   Interesting Chemical Experiment (II)

Objectives

(1) To make students realize that there is chemistry everywhere, stimulate students' interest in learning and train students' ability of using chemical knowledge to analyze and solve problems.

(2) To master the experiment principles and test methods of blue bottle, dripping to produce smoke, ballet in water, karst cave wonders, gold in water, etc.

Principles

1. Blue Bottle Experiment

Methylene blue is a kind of dark green crystal, which can be dissolved in water and ethanol. In alkaline solution, blue methylene blue is easily reduced to colorless methylene white by glucose. When oscillating the colorless solution, the contact area between solution and air increases, leading to the increased amount of oxygen dissolved in the solution. In this case, oxygen oxidizes methylene white to methylene blue and the solution becomes blue again. When the solution is let stand, some of the dissolved oxygen escapes, and methylene blue is again reduced to methylene white by glucose. If repeatedly oscillating and resting the solution, we can find that the color alternates as "blue—colorless—blue—colorless", which is called the "chemical oscillation" of methylene blue.

2. Dripping to Produce Smoke

The lumpish metal aluminum has a compact oxide film on its surface, which contributes to its high stability at room temperature so that it can be widely used to make daily utensils. However, when the metal is in powder form, its activity is greatly increased. Under the condition of water as a catalyst, aluminum powder can react with elementary iodine at room temperature to form solid aluminum iodide, and a large amount of heat is released to make excess iodine sublimate into iodine vapor and present beautiful purple smoke. The metal powder used in the "dripping to produce smoke" experiment can be not only aluminum, but also magnesium, zinc, iron, copper, etc.

$$2Al + 3I_2 \xrightarrow{H_2O} 2AlI_3 + Q$$

3. Ballet in Water

Sodium and potassium react violently in water to generate hydrogen and give out heat, so the metal melts into a pellet which rapidly rotates on the surface of the water driven by the gas. Potassium can even cause spontaneous combustion and produce a purple flame. The produced MOH(M=Na, K) solution is alkaline, which can make the phenolphthalein indicator turn red.

$$2Na + 2H_2O = 2NaOH + H_2\uparrow$$
$$2K + 2H_2O = 2KOH + H_2\uparrow$$

4. Karst Cave Wonders

When adding a variety of metal salt particles that can be dissolved in water to sodium silicate solution, a thin film of insoluble silicate forms on the surface. Due to the semi-permeable membrane property of this film, the solvent water molecules outside the film continuously permeate into the inside under the driving of osmotic pressure, causing the film to expand. When a certain degree of pressure difference is reached, the thin film is broken, and the metal salt solution flows out of the fissure, which can react with the sodium silicate outside the film to produce new silicate. The process repeats itself over and over, just as plants grow. A colorful "garden in water" can be observed due to the different colors from insoluble silicates of different metals.

$$M^{2+} + SiO_3^{2-} = MSiO_3\downarrow$$

5. Gold in Water

The reaction of KI with soluble lead salt results in the formation of PbI_2, which is a filamentous precipitate with golden yellow sheen and resembles "gold" in morphology.

$$2KI + Pb(NO_3)_2 = PbI_2\downarrow + 2KNO_3$$

Therefore, the above reaction can be used to prepare "gold" in water. However, the salt is readily soluble in boiling water or excess KI solution.

Experimental Items

Apparatuses and materials: measuring cylinder, watch glass, flask, rubber stopper, test tube, mortar, evaporating dish, tweezers, beaker, funnel, platform scale, alcohol burner.

Reagents: 1% methylene blue, 20% Na_2SiO_3, KI (s), I_2 (s), NaOH (s), glucose (s), aluminum powder, Na (s), K (s), $CaCl_2$ (s), $Co(NO_3)_3$ (s), $CuSO_4$ (s), $NiSO_4$ (s), $MnSO_4$ (s), $FeSO_4$ (s), $FeCl_3$ (s), $Pb(NO_3)_2$ (s).

Procedures

1. Blue Bottle Experiment

Add 50 mL of water in a beaker (the stopper is additional) of 50 mL, add 1 g of NaOH and 1 g of glucose to water to dissolve them, and then add 5 drops of 1% methylene blue solution. After shaking it well, stop the opening to observe the solution turns colorless. Then remove the stopper and shake the bottle to observe the solution quickly turns blue. After being placed for a while, the solution becomes colorless again. Repeat the operations and observe the color change.

2. Dripping to Produce Smoke

Take 10 iodine particles and grind them in a mortar, then add 1 teaspoon of aluminum powder (about 1/10 of the amount of iodine), grind them together and mix them well. Pour the mixture into the center of an evaporating dish and add 1~2 drops of water to the mixture, and then immediately cover the evaporating dish with a large beaker (note that the apparatuses and reagents must be dry). A moment later, a thick and beautiful smoke appears. Observe the color.

3. Ballet in Water

Carefully cut a small piece (the size of mung bean) of metallic sodium and potassium with tweezers and a small knife and use filter paper to drain the kerosene on the surface. Put them into a beaker half full with water and immediately cover the beaker with an inverted funnel to prevent the metal sodium and potassium from popping up, which may cause spontaneous combustion. Observe the reaction. After the reaction, add 1~2 drops of phenolphthalein reagent and observe the color change of the solution.

4. Karst Cave Wonders

Add about 30 mL of 20% sodium silicate solution into a 100 mL beaker, and then add a small grain of calcium chloride, cobalt nitrate, copper sulfate, nickel sulfate, manganese sulfate, ferrous sulfate and ferric trichloride into the bottom of the beaker (note that keep a certain interval between each solid). Write down the location of each metal salt, and wait for a while to observe what happens.

5. Gold in Water

Take two beakers of 50 mL and add 20 mL of water, followed by adding 3.5 g of KI and 3.5 g of $Pb(NO_3)_2$, respectively, stir the solution and dissolve the solids, and then obtain 1 mol/L KI solution and 0.5 mol/L $Pb(NO_3)_2$ solution. Mix and heat the two, when the mixed solution is going to boiling, cool it down rapidly and put it in cold water, then observe the phenomenon.

Notes

(1) Iodine vapor is highly toxic, and experiments should be carried out in a fume hood.

(2) The solution of the blue bottle experiment must be alkaline, but alkali should not be excessive. The best temperature for a blue bottle experiment is 25℃, so the experiment needs a little heating up in the winter, but the temperature should not be too high.

(3) Sodium and potassium cannot be taken too much, otherwise the reaction is too violent. Excess solid sodium must be returned to the reagent bottle, and the used filter paper should not be discarded everywhere to avoid spontaneous combustion of residual solids, which can be treated in the water.

(4) After the cave wonder experiment, the beaker should be promptly cleaned to avoid the corrosion of alkaline silicate.

Exp. 42   Identification of Salt and Sodium Nitrite

Objectives

To learn the properties of sodium nitrite and identify sodium nitrite and salt through experiments.

Principles

Sodium nitrite is a kind of white to light yellow powder or granule which is similar in appearance to salt, salty and easily soluble in water. It is commonly known as "industrial salt", widely used in the industry and construction industry, and at the same time, it is also allowed as a colorant for limited use in meat products. The chance of food poisoning resulting from sodium nitrite is higher that only 0.3~0.5 g of ingestion can cause poisoning and even death. The market price for salt is about 2000 yuan per ton while industrial salt is about 230 yuan per ton, so some unscrupulous salt manufacturers are making high profits by adding sodium nitrite to salt.

Sodium nitrite and salt can be identified from two aspects, physical properties and chemical properties, including colorimetry, heating and melting, water dissolution, acidic potassium permanganate titration, adding acid, KI-starch test paper, silver nitrate, ferrous sulfate, phenolphthalein, etc.

Physical methods:

(1) Colorimetry.

Sodium chloride is a kind of cubic crystal or small crystalline powder, which is white; sodium nitrite is a kind of orthorhombic crystal, which is slightly pale yellow.

(2) Heating and melting.

The melting point of sodium chloride is 801℃, while the melting point of sodium nitrite is 271℃. Place sodium nitrite and sodium chloride on the same small sheet metal and heat it with an alcohol burner, the one which melts is sodium nitrite while the one which doesn't melt is sodium chloride.

(3) Water dissolution.

Sodium chloride and sodium nitrite are both soluble in water, but the solubility of both is very different. The effect of temperature on the dissolution of sodium chloride is very small; while sodium nitrite is easily soluble in water, which is endothermic in the dissolution process and dissolved faster in hot water.

Chemical methods:

(1) Generation and instability of sodium nitrite (by adding acid):

$$NaNO_2 + H_2SO_4 \text{ (dilute)} == NaHSO_4 + HNO_2$$
$$2HNO_2 == NO\uparrow + NO_2\uparrow \text{ (reddish brown)} + H_2O$$

Salt has no similar reactions or phenomena.

(2) Oxidability of sodium nitrite (by ferrous sulfate and KI-starch test paper):

$$2NaNO_2 \text{ (colorless)} + 2FeSO_4 + 2H_2SO_4 == 2NO\uparrow + Fe_2(SO_4)_3 \text{ (tawny)} + Na_2SO_4 + 2H_2O$$
$$2NaNO_2 \text{ (colorless)} + 2KI + 2H_2SO_4 == 2NO\uparrow + I_2 \text{ (brown)} + 2H_2O + Na_2SO_4 + K_2SO_4$$

Salt has no similar reactions or phenomena.

(3) Reducibility of sodium nitrite (by acidic potassium permanganate titration):

$$2KMnO_4 \text{ (fuchsia)} + 5NaNO_2 + 3H_2SO_4 == K_2SO_4 + 2MnSO_4 \text{ (colorless)} + 5NaNO_3 + 3H_2O$$

Salt has no similar reaction or phenomenon.

(4) Identification of sodium nitrite (by silver nitrate):

$$NaNO_2 + AgNO_3 == AgNO_2\downarrow \text{ (white)} + NaNO_3$$
$$NaCl + AgNO_3 == AgCl\downarrow \text{ (white)} + NaNO_3$$
$$AgNO_2 + HNO_3 == AgNO_3 + HNO_2 \text{ (the white precipitate is dissolved)}$$

AgCl does not react with dilute nitric acid.

(5) Acid-base property of sodium nitrite solution (by phenolphthalein).

$NaNO_2$ is a kind of strong base and weak acid salt. Its aqueous solution is alkaline and the alkalinity enhances with the water evaporation after heating, which makes the phenolphthalein test paper turn red. Table salt is a kind of strong acid and strong base salt. Its aqueous solution is neutral and the acid-base property does not change after heating, so does the color of the phenolphthalein test paper.

Experimental Items

Apparatuses and materials: platform scale, alcohol burner, test tube.

Reagents: sodium nitrite, sodium chloride, phenolphthalein indicator, dilute sulfuric acid, 0.1 mol/L $FeSO_4$, KSCN saturated solution, 0.1 mol/L KI, starch solution, 0.01 mol/L $KMnO_4$, 0.1 mol/L $AgNO_3$, dilute nitric acid.

Procedures

At room temperature, weigh 5 g of sodium nitrite and sodium chloride and place them in different test tubes respectively, then add 10 mL of water to dissolve them and observe the change of the solution temperature. Afterwards, divide the two solutions into six test tubes respectively and carry out the following experiments.

(1) Add the phenolphthalein indicator to the two solutions, heat the solutions and observe their color changes.

(2) Add icy dilute sulfuric acid to the two solutions, let them stand for a moment, oscillate the test tubes, and then observe the color change of solutions and if there is gas being released.

(3) Add 5 drops of dilute sulfuric acid to the two solutions for acidification, and then add 5 drops of 0.1 mol/L $FeSO_4$. Oscillate the test tubes, and then let the solutions stand and observe their color changes. Then add 5 drops of KSCN saturated solution and observe the color change.

(4) Add 5 drops of dilute sulfuric acid to the two solutions for acidification, and then add 5 drops of 0.1 mol/L KI. Oscillate the test tubes, and then let the solutions stand and observe their color changes. Then add 5 drops of starch solution and observe the color change.

(5) Add 5 drops of dilute sulfuric acid to the two solutions for acidification, and then add 5 drops of 0.01 mol/L $KMnO_4$. Oscillate the test tubes, and then let the solutions stand and observe their color changes.

(6) Add 5 drops of 0.1 mol/L $AgNO_3$ to the two solutions, oscillate the test tubes, then let the solutions stand and observe the formation of precipitation. Afterwards, add dilute nitric acid to the precipitates from centrifugal separation and observe whether the precipitates are dissolved.

第1章 绪　　论

1.1　化学实验的重要意义

化学是一门创造新物质的学科。当今，无论从自然科学的分类、研究对象的层次还是与新兴学科的交叉和融合来看，化学正处于一个多边关系的中心。因此，化学是一门中心学科，又是一门实用性很强的科学。化学为人类的生产和生活提供了大量的物质基础，并通过实验研究不断创造出惊人的成果。化学家不仅发现和合成了千百万种天然的和非天然的化合物，而且已能通过分子设计来合成新化合物。现代化学虽然已进入理论与实践并重的阶段，但是它仍是以实验为基础的学科，因为化学理论和规律的产生是建立在实验基础之上，实验又是验证理论的唯一标准。实验是化学与生产力发展的桥梁，化学实验推动化学科学的飞速发展，引领人类进入崭新的物质世界。

1.2　无机化学实验课的教学目的

对于即将从事化学及相关专业的人员来说，如不经过严格的实验训练、不具备独立的实验操作能力，就不可能真正掌握和理解化学专业知识，今后也无法通过实验进行探索、发明和创造。无机化学实验是高等学校化学、应用化学以及其他化学相关专业学生的第一门重要的专业基础课。学生通过观察实验现象和性质变化了解和认识无机化学反应的事实及其规律，加深对无机化学基本原理和基础知识的理解；学生通过实验训练培养熟练的无机化学实验的基本操作技能，掌握无机化合物的一般制备方法。实验课是由教师组织指导并由学生独立完成的课程。它承担着提高学生的观察、分析和解决问题能力以及书面表达能力；培养学生实事求是、独立思考、坚韧执著的科学态度和创新能力；培养学生规范整洁有序的实验室工作习惯和良好的科学素养。因此，实验教学具有理论教学无法替代的特殊目的和作用。

1.3　实验课的学习要求和方法

本书将实验内容分为以下几个模块：基础知识和基本操作实验、无机化学原理实验、无机化合物性质实验、无机化合物制备实验、综合和设计实验等。每个实验项目包括目的、原理、内容、基本操作、注意事项及其他要求。要达到实验课的学习目的和要求，必须有正确的学习方法。我国著名化学家卢嘉锡曾对如何做好化学实验有过精辟的论述：聪明的头脑、灵巧的双手和整洁的习惯。具体来说，即具备丰富的化学理论基础知识，善于用理论指导实验、分析和解决问题，有扎实规范的实验操作技能，有整洁、细致、实事求是的实验习惯和良好的科学素养。无机化学实验的学习方法大致可以分为下列三个步骤。

1.3.1 预习

为了获得良好的实验效果，及时发现问题并予以纠正，规避实验室安全的风险，要求每次实验前必须认真预习。通过阅读实验教材、教科书和相关的参考资料，明确本次实验目的，理解实验原理，细读实验内容，了解基本操作要点及注意事项等。在此基础上将实验内容中的每个步骤进行分解细化和梳理，用简洁明了有条理的文字和符号来表达实验步骤，写好预习笔记。可以用填空方式写出实验内容，包括每个步骤的试剂名称、浓度和用量，所用仪器，操作名称等，且在每步后的实验现象或数据处留出空格供实验时记录用。实验预习笔记本又是实验记录本。反之，若对实验内容心中无数，手忙脚乱，不仅影响实验进度，还可能会遗漏内容，加错试剂或用错仪器，甚至发生安全事故。因此，要求每位学生必须准备一本预习记录本，未做预习者不得进入实验室。

1.3.2 实验

根据实验教材上规定的方法、步骤和试剂用量进行操作，并做到以下几点：
(1)认真操作，仔细观察，及时如实地将实验现象和数据记录在实验记录本上。
(2)在实验中严格按照实验操作规程和仪器操作规范进行实验。
(3)严格遵守实验室规则，保持台面和仪器的整洁，在实验进度上做到合理穿插、统筹安排，不拖拉。
(4)在实验过程中勤于思考和分析，力争自己解决问题。如果发现实验现象和结果与理论不相符合，应首先尊重实验事实，认真分析并检查实验操作上的原因，进行及时纠正。也可以做对照试验、空白试验或经教师同意后做自行设计的实验来核对。对于耗时较长的实验应报告教师，经同意后方可重做。

1.3.3 实验报告

实验完成后应对实验现象进行解释并作讨论，或对实验数据进行处理和计算，得出结果并进行讨论，写出实验报告，并交指导教师批阅。实验报告要求：内容完整、文字和符号表达专业、书写规范、字迹端正、现象和结果准确、解释和讨论合理、条理性强、整齐清洁。

实验报告内容包括：
(1)实验名称、实验目的和实验原理。
(2)实验内容：列出实验步骤(试剂、浓度和用量、所用仪器、操作方法等)、实验现象(物质状态、颜色)和实验数据(表格形式)。
(3)实验现象和解释：用化学反应式或化学原理对实验现象进行解释和归纳。
(4)实验数据和结果处理：用化学公式或作图法进行分析和计算，得出结果。
(5)问题和讨论：对所学的新操作或本实验操作的关键点进行简述；或对实验中遇到的问题和体会酌情加以分析讨论；或回答书中思考题。

下面列出几种类型的实验报告格式，以供参考。

无机化学性质实验报告

实验名称 _____

姓名_____ 专业_____ 班级_____ 编号_____ 指导教师_____ 日期_____

一、实验目的：

二、实验内容：

实验步骤	实验现象	解释和反应方程式

三、问题和讨论：

无机化学测定实验报告

实验名称 _____

姓名_____ 专业_____ 班级_____ 编号_____ 指导教师_____ 日期_____

一、实验目的：

二、实验原理：

三、实验步骤：

四、数据记录与处理：

五、问题和讨论：

无机化学制备实验报告

实验名称 _____

姓名_____ 专业_____ 班级_____ 编号_____ 指导教师_____ 日期_____

一、实验目的：

二、实验原理：

三、实验步骤：

四、实验现象和反应方程式：

五、实验结果（产品颜色、状态、产量、产率）

六、问题和讨论：

第 2 章 实验室基础知识

2.1 实验室安全知识、纪律及注意事项

实验课是由教师组织指导并由学生独立完成的教学活动。它承担着观察实验现象、巩固理论知识、学习和掌握实验技能和方法、养成科学严谨和良好的实验室工作习惯、培养创新意识和提高科学素养的重要功能。

学生对于实验室工作方面的知识知之甚少,进入大学后无机化学实验是所学的第一门基础化学实验课程。为了使学生尽快熟悉实验教学方式、规范教学秩序、掌握一些实验室工作知识,必须制定相关的规章制度,同时积极进行这方面的教育和培养。但是,这种培养和锻炼不可能一蹴而就,而是逐步形成的,需要贯穿在今后各学科的化学实验教学之中。

2.1.1 实验室的基本布局和常用设备

1. 基本布局

无机化学实验室必须有水、电、通风等系统,其基本布局有:

(1) 学生实验台和仪器柜;放置公用试剂、材料和仪器的桌子和柜子;台间距离既是人员来往通道又是安全逃生通道,故不得放置凳子等杂物。

(2) 实验台两侧及通风橱中装有水龙头,下方有水槽,并通入废水池。

(3) 实验台、墙面及通风橱中装有多种插座,并配置电总闸。

(4) 实验室一角安装有公用的通风橱,学生实验台上方装有抽风口,并直排楼顶。易挥发试剂、有机试剂等置于通风橱中。

2. 常用仪器和设备

常用仪器和设备包括台秤、分析天平、气压计、离心机、真空泵、电炉、恒温水浴箱、电热干燥箱等。其他仪器和材料均分门别类放在指定的地点:铁架台、三脚架等放在指定的柜子中;铁夹、铁圈、橡皮塞、橡皮管等放在指定的抽屉中;去离子水、燃料酒精、各种公用试剂放在公用台上;常用的少量试剂分装在试剂瓶或滴瓶中,整齐有序地放在学生实验台上,用完后应及时放回原处。

2.1.2 实验室规则

实验是人们运用或探索科学知识和规律的一种实践活动。它不同于听课或看书,有自身的特点和规律,有一定的工作程序和规则。实验室规则、安全守则以及学校有关的制度和规定是人们在长期的实验室工作中归纳和总结出来的,它是保持正常的实验工作秩序和良好的实验环境,防止意外事故的发生和做好实验的一个重要前提和保障,因此必须人人遵守。

(1) 认真预习:实验前必须认真预习,理解实验原理和要求,拟订实验步骤,做到胸有

成竹。否则不能进入实验室做实验。做规定以外的实验，应经教师允许。

(2) 遵守操作规程：为了保证人身、财产和实验室安全，提高实验效率，在实验时必须集中精力，认真严格地按照仪器操作规范和实验操作规程进行实验。仔细观察、独立思考、独立完成，在实验进度上做到合理穿插、统筹安排、不拖拉。应及时如实详细地将实验现象和数据记录在记录本上。发现错误应重做，对于耗时较长的实验应报告教师，经同意后方可重做。实验后，根据原始记录认真地分析问题，处理数据，写出实验报告。

(3) 遵守实验室纪律：实验室应保持肃静，不准大声喧哗、随便走动、打电话、发短信等。实验课不得迟到或未经允许而早退，未经教师同意不得缺课、调课，因故缺席未做的实验应及时补做。

(4) 爱护国家财产：实验时应小心使用仪器和设备，精密仪器用完后应及时复原并登记。仪器和设备有异常或损坏必须立即停止使用并报告教师。公用仪器用毕应擦净后送回原处。不得随意动用他人的仪器。

(5) 厉行节约：在不影响实验效果的前提下，节约水、电、试剂和耗材等。按规定的等级和用量取用试剂。用水刷洗或润洗仪器时应按少量多次方法洗净仪器。仪器、试剂及实验产品不得带出实验室，培养良好的实验习惯。

(6) 减少污染：减少毒气、废液和废物等对实验室环境的污染。剧毒药品必须按照严格的管理和使用制度执行，包括领用时要登记，用完后要回收或销毁，扫扫实验室，并洗净双手。产生有毒气体的实验放在通风橱中进行。每人准备一个废品杯和一个废液杯，将实验中的废纸、火柴梗等随时放入废品杯中，将废液随时倒入废液杯中，待实验结束后，分别倒入公用的垃圾筒和废液桶中。切勿将废物和酸性废液直接倒入水槽，以免堵塞和腐蚀下水道。

(7) 保持实验室整洁有序：实验试剂应整齐摆放在实验台上，用完及时放回原处。随时保持实验台面的干燥整洁，实验时取用需要的仪器，并整齐地放在实验台的前方。实验后将仪器洗净并按内高外低整齐地放回柜中。实验台和水槽用湿抹布擦干净。

(8) 实行轮流值日生制度：每次实验结束后，值日生负责打扫实验室，包括检查试剂瓶、仪器设备等是否整齐完好，关好水龙头、拔掉电器插头，拖地，整理和擦净公用仪器和公用台面；倒废物桶；关好门窗等。经教师检查合格后方可离开实验室。

(9) 发生意外事故立即报告：发生意外事故应保持镇静，立即报告教师并采取措施紧急处理。

2.1.3 实验室安全守则

化学实验室有水、电、各种仪器设备和许多试剂，化学试剂中有很多是易燃、易爆、有腐蚀性和有毒的，所以必须十分重视安全问题。每次实验前充分了解实验中的安全问题和注意事项，实验时必须严格遵守实验室安全守则、仪器操作规范和实验操作规程。同时学习一些安全和紧急救护知识非常必要。万一发生事故，要立即紧急处理。具体应做到：

(1) 不要用湿的物品或手接触电源。水、电、酒精灯及电炉等用完毕立即关闭。点燃的火柴用后立即熄灭，不得乱扔。

(2) 涉及易燃、易爆物质的实验要在远离火源及热源处进行。

(3) 涉及有毒和刺激性气体的实验要在通风橱中进行。绝不能用鼻子直接对着瓶口或管口嗅闻气体。

(4)加热时不能从容器上方俯视，以免反应物溅到脸部。加热试管中液体时，管口不能对人。加热固体或浓缩液体时，要不停地搅拌，以免固体飞溅伤人。可戴上护目镜。

(5)浓酸、浓碱具有强腐蚀性、使用时要小心，不能让它溅在皮肤和衣服上，更应注意保护眼睛。稀释浓硫酸时，要把酸注入水中。移取试剂瓶时，须将手心对准瓶体标签处而不能仅用手指抓住瓶盖。

(6)严禁在实验室内饮食或把食品带入实验室。进实验室必须穿实验服，长头发必须扎起来。严禁留长指甲、戴戒指或手镯，严禁穿背心、短裤或拖鞋进实验室。不要低头、趴着或靠着实验台做实验。

(7)严禁皮肤直接接触化学试剂，严禁化学试剂入口或接触伤口。实验剩余的废物和废液严禁倒入水槽，实验后应回收集中处理。

(8)实验结束后，应将手洗干净后离开实验室。

2.1.4 常用试剂的保管

保管化学试剂，要注意防火、防水、防挥发、防曝光和防变质。若保管不当，不仅会造成国家财产损失，还可能对环境造成污染，人员受到伤害。化学试剂的保存应根据试剂的易燃性、易分解性、毒性、腐蚀性和潮解性等各不相同的特点，采用不同的方式。危险药品是指受光、热、空气、水或撞击等外界因素的影响，可能引起燃烧、爆炸的药品，或具有剧毒性、强腐蚀性的药品。

一般性质稳定的单质和无机盐类的固体应放在试剂柜内，无机试剂和有机试剂应分开存放。危险性的试剂必须严格管理，分类存放。

(1)易燃试剂：

(i)易燃液体主要是有机试剂，极易挥发，遇明火即燃烧，应单独存放，注意阴凉通风且远离火源和热源。

(ii)易燃固体有硫磺、红磷、镁粉和铝粉等，着火点都很低，应单独存放，注意阴凉通风干燥且远离火源。白磷在空气中可自燃，应保存在水中，并放于避光处。

(iii)遇水易燃的固体有金属锂、钠、钾、电石、锌粉等，可与水剧烈反应放出可燃性气体，必须存放在阴凉干燥处，同时锂要用石蜡密封，钠和钾应保存在煤油中。

(2)爆炸品：爆炸品指遇热、摩擦或撞击即可引起剧烈分解反应，放出大量气体而产生爆炸的试剂，如硝酸铵、苦味酸、三硝基甲苯等，应存放在阴凉低温处，用时必须轻拿轻放。

(3)强氧化试剂：强氧化试剂如氯酸钾、硝酸钾、过氧化氢、过氧化钠、高锰酸盐和重铬酸盐等都易分解，当遇酸或加入还原性物质、受高热、撞击时，就可能引起爆炸。保存这类物质，一定不能与可燃性物质或还原性物质放在一起，要放在阴凉通风处。

(4)剧毒品：氰化物、砷的化合物、汞及汞的化合物等均为剧毒化学品，应锁在固定的铁柜中，并由双人负责保管。可溶性的铬、镉、钡、铅、锑、铊盐等也是有毒化学品，也应妥善保管。要做到现用现领，用后将剩余物及时交回，并应有使用登记制度。

(5)腐蚀性试剂：强酸、氟化氢、强碱、溴、酚及其他具有强腐蚀性的试剂不要与强氧化性、易燃品、易爆品放在一起。

(6)其他试剂：

(i)易潮解试剂：高氧化态金属氯化物如三氯化锑、氯化(亚)锡等、硫化物、氢氧化钠、氢氧化钾及其他易潮解试剂必须放置在干燥处。

(ii) 见光易分解或变质的试剂：硝酸、硝酸盐、碘化物、过氧化氢、亚铁盐和亚硝酸盐等都应储存于棕色瓶中，避光保存。

2.1.5 实验室事故的紧急处理

如果在实验过程中发生了意外事故，可以采取以下应急措施进行紧急处理。

(1) 割伤：先用消毒棉把伤口清理干净，然后挑出异物，再涂上碘酒或用止血贴包扎，必要时送医院治疗。

(2) 烫伤：切勿用水冲洗，可在烫伤处涂烫伤膏或万花油。

(3) 化学灼伤：

(i) 酸或碱腐蚀：先用干净的布或吸水纸揩去皮肤上的化学试剂，再用大量清水冲洗。若被酸腐蚀致伤，可用饱和碳酸氢钠溶液或稀氨水冲洗；若被碱腐蚀致伤，可用 3%~5%乙酸或 3%硼酸溶液冲洗，最后用水冲洗，必要时送医院治疗。

(ii) 酸或碱溅入眼睛：应立即用大量流水冲洗，再用 3%~5%碳酸氢钠或 3%硼酸溶液冲洗，然后送医院治疗。

(4) 吸入刺激性或有毒气体：若吸入氯气、氯化氢，可吸入乙醇和乙醚的混合蒸气解毒；若吸入硫化氢或一氧化碳气体而感到不适，应立即到室外呼吸新鲜空气。

(5) 毒物误入口内：可口服一杯含 5~10 mL 稀硫酸铜溶液的温水，再用手指伸入咽喉促使呕吐，并立即送医院治疗。

(6) 触电：应立即切断电源，必要时进行人工呼吸，找医生抢救。

(7) 灭火：物质的燃烧需要空气和温度，所以灭火的方法就是隔绝空气或降温。若实验中不慎起火，应立即灭火，并马上采取切断电源、移走易燃药品、停止通风等措施防止火势蔓延。灭火方法和灭火设备要根据起火的原因选择。

(i) 一般性着火，若火势小，可用湿布、石棉布或沙子盖灭；若火势大，可用水、泡沫灭火器等扑灭。

(ii) 当身上衣服着火时，切勿惊慌乱跑，以免风助火势。应立即脱下衣服或用湿布或石棉布盖灭着火处。如果燃烧面积较大，可就地打滚起灭火作用。

(iii) 活泼金属如 Na、K、Mg、Al 等引起的着火不能用水、泡沫灭火器灭火，以免产生可燃气体助长火势，只能用沙子、干粉灭火器等灭火。

(iv) 有机溶剂着火时，切勿使用水、泡沫灭火器灭火，以免增加流动性使火势蔓延，应用二氧化碳灭火器、1211 灭火器、专用防火布、沙子或干粉灭火器等灭火。

(v) 电器着火时，首先关闭电源，再用防火布、干粉灭火器等灭火。不能用水、泡沫灭火器灭火，以免触电。

2.2　无机化学常用仪器的使用、洗涤与干燥

2.2.1　常用仪器

常用仪器见图 2.1。

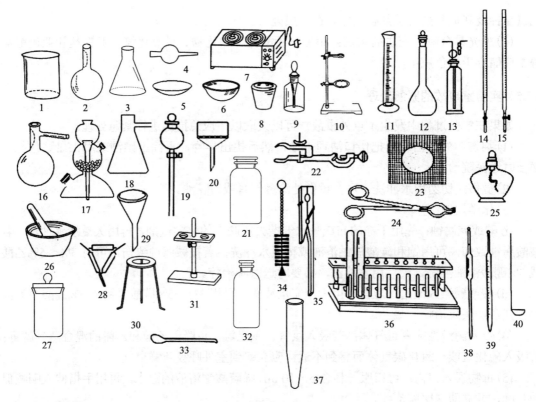

图 2.1 常用仪器

1. 烧杯；2. 烧瓶；3. 锥形瓶；4. 干燥管；5. 表面皿；6. 蒸发皿；7. 水浴锅；8. 坩埚；9. 滴瓶；10. 铁架和铁圈；11. 量筒；12. 容量瓶；13. 洗气瓶；14. 碱式滴定管；15. 酸式滴定管；16. 蒸馏烧瓶；17. 启普发生器；18. 抽滤瓶；19. 分液漏斗；20. 布氏漏斗；21. 广口瓶；22. 铁夹；23. 石棉网；24. 坩埚钳；25. 酒精灯；26. 研钵；27. 称量瓶；28. 泥三角；29. 漏斗；30. 三脚架；31. 漏斗架；32. 细口瓶；33. 药匙；34. 试管刷；35. 试管夹；36. 试管架和试管；37. 离心试管；38. 移液管；39. 吸量管；40. 燃烧匙

2.2.2 常用仪器的使用方法与注意事项

1. 试管、离心试管

规格：分硬质试管、软质试管，有刻度、无刻度，有支管、无支管等。按容量(mL)分有 5、10、15、20、25、50 等。无刻度试管按管外径×管长(mm)，如 15×75 等。

主要用途：常温或加热条件下用作少量试剂反应容器，便于操作和观察。用于收集少量气体。具支试管可检验气体产物。离心试管可用于沉淀分离。

使用方法和注意事项：反应液不超过 1/2；加热时不超过 1/3，以防溅出。管外水应擦干，以防受热不均，导致试管破裂和烫手。加热液体时，倾斜 45°，以防暴沸，管口不对人。加热固体管口略下倾，增大受热面，避免管内冷凝水回流。离心试管不可直接加热。

2. 烧杯

规格：硬质、软质，一般按容量(mL)分有 25、50、100、150、200、250 等。

主要用途：常温或加热条件下用作大量物质反应器，反应物易混合均匀。用于配制溶液。

使用方法和注意事项：反应液不超过2/3，以防液体溅出，加热时反应液不超过1/2。外壁擦干，烧杯底要垫石棉网，以防受热不均。

3. 锥形瓶

规格：硬质、软质，有刻度、无刻度、广口、细口、微型等，按容量(mL)分有50、100、150、200等。

主要用途：用作反应容器，振荡方便，适用于滴定。

使用方法和注意事项：盛液不能太多，防止液体溅出。实验时应下垫石棉网或置于水槽中，以防受热不均。

4. 滴瓶

规格：玻璃质，分棕色、无色两种，滴管上带有橡皮胶头。按容量(mL)分有 15、30、60、125等。

主要用途：用于盛放少量液体试剂或溶液，便于取用。

使用方法和注意事项：棕色瓶装见光易分解或不太稳定的物质，防止物质分解或变质。滴管不能吸得太满，也不能倒置，防止试剂侵蚀橡皮胶头。滴管不得弄乱，以防污染试剂。

5. 漏斗

规格：长颈、短颈，按斗径分(mm)：60、100等。热漏斗用于热过滤。

主要用途：用于过滤或倾注液体。长颈漏斗常装配气体发生器，用于加液。

使用方法和注意事项：不可直接加热。过滤时漏斗颈尖端必须紧靠盛接滤液的容器壁。用长颈漏斗加液时，漏斗颈应插入液面下。

6. 称量瓶

规格：按容量(mL)分。高型：10、20等；矮型：5、10、15等。

主要用途：用于准确称取一定量固体药品。

使用方法和注意事项：不能加热，盖子是磨口配套的，不得丢失、弄乱，不用时应洗净。磨口处垫纸条，防止粘连。

7. 量筒

规格：按容量分(mL)有5、10、20、25、50、100、200等。上部大下部小的称为量杯。

主要用途：用于量取一定体积的液体。

使用方法和注意事项：读数时视线与弯月面相切。不可加热，不可作为实验容器(溶解、稀释等)，不可量热溶液或液体。

8. 移液管

规格：刻度管型和单刻度胖肚型，此外还有自动移液管。按最大标度(mL)有1、2、5、10、25等，微量0.1、0.2、0.25。

主要用途：用于精确移取一定体积的液体。

使用方法和注意事项：吸入液体液面超过刻度时，用食指按住管口，轻轻放气，使液面低于刻度。食指按住管口，移往指定容器上，放开食指，使液体注入。用时先用待移液润洗三次。最后一滴残留液不要吹出（有"吹"字除外）。

9. 容量瓶

规格：按刻度以下的容量(mL)分有 25、50、100、250、500、1000 等。现在也有塑料塞的。

主要用途：用于配制准确浓度溶液。

使用方法和注意事项：溶质先在烧杯内全部溶解，冷却后移入容量瓶。不能加热，以免影响容量瓶的精确度。不能代替试剂瓶存放溶液。

10. 抽滤瓶和布氏漏斗

规格：布氏漏斗为瓷质，以口径大小表示。抽滤瓶为玻璃质，规格以容量(mL)表示。

主要用途：两者配套使用，用于无机制备实验中晶体或沉淀的减压过滤。

使用方法和注意事项：不能直接加热。滤纸要略小于漏斗内径，又要把小孔全部盖住，以免漏滤。使用时先抽气，再过滤，停止过滤时，要先放气，后关泵。

11. 蒸发皿

规格：平底、圆底等。规格按上口径(mm)分。

主要用途：用于蒸发、浓缩溶液，视液体性质不同选不同质地的蒸发皿。

使用方法和注意事项：能耐高温，但不宜骤冷。

12. 坩埚

规格：瓷质，也有石英或金属制成。规格以口径(mm)表示。

主要用途：用于强热、煅烧固体，视固体性质不同选不同质地的坩埚。

使用方法和注意事项：放在泥三角上直接强热或煅烧。加热或反应完毕后，用坩埚钳取下时，坩埚钳应预热，取下后应放置在石棉网上。

13. 细口瓶

规格：玻璃质，有磨口和不磨口，无色、棕色和蓝色。按容量(mL)分有 100、125、250、500、1000 等。细口瓶又称试剂瓶。

主要用途：用于储存溶液和液体药品。

使用方法和注意事项：不能直接加热。瓶塞不能弄脏、弄乱。盛放碱液应改用胶塞。有磨口塞的细口瓶不用时应洗净，并在磨口处垫上纸条。有色瓶盛见光易分解或不太稳定的物质的溶液或液体。

14. 广口瓶

规格：玻璃质，有无色和棕色，有磨口和不磨口。磨口有塞，若无塞的口上是磨砂的则为集气瓶。按容量(mL)分有 30、60、125、250、500 等。

主要用途：用于储存固体药品，收集气体。

使用方法和注意事项：不能直接加热。不能放碱，瓶塞不得弄脏、弄乱。做气体燃烧实验时瓶底应放少许沙子或水。收集气体后，要用毛玻璃片盖住瓶口。

15. 酸式滴定管/碱式滴定管

规格：玻璃质，分酸式(具玻璃旋塞)和碱式(具乳胶管连接的玻璃尖嘴)两种。按刻度最大标度(mL)分有 25、50、100。微量的有 1、2、3、4、5、10 等。

主要用途：用于滴定，或量取校准体积的液体。

使用方法和注意事项：用前洗净，装液前用预装溶液淋洗三次。使用酸式滴定管时，用左手开启旋塞。使用碱式滴定管时，左手轻捏乳胶管内玻璃珠，溶液即可放出。碱管要注意赶尽气泡；酸管旋塞应涂凡士林，碱管下端乳胶管不能用洗液洗。酸管、碱管不能对调使用。

16. 干燥管

规格：玻璃质，还有其他形状的，以大小表示。

主要用途：用于干燥气体。

使用方法和注意事项：干燥剂颗粒要大小适中，填充时松紧要适中不与气体反应。两端要用棉花团。干燥剂变潮后应立即换干燥剂，用后应清洗。两头要接对(大头进气，小头出气)，并固定在铁架台上使用。

17. 洗气瓶

规格：玻璃质，形状有多种。按容量(mL)分有 125、250、500、1000 等。

主要用途：用于净化气体，反接也可用作安全瓶(或缓冲瓶)。

使用方法和注意事项：接法要正确(进气管通入液体中)，洗涤液注入容器高度的 1/3，不得超过 1/2。

18. 铁架台

规格：铁制品，铁夹现在有铝制的。铁架台有圆形和长方形。

主要用途：用于固定或放置反应容器。铁圈还可代替漏斗夹使用。

使用方法和注意事项：仪器固定在铁架台上时，仪器和铁架的重心应落在铁架台底盘中部。用铁夹夹持仪器时，不能过紧或过松，应以仪器不能转动为宜。加热后的铁圈不能撞击或摔落在地。

19. 毛刷

规格：以大小或用途表示，如试管刷、滴定管刷等。

主要用途：用于洗刷玻璃仪器。

使用方法和注意事项：洗涤时手持刷子的部位要合适。注意毛刷顶部竖毛的完整程度。

20. 研钵

规格：瓷质，也有玻璃、玛瑙或铁制品。以口径大小表示。

主要用途：用于研碎或混合固体物质。按固体的性质和硬度选用不同的研钵。

使用方法和注意事项：大块物质只能压碎，不能舂碎。放入量不宜超过研钵容积的 1/3。易爆物质只能轻轻压碎，不能研磨。

21. 试管架

规格：有木质和铝质，有不同的形状和大小。
主要用途：用于放置试管。
使用方法和注意事项：加热后的试管应用试管夹夹住悬放架上。

22. 试管夹

规格：有木制、竹制，也有金属丝(钢或铜)制品，有不同的形状。
主要用途：用于夹持试管。
使用方法和注意事项：夹在试管上端。不要把拇指按在夹的活动部分，一定要从试管底部套上和取下试管夹。

23. 三脚架

规格：铁制品，有大小、高低之分，比较牢固。
主要用途：用于放置较大或较重的加热容器。
使用方法和注意事项：放置加热容器(除水浴锅外)应先放石棉网。下面加热灯焰的位置要合适，一般用氧化焰加热。

24. 燃烧匙

规格：匙头铜质，也有铁制品。
主要用途：检验可燃性，用于固气燃烧反应。
使用方法和注意事项：放入集气瓶时应由上而下慢慢放入，且不要触及瓶壁。硫磺、钾、钠燃烧实验，应在匙底垫上少许石棉或沙子。用完立即洗净匙头并干燥。

25. 泥三角

规格：由铁丝扭成，套有瓷管。有大小之分。
主要用途：用于灼烧坩埚时放置坩埚。
使用方法和注意事项：使用前应检查铁丝是否断裂，断裂的不能使。坩埚放置要正确，坩埚底应横着斜放在三个瓷管中的一个瓷管上。灼烧后小心取下，不要摔落。

26. 药匙

规格：由牛角、瓷或塑料制成。现多数是塑料的。
主要用途：用于拿取固体药品。药匙两端各有一个勺，一大一小。根据药品用量分别选用。
使用方法和注意事项：取用一种药品后，药匙必须洗净，并用滤纸擦干后，才能取用另一种药品。

27. 石棉网

规格：由铁丝编成，中间涂有石棉。有大小之分。

主要用途：石棉是一种不良导热体，能使物体均匀受热。

使用方法和注意事项：使用前应先检查，石棉脱落的不能用。不能与水接触，不可卷折。

28. 坩埚钳

规格：铁制品，有大小、长短的不同(打开或关闭钳子时不要太紧和太松)。

主要用途：夹持坩埚加热或往高温电炉(马弗炉)中放、取(也可用于夹取热的蒸发皿)。

使用方法和注意事项：使用时必须用干净的坩埚钳。坩埚钳用后，应尖端向上平放在实验台上(如温度很高，则应放在石棉网上)。实验完毕后，应将坩埚钳擦干净，放入实验柜中，干燥放置。

2.2.3 常用玻璃仪器的洗涤

玻璃仪器的洗涤方法很多，应根据实验的要求、污物的性质、沾污的程度选用。常用的洗涤方法如下：

(1) 水洗：用水和毛刷刷洗，除去仪器上的尘土、其他不溶性杂质和可溶性杂质。

(2) 用去污粉、肥皂或洗涤剂洗：洗去油污和有机物质，若油污和有机物质仍洗不干净，可用热的碱液洗。

(3) 用铬酸洗液洗：在进行精确的定量实验时，对仪器的洁净程度要求很高，所用仪器形状特殊，这时要用洗液洗。洗涤方法如下：

(i) 将玻璃器皿用水或洗衣粉洗刷一遍，尽量把器皿内的水去掉，以免冲稀洗液。

(ii) 将洗液倒入待洗容器，反复浸润内壁，使污物被氧化溶解。

(iii) 用毕将洗液倒回原瓶内，以便重复使用。

(iv) 洗液瓶的瓶塞要塞紧，以防洗液吸水失效。

铬酸洗液有强酸性和强氧化性，去污能力强，适用于洗涤油污及有机物。洗液有强腐蚀性，勿溅在衣物、皮肤上。当洗液的颜色由原来的深棕色变为绿色，即重铬酸钾被还原为硫酸铬时洗涤效能下降，应重新配制。比色皿应避免使用毛刷和铬酸洗液。

(4) 用浓 HCl 洗：可以洗去附着在器壁上的氧化剂如二氧化锰，大多数不溶于水的无机物都可洗去。例如，灼烧过沉淀物的瓷坩埚可先用热 HCl(1:1)洗涤，再用洗液洗。

(5) 用氢氧化钠的高锰酸钾洗液洗：可以洗去油污和有机物，洗后在器壁上留下的二氧化锰沉淀可再用盐酸洗。

(6) 其他洗涤方法：除上述方法外，还可根据污物的性质选用适当试剂，如 AgCl 沉淀可用氨水洗涤，硫化物沉淀可用硝酸加盐酸洗涤。

用以上各种方法洗涤后，玻璃仪器上往往还留有 Ca^{2+}、Mg^{2+}、Cl^- 等离子。如果实验中不允许这些离子存在，应再用蒸馏水把它们洗去。使用蒸馏水的目的只是洗去附在仪器壁上的自来水，所以应该尽量少用，符合少量(每次用量少)多次(一般洗 3 次)的原则。

洗涤玻璃仪器的基本要求：

(1) 洗净的玻璃仪器壁上不应附着不溶物、油污。检查方法是将玻璃仪器用水完全湿润，把仪器倒过来，水即顺器壁流下，器壁上只留下一层薄而均匀的水膜，不挂水珠，表示仪器已洗净。

(2) 在定性、定量实验中，由于杂质的引进会影响实验的准确性，对玻璃仪器的要求比较高。但在有些情况下，如一般的无机制备、性质实验，对玻璃仪器的洁净程度要求不是很高，只要刷洗干净。工作中应根据实际情况决定洗涤的程度。

(3) 已洗净的玻璃仪器不能用布或纸擦，因为布和纸的纤维会留在器壁上。

2.2.4 仪器的干燥

常用仪器可用以下方法干燥（图 2.2）：

(1) 晾干：不急用的仪器洗净后可倒挂在干净的实验柜内或仪器架上，任其自然干燥。

(2) 烘箱烘干：将洗净的仪器尽量倒干水后，放进烘箱内，放时应使仪器口朝下，并在烘箱最下层放一个搪瓷盘，盛接从仪器上滴下来的水，以免水滴到电热丝上，损坏电热丝。

(3) 烤干：一些常用的烧杯、蒸发皿等可放在石棉网上用小火烤干。试管可用试管夹夹住，在火焰上来回移动，直至烤干，但必须使管口朝下，以免水珠倒流至试管灼热部分，使试管炸裂。

(4) 气流烘干：试管、量筒等适合在气流烘干器上烘干。

(5) 电吹风吹干。

(6) 有机溶剂挥发带走水汽。

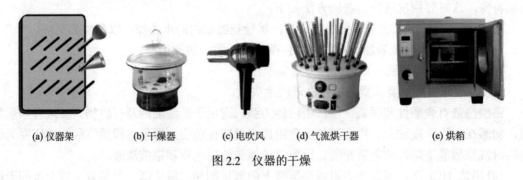

(a) 仪器架　　(b) 干燥器　　(c) 电吹风　　(d) 气流烘干器　　(e) 烘箱

图 2.2　仪器的干燥

2.3　常用加热器具的介绍与使用

2.3.1 酒精灯与酒精喷灯

1. 酒精灯

酒精灯（加热温度通常为 400～500℃）由灯罩、灯芯和灯壶三部分组成，如图 2.3(a) 所示。

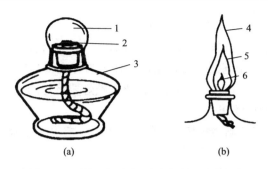

图 2.3　酒精灯的结构(a)和灯焰示意图(b)

1. 灯罩；2. 灯芯；3. 灯壶；4. 外焰；5. 内焰；6. 焰芯

使用酒精灯时，先检查灯芯。如果灯芯顶端不平或已烧焦，需要剪去使其平整。然后检查灯内有无酒精。添加酒精应在灯熄灭的状态下，取出灯芯，借助漏斗将酒精注入。灯内酒精的体积应为酒精灯容积的 1/2～2/3。用火柴点燃，绝对禁止用酒精灯引燃另一盏酒精灯。熄灭火时，用灯罩盖灭，不可用嘴吹灭。正确使用如图 2.4 所示。

图 2.4　酒精灯使用示意图

安全要求：酒精是易燃品，使用时要严格按照规范操作。不要碰倒酒精灯或将酒精洒在容器外面，以免引起火灾。万一洒出的酒精在桌上燃烧起来，不要惊慌，应立即用湿抹布扑盖。

2. 酒精喷灯

酒精喷灯(加热温度通常为 800～900℃)也是用酒精作燃料，但它是将酒精气化并与空气混合后燃烧。酒精喷灯的火焰温度高而稳定，可达 800～900℃。酒精喷灯有座式和挂式两种，如图 2.5 所示。

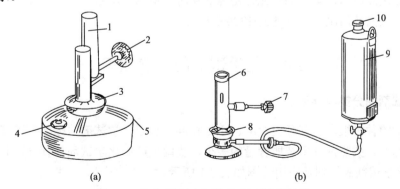

图 2.5　座式喷灯(a)和挂式喷灯(b)的结构

1. 灯管；2. 空气调节器；3. 预热盘；4. 铜帽；5. 酒精壶；6. 灯管；7. 空气调节器；8. 预热盘；9. 酒精储罐；10. 盖子

座式喷灯和挂式喷灯使用方法基本相同。

(1) 添加酒精：在酒精壶或酒精储罐内加入酒精，切忌在使用过程中续加，以免着火。

(2) 预热：在预热盘中加满酒精点燃预热。挂式喷灯应将酒精储罐下面的开关打开，从灯管口冒出酒精后再关上，再点燃前再打开；座式喷灯要将灯的空气调节器关上，预热一会儿后再缓缓调整，防止酒精喷出。

(3) 点燃：等预热盘中的酒精燃烧完将灯管灼热后，打开空气调节器并用火柴将灯点燃，调节火焰平稳即可使用。

(4) 关火：用完后关闭空气调节器或用石棉网盖住灯管口即可将灯熄灭。

挂式喷灯不用时，要将酒精储罐下面的开关关闭，长时间不用应将酒精倒出。

安全要求：酒精是易燃品，使用时要严格按照规范操作。使用酒精喷灯时灯管必须灼热后再点燃，否则易造成液体酒精喷出引起火灾。座式喷灯最多使用 30 min，挂式喷灯也要注意使用时间不要太久。若需继续使用，应到时将喷灯熄灭冷却，添加酒精后再次点燃。

2.3.2 电加热设备

电加热设备包括电炉、电热板、电热套、高温炉等，如图 2.6 所示。

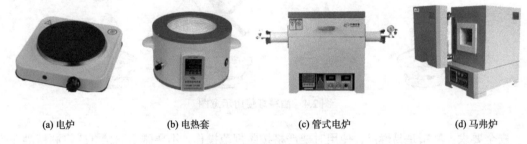

(a) 电炉　　(b) 电热套　　(c) 管式电炉　　(d) 马弗炉

图 2.6　常用的电加热设备

(1) 电炉和电热套的温度可通过外接变压器调节，出于安全考虑实验一般使用封闭式电炉。使用电炉时，容器(如烧杯等)和电炉之间要隔一块石棉网。电热板的加热面积比电炉大，用于加热体积较大或数量较多的试样。

(2) 常用的高温炉有管式电炉和马弗炉，主要用于高温灼烧或进行高温反应。其温度用温度控制仪连接热电偶控制。加热元件是电热丝时，最高使用温度可达 950℃左右；如果用硅碳棒加热，最高使用温度可以达到 1300℃左右。

2.4　化学试剂的一般知识

2.4.1　化学试剂的分类与存放

化学试剂的种类很多，其分类标准也不尽相同。我国化学试剂的标准有国家标准、化工标准及企业标准。按照试剂中杂质含量的多少，我国生产的化学试剂(通用试剂)基本上分为 4 个级别。化学试剂的级别及应用范围如表 2.1 所示。

表 2.1　化学试剂的级别及应用范围

级别	名称	英文符号	标签颜色	应用范围
一级	优级纯(保证试剂)	GR	绿	精密分析研究工作
二级	分析纯(分析试剂)	AR	红	分析实验
三级	化学纯	CP	蓝	一级化学实验
四级	实验试剂	LR	黄	工业或化学制备
生化试剂	生化试剂(生物染色剂)	BR	咖啡或玫红	生化实验

根据规定，试剂瓶的标签上应标出试剂名称、化学式、分子量、级别、技术规格、产品标准号、生产许可证、生产批号、生产厂家等，危险品和剧毒品还应给出相应的标志。一般来说，无机化学实验中常使用化学纯或分析纯级别的试剂。

存放要做到试剂分开存放、取用方便、注意安全、保证质量。一般的化学试剂均应密封，并分门别类地存放在低温、干燥、通风处。特殊化学试剂有特殊的保存方法。白磷要保存在水中，金属钾、钠通常保存在煤油中。化学试剂分装时，一般把固体试剂装在广口瓶中，液体试剂或配制的溶液盛放在细口瓶或带有滴管的滴瓶中，见光易分解的试剂或溶液如硝酸银等盛放在棕色瓶中，每个试剂瓶都应贴上标签，写上试剂名称、日期、规格或浓度。

2.4.2　化学试剂的取用

取用试剂前，要先看清试剂瓶的标签。取用时，先打开瓶塞，将瓶塞反放在实验台上，不可将瓶塞横置桌上而污染，试剂用多少取多少。取完后及时盖好瓶盖，决不允许张冠李戴，然后把试剂瓶放回原处，养成良好的实验习惯。

1. 试剂瓶打开方法

(1) 软木塞：欲打开固体试剂瓶上的软木塞时，可手持瓶子，将瓶斜放在实验台上，然后用锥子斜着插入软木塞将塞子取出。即使软木塞渣附在瓶口，渣也不会落入瓶中，可用卫生纸轻轻擦掉。

(2) 塑料塞：盐酸、硫酸、硝酸等液体试剂瓶多用塑料瓶塞。如果塞子打不开，可用热水浸过的抹布裹上塞子的头部，然后用力拧，一旦松动就能拧开。

(3) 玻璃瓶塞：可在水平方向用力转动塞子，或左右交替横向用力摇动塞子，轻轻打开。如果打不开，可紧握瓶的上部，用木柄或木槌敲打塞子，也可在桌端轻轻叩敲。注意：绝不能用铁锤敲打。

2. 固体试剂的取用方法

(1) 用清洁、干燥的药匙取试剂。取大量的固体时用药匙的大头，取少量的固体时用药匙的小头。专匙专用，用过的药匙必须洗净擦干才能再用。

(2) 要求取用一定质量的固体试剂时，可把固体放在干燥的称量纸上称量，有腐蚀性或易潮解的固体应放在表面皿上或玻璃容器内称量。

(3) 往试管(特别是湿试管)中加入试剂时，可用药匙或将取出的药品放在对折的纸片上，伸进试管约 2/3 处[图 2.7(a)和(b)]。加入块状固体时，应将试管倾斜，使其沿试管壁慢慢滑下[图2.7(c)],以免碰破试管壁。

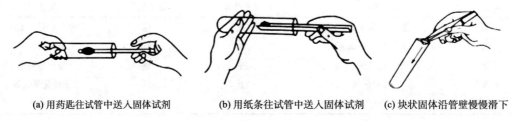

(a) 用药匙往试管中送入固体试剂　　(b) 用纸条往试管中送入固体试剂　　(c) 块状固体沿管壁慢慢滑下

图 2.7　固体试剂的取用方法

注意

(1) 取试剂不要超出所需剂量。多取的试剂不能倒回原瓶,可放在指定的容器中供他人使用。

(2) 固体颗粒较大时,可在清洁、干燥的研钵中研碎。研钵中所盛固体的量不要超过其容积的 1/3。

(3) 有毒试剂要在教师的指导下取用。

3. 液体试剂的取用方法

(1) 从滴瓶中取用液体试剂时,用滴瓶中的滴管。滴管绝不能伸进所用的实验容器中,以免接触器壁而污染试剂[图 2.8(a)]。如果从试剂瓶中取出少量液体试剂,则需用附于该试剂瓶的专用滴管取用。装有试剂的滴管不要横置或滴管口向上斜放,以免液体流入滴管的乳胶头中。

(2) 从细口瓶中取用液体试剂时,用倾注法。先将瓶塞取下反放在桌面上,手握住试剂瓶上贴标签的一面,慢慢倾斜瓶子,让试剂沿着试管壁流入试管或沿着玻璃棒注入烧杯中[图 2.8(b)和(c)]。倒出所需量后,将试剂瓶口在容器上靠一下,再逐渐立起瓶子,以免遗留在瓶口的液滴流到瓶的外壁。

(3) 定量取用液体时,用量筒或移液管。量筒用于量度一定体积的液体,可根据需要选用不同容量的量筒。量取液体时,要按图 2.9 所示,使视线与量筒内液体弯月面的最低处保持水平,偏高或偏低都会读不准读数而造成较大的误差。

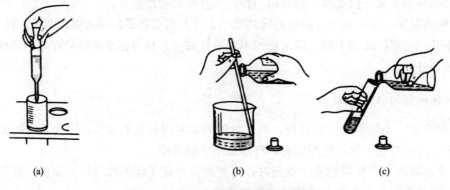

(a)　　(b)　　(c)

图 2.8　滴液入试管(a)和倾注法(b 和 c)

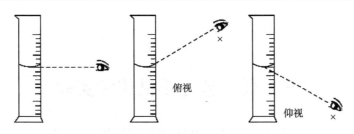

图 2.9　读取量筒内液体的容积

注意

(1) 按需取用液体试剂，多取的试剂不能倒回原试剂瓶中。

(2) 用滴管取用试剂时，学会估计取用液体的量。1 mL 相当于多少滴？5 mL 占一支试管容量的多少？建立"少量"的概念。

(3) 在试管中进行实验时，倒入试管中溶液的量一般不要超过其容积的 1/3。

参考文献

赵新华. 2014. 无机化学实验. 4 版. 北京: 高等教育出版社

第3章 基础无机化学实验

实验1 仪器的认领、洗涤与干燥

实验目的

(1) 熟悉无机化学实验室的规则和要求。
(2) 领取无机化学实验常用仪器，熟悉其名称、规格及使用注意事项。
(3) 学习并练习常用仪器的洗涤和干燥方法。

实验步骤

(1) 按表 3.1 中仪器数量、规格、型号逐个认领无机化学实验中常用仪器。

表 3.1　无机化学实验常用仪器

名称	规格	数量	备注	名称	规格	数量	备注
试管架		1		瓷坩埚	有盖	1	
试管夹		1		点滴板	白	1	
试管刷		1	公用	玻璃片		1	
试管	10 mm×150 mm	10		酒精灯		1	
离心试管		6		镊子		1	
烧杯	50 mL	2		石棉网		1	
烧杯	100 mL	2		泥三角		1	
烧杯	250 mL	2		洗耳球		1	
烧杯	≥400 mL	2		燃烧匙		1	
量筒	10~20 mL	1		坩埚钳		1	
量筒	50 mL	1		蒸发皿		1	
集气瓶		1		塑料洗瓶		1	
抽滤瓶		1		布氏漏斗		1	
烧瓶		1		玻璃漏斗		1	
锥形瓶	150 mL	2		表面皿		1	
锥形瓶	250 mL	3					
容量瓶	50 mL，有盖	1					
容量瓶	100 mL，有盖	1					

(2) 洗涤：根据不同的仪器选择正确的洗涤方法进行洗涤，抽取两件洗干净的仪器交教师检查，并将洗净的仪器合理有序地摆放在柜子中。

(3) 干燥。

注意事项

(1) 试管内废液必须先倒入废液缸中，后注水洗涤，洗涤时应做到少量多次。
(2) 不能一手同时握拿多支试管。刷洗时用力适度。实验柜内仪器摆放应便于取用。
(3) 若用王水洗涤，王水必须用新配制的(王水不稳定)。
(4) 铬酸洗液常放置在通风橱内，用后要回收。
(5) 无机化学实验主要采用微量化。所有实验中涉及的试剂，在药品箱二级试剂瓶中应备有，若无，则在公共药品台上。二级试剂仅限当天实验用，不必多取。二级试剂瓶不要混装，瓶盖也不可任意调换。

思考题

(1) 烤干试管时为什么管口略向下倾斜？
(2) 什么仪器不能用加热的方法进行干燥，为什么？
(3) 画出离心试管、多用滴管、烧杯、量筒、容量瓶简图，说明其主要用途和注意事项。

实验 2　试剂的取用和试管操作

实验目的

(1) 了解化学试剂纯度的等级标准和应用范围，以及试剂瓶的种类。
(2) 学习并掌握固体样品的称取、液体试剂的量取及使用方法。
(3) 练习并掌握试管操作及固体或液体的加热方法。
(4) 练习搅拌、溶解、研磨、仪器洗涤等基本操作。
(5) 了解实验的化学原理。

实验原理

1. "三色杯"实验

非极性的 I_2 分子易溶于许多有机溶剂中而呈现出不同的颜色。碘在乙醇或异戊醇中生成的溶液显棕色，这是由于生成了溶剂合物；碘在介电常数较小的溶剂如二硫化碳、四氯化碳中生成紫色溶液，这是因为在溶剂中碘以分子状态存在。碘在水中的溶解度虽然小，但是在碘化钾溶液中溶解度明显增大。碘盐的浓度越大、溶解的碘越多，则生成的溶液颜色越深。在溶液中形成了 I_3^-：$I_2 + I^- \rightleftharpoons I_3^-$。从此平衡可见，溶液中总有碘分子存在，使碘的水溶液呈现颜色。

2. "硫酸铜变色"实验

$CuSO_4 \cdot 5H_2O$ 俗称胆矾，是最常见的铜盐。它是蓝色斜方晶体，在不同温度下可以逐步失水：

$$CuSO_4 \cdot 5H_2O \xrightarrow{375\,K} CuSO_4 \cdot 3H_2O \xrightarrow{423\,K} CuSO_4 \cdot H_2O \xrightarrow{523\,K} CuSO_4$$

若继续加热至高于 873 K，白色的 $CuSO_4$ 将分解为 CuO、SO_2、SO_3 和 O_2。可见，各个

水分子的结合力不完全一样。实验证明，四个水分子与 Cu^{2+} 以配位键结合，第五个水分子以氢键与两个配位水分子和 SO_4^{2-} 结合。$CuSO_4 \cdot 5H_2O$ 的简单平面结构如下：

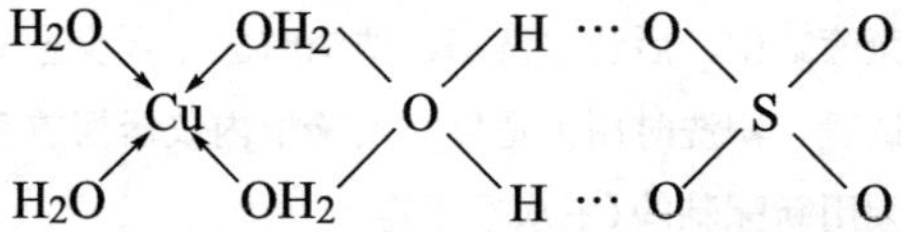

$$\begin{array}{c} H_2O \quad OH_2 \quad \quad H \cdots O \quad \quad O \\ \diagdown \quad \diagup \quad \diagdown \quad \diagup \quad \diagdown \quad \diagup \\ Cu \quad \quad \quad O \quad \quad \quad S \\ \diagup \quad \diagdown \quad \diagup \quad \diagdown \quad \diagup \quad \diagdown \\ H_2O \quad OH_2 \quad \quad H \cdots O \quad \quad O \end{array}$$

加热失水时，先失去 Cu^{2+} 左边的两个水分子，再失去 Cu^{2+} 右边的两个水分子，最后失去以氢键与 SO_4^{2-} 结合的水分子。

3. "五色管"实验

在水溶液中，Ni^{2+} 与乙二胺(en)能形成配位比为 1∶1、1∶2 和 1∶3 的配合物，它们的稳定性不同并呈现不同的颜色：$[Ni(H_2O)_4(en)]^{2+}$（浅蓝色）、$[Ni(H_2O)_2(en)_2]^{2+}$（蓝色）、$[Ni(en)_3]^{2+}$（紫色）。Ni^{2+} 与丁二酮肟(dmg)在水溶液中能生成螯合物二丁二酮肟合镍(Ⅱ) $[Ni(dmg)_2]$，它是一种鲜红色沉淀。

$$[Ni(H_2O)_6]^{2+}(绿色) + en \longrightarrow [Ni(H_2O)_4(en)]^{2+}(浅蓝色) + 2H_2O$$

$$[Ni(H_2O)_6]^{2+}(绿色) + 2en \longrightarrow [Ni(H_2O)_2(en)_2]^{2+}(蓝色) + 4H_2O$$

$$[Ni(H_2O)_6]^{2+}(绿色) + 3en \longrightarrow [Ni(en)_3]^{2+}(紫色) + 6H_2O$$

$$[Ni(H_2O)_6]^{2+}(绿色) + 2dmg \longrightarrow Ni(dmg)_2(红色) + 6H_2O + 2H^+$$

4. 空气中氧气的体积

红磷是单质磷的一种同素异形体。磷的同素异形体在不同的条件下可相互转化：

$$黑磷 \xleftarrow{高温，高压} 白磷 \xrightarrow{400℃隔绝空气} 红磷$$

红磷的化学性质较稳定。它在 400℃以上燃烧，不溶于有机溶剂。红磷在空气中燃烧生成磷的氧化物，均为易溶于水的白色固体。

$$4P \xrightarrow{O_2(不足)} P_4O_6 \xrightarrow[\triangle]{O_2} P_4O_{10}$$

实验用品

仪器：试管、试管夹、烧杯、洗瓶、量筒、研钵、药匙、台秤、酒精灯、铁架台、铁夹、玻璃棒、滴管、橡皮塞。

试剂：碘、碘化钾、红磷、硫酸铜、四氯化碳、异戊醇、硫酸镍(0.1 mol/L)、乙二胺(25%, V/V)、丁二酮肟(1%)。

实验步骤

1. "三色杯"实验

取一个 10 mL 量杯(筒)沿杯壁注入 2 mL 四氯化碳，再依次注入 5 mL 水和 2 mL 异戊醇。用药匙的小头分别取一小匙 KI 固体和一小匙已在研钵中研细的固体碘置于表面皿上，并将它们混合均匀。用一支已用水润湿的玻璃棒粘起表面皿上的混合物，插入装有上述溶液的量杯中，并轻轻地搅动，观察量杯中三层溶液的颜色。

2. "硫酸铜变色"实验

在试管内放入几粒 $CuSO_4 \cdot 5H_2O$ 晶体，按固体试剂的加热方法实验。当所有晶体变为白色时，停止加热，当试管冷却至室温后加入 3～5 滴水。注意观察颜色的变化，用手摸一下试管壁有什么感觉。

3. "五色管"实验

取 5 支试管，在每支试管中加入 10 滴 1 mol/L $NiSO_4$ 溶液。第一支试管中加入 1～2 滴 25%乙二胺(en)溶液，第二支试管中加入 3～4 滴 25%乙二胺溶液，第三支试管中加入 5～6 滴 25%乙二胺溶液，第四支试管中加入 9～10 滴 1%丁二酮肟(dmg)溶液，第五支试管用作对比颜色。充分振荡试管后，观察并比较五支试管中配合物的不同颜色。

4. 空气中氧气的体积

取一支干燥、洁净的硬质试管，将约 0.2 g 红磷放入试管底部，用橡皮塞盖紧。在酒精灯上加热试管(注意安全！)，待红磷燃烧的火焰熄灭后，冷至室温。将试管倒置于盛有水的大烧杯中，在水下拔掉橡皮塞(注意：管口不能露出水面)，观察试管中液面上升的现象。待水面不再上升时，在水下盖上橡皮塞。取出试管，观察试管中水的体积约占试管总体积的几分之几，从而说明空气中大约含有多少体积的氧气。

注意事项

(1) "三色杯"实验中，玻璃棒不要上下搅动，以免两种有机层混合。
(2) 由于有机溶剂挥发性强且毒性大，所以"三色杯"实验后应及时处理废液，以免污染环境。

思考题

在试管中加热液体或固体试剂应注意什么？

参考文献

北京师范大学无机化学教研室等. 1991. 无机化学实验. 2 版. 北京：高等教育出版社

实验 3　溶液的配制

实验目的

(1) 掌握溶液的质量分数、质量摩尔浓度、物质的量浓度、体积比浓度等一般配制方法。
(2) 学习比重计、移液管、容量瓶、台秤和分析天平等的使用方法。
(3) 了解特殊溶液的配制。

实验原理

1. 溶液浓度的几种表示方法

1)质量分数

质量分数(w)表示溶液中所溶解溶质的质量分数,习惯上称为质量百分比浓度。

$$w = \frac{m_{溶质}}{m_{溶液}}$$

所以

$$m_{溶质} = \frac{w\rho_{溶剂}V_{溶剂}}{1-w}$$

式中,w 为溶质的质量分数;$m_{溶质}$ 为固体试剂的质量(g);$m_{溶液}$ 为溶液的质量(g);$\rho_{溶剂}$ 为溶剂的密度(g/mL);$V_{溶剂}$ 为溶剂的体积(mL)。

2)质量摩尔浓度

质量摩尔浓度(b)是指 1000 g 溶剂中溶解溶质的物质的量(mol/kg),一般以水为溶剂。

$$b = \frac{n}{1000 \text{ g 溶剂}}$$

所以

$$x_{溶质} = \frac{Mbm_{溶剂}}{1000} = \frac{Mb\rho_{溶剂}V_{溶剂}}{1000}$$

式中,b 为质量摩尔浓度(mol/kg);$x_{溶质}$ 为溶质的质量(g);M 为固体试剂的摩尔质量(g/mol);$m_{溶剂}$ 为溶剂的质量(g);其他符号的说明同前。

3)物质的量浓度

物质的量浓度(c)是指 1 L 溶液中溶解溶质的物质的量。

$$c = \frac{m_{溶质}}{VM}$$

所以

$$m_{溶剂} = cVM \qquad c_1V_1 = c_2V_2$$

式中,c 为物质的量浓度(mol/L);其他符号的说明同前。

4)体积比浓度

体积比浓度是指一定体积的溶质溶解在一定体积的溶剂中(V/V)。

2. 溶液配制的基本方法

无机化学实验通常配制的溶液有一般溶液、标准溶液和特殊溶液。

1)一般溶液的配制

一般溶液是指实验对溶液浓度的准确度要求不高,通常利用台秤、量筒或带刻度的烧杯等低准确度的仪器就能满足要求。该配制过程又称粗略配制。

配制一般溶液常用以下几种方法。

a. 直接水溶法

这种方法适用于易溶于水但不发生水解的固体试剂，如 NaOH、NaCl、KNO$_3$ 等。配制溶液时，可用台秤称取一定量的固体于烧杯中，加适量蒸馏水搅拌溶解，稀释至所需体积，再转入试剂瓶中。

b. 稀释法

对液态试剂，如浓酸、浓碱或一定浓度的电解质溶液，配制其稀溶液时，先用量筒量取所需体积的浓溶液，然后用适量的蒸馏水稀释。特别注意的是，用浓硫酸配制其稀溶液时，应在不断搅拌下将浓硫酸缓慢地倒入盛水的容器中，切不可颠倒顺序。

2) 标准溶液的配制

标准溶液是指已知准确浓度的溶液，即实验对溶液浓度的准确度要求高。通常利用分析天平、移液管、容量瓶等高准确度的仪器配制溶液。该配制过程又称准确配制。配制标准溶液的方法有两种。

a. 直接法

用分析天平准确称取一定量的基准试剂于烧杯中，加入适量纯水(离子交换水或二次蒸馏水)溶解后，转入容量瓶，再用纯水稀释至刻度，摇匀。其准确浓度可由称量数据及稀释体积求出。

b. 标定法

对于非基准试剂，不能用直接法配制标准溶液，但可先配成近似于所需浓度的溶液，然后用基准试剂或已知准确浓度的标准溶液标定它的浓度。

用稀释法配制较稀的标准溶液时，可用移液管吸取适当体积的浓溶液至一定体积的容量瓶中配制。

3) 特殊溶液的配制

特殊溶液是指某些溶质在溶解过程中易发生水解、见光分解或发生氧化还原反应的溶液。在配制的过程中必须加一些物质或采取一定的措施防止它们在保存期间失效。

当配制 FeCl$_3$、SnCl$_2$、SbCl$_3$ 和 BiCl$_3$ 等溶液时，加入一定量的稀盐酸使其溶解，再用蒸馏水稀释至所需体积，然后转入试剂瓶中保存。对特别易水解的物质，如 SnCl$_4$ 可先溶解在浓盐酸中，然后用水稀释。对容易发生氧化还原反应的溶液，要防止在保存期间被氧化失效。例如，Sn^{2+}、Fe^{2+} 溶液应分别放入一些 Sn 粒和 Fe 屑。AgNO$_3$、KMnO$_4$、KI 等易见光分解的溶液应储于洁净的棕色瓶中。容易发生化学腐蚀的溶液应储于合适的容器中。

对在水中溶解度较小的物质，如 I$_2$ 水的配制，可将 I$_2$ 用 KI 溶液溶解，然后用水稀释至所需体积，摇匀后转入试剂瓶中。

实验用品

仪器与材料：烧杯、移液管、吸量管、容量瓶、滴瓶、量筒、试剂瓶、称量瓶、台秤、分析天平等。

试剂：NaCl、KCl、CaCl$_2$、NaHCO$_3$、SnCl$_2$、锡粒、浓硫酸、浓盐酸、HAc 溶液(2.00 mol/L)、Na$_2$CO$_3$(AR)、草酸二水合物(AR)。

实验步骤

1. 配制一般溶液

(1) 粗略配制 50 mL 3 mol/L H_2SO_4 溶液[1]。

(2) 粗略配制 20 mL 0.2 mol/L Na_2CO_3 溶液。

(3) 由已知浓度为 2.00 mol/L HAc 溶液准确配制 50 mL 0.200 mol/L HAc 溶液。

(4) 准确配制 100 mL 质量分数为 0.90%的生理盐水,按质量比 $NaCl:KCl:CaCl_2:NaHCO_3 = 45:2:1.2:1$ 配制。

2. 配制 $H_2C_2O_4 \cdot 2H_2O$ 标准溶液

准确称取 1.5000~1.6000 g $H_2C_2O_4 \cdot 2H_2O$[2]固体于烧杯中,加入少量纯水(二次蒸馏水或去离子水)使其完全溶解,转移至容量瓶中。再用少量纯水淋洗烧杯及玻璃棒数次,并将每次淋洗的水全部转入 250 mL 容量瓶。最后用纯水稀释至刻度,摇匀。计算溶液的准确浓度。

3. 配制特殊溶液

配制 100 mL 0.1 mol/L $SnCl_2$ 溶液(注意配制过程中防止水解和储存过程中防止氧化)。

附注

[1] 浓硫酸的相对密度与质量分数对照表

d_4^{20}	1.8144	1.8195	1.8240	1.8279	1.8312	1.8337	1.8355	1.8364	1.8361
$x/\%$	90	91	92	93	94	95	96	97	98

[2] 基准物质草酸二水合物($H_2C_2O_4 \cdot 2H_2O$)分子量为 126.03,含两个可与 OH^- 作用的 H^+,用于标定碱的浓度。

思考题

(1) 配制 $SnCl_2$ 溶液时,如何防止其氧化和水解?

(2) 精确配制溶液时,容量瓶是否需要干燥?用浓溶液配制稀溶液时,容量瓶是否要用被稀释液润洗?

(3) 洗净后的移液管是否还要用吸取的溶液润洗?为什么?

实验4 二氧化碳分子量的测定

实验目的

(1) 掌握相对密度法测定气体分子量的原理和方法。

(2) 熟悉启普发生器的使用以及气体洗涤、干燥等基本操作。
(3) 巩固分析天平的使用。

实验原理

根据阿伏伽德罗定律，在相同温度、压力下，相同体积的两种理想气体的质量之比等于其分子量之比。

$$\frac{m_A}{m_B} = \frac{M_A}{M_B}$$

式中，m_A、m_B 分别为 A、B 两种气体的质量，M_A、M_B 分别为 A、B 两种气体的分子量。因此，只要在相同温度、压力下测定相同体积的两种气体的质量，其中一种气体的分子量已知，即可以求出另一种气体的分子量。

因为已知空气的平均分子量为 29.0，所以只要测得 CO_2 与空气在相同条件（温度、压力、体积）下的质量，便可求出 CO_2 的分子量，即

$$M_{CO_2} = \frac{m_{CO_2}}{m_{空气}} \times 29.0$$

式中，CO_2 的质量 m_{CO_2} 可直接通过两次称量求得。

第一次称量充满空气的容器的质量为

$$m_1 = m_{空气} + m_{瓶} + m_{塞}$$

第二次称量充满 CO_2 的容器的质量为

$$m_2 = m_{CO_2} + m_{瓶} + m_{塞}$$

$$m_2 - m_1 = m_{CO_2} - m_{空气}$$

$$m_{CO_2} = m_2 - m_1 + m_{空气}$$

而同体积的空气质量 $m_{空气}$ 可根据实验时测得的大气压（p）和热力学温度（T），利用理想气体状态方程计算得到。

$$m_{空气} = \frac{pV}{RT} \times 29.0$$

式中，体积 V 可以通过称量充满水的容器的质量来计算

$$m_3 = m_{H_2O} + m_{瓶} + m_{塞}$$

$$m_3 - m_1 = m_{H_2O} - m_{空气} \approx m_{H_2O}$$

$$V = m_{H_2O} / \rho_{H_2O}$$

实验用品

仪器与材料：分析天平、台秤、启普发生器、洗气瓶、干燥管、锥形瓶、玻璃管、橡皮管、温度计、气压计。

试剂：石灰石、HCl（6 mol/L）、$NaHCO_3$（1 mol/L）、浓硫酸。

实验步骤

1. CO_2 的制备

按图 3.1 所示装配制取 CO_2 的实验装置。从启普发生器出来的 CO_2 气体经过洗气瓶 2 中的 $NaHCO_3$ 溶液，除去其中少量的 H_2S、HCl 等酸性气体杂质，再经过洗气瓶 3 中的浓硫酸除水，使导出的气体为干燥、纯净的 CO_2 气体。

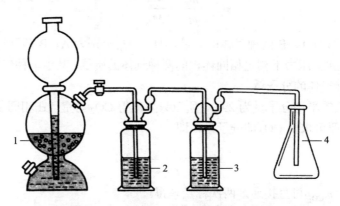

图 3.1　CO_2 制取、净化、收集实验装置
1. 启普发生器；2. 洗气瓶($1\ mol/L\ NaHCO_3$ 溶液)；3. 洗气瓶(浓 H_2SO_4)；4. 锥形瓶

2. 称量

(1) 充满空气的瓶和塞子的称量：取一个干燥、洁净的锥形瓶，用一个合适的橡皮塞塞住瓶口，在橡皮塞上做一记号，以固定橡皮塞塞入瓶口的位置。然后在分析天平上称量，此质量为(空气+瓶+塞)的总质量，记为 m_1。

(2) 充满 CO_2 的瓶和塞子的称量：从启普发生器出来的 CO_2 经纯化和干燥后，导入锥形瓶底部。待 CO_2 充满后，缓慢取出导气管，用橡皮塞塞入瓶口至原记号位置，在同一分析天平上进行称量，此质量为(CO_2+瓶+塞)的总质量，记为 m_2。再重复充 CO_2 的操作，直到前后两次称量的质量只相差 $1\sim 2\ mg$ 为止。

(3) 充满水的瓶和塞子的称量：往锥形瓶中加满水，用橡皮塞塞入瓶口至原记号位置，用吸水纸擦干锥形瓶外壁的水珠，在台秤上称量，此质量为(H_2O+瓶+塞)的总质量，记为 m_3。记录实验时的温度 T 和大气压 p。

数据记录与处理

室温：$t=$ ＿＿＿＿℃；大气压：$p=$ ＿＿＿＿mmHg

(空气+瓶+塞)的总质量 $m_1=$ ＿＿＿＿ g

第一次(CO_2+瓶+塞)的总质量 ＝ ＿＿＿＿ g

第二次(CO_2+瓶+塞)的总质量 ＝ ＿＿＿＿ g

(CO_2+瓶+塞)的总质量 $m_2=$ ＿＿＿＿ g

(水+瓶+塞)的总质量 $m_3=$ ＿＿＿＿ g

瓶的体积 $V = (m_3 - m_1)/1.00 =$ _____ mL

瓶内空气的质量 $m_{空气} =$ _____ g

CO_2 的质量 $m_{CO_2} = m_2 - m_1 + m_{空气} =$ _____ g

CO_2 的分子量 M_{CO_2} _____

误差 _____

思考题

(1) 完成数据记录和结果处理，并分析误差产生的原因。

(2) 指出实验装置图中各部分的作用，并写出有关反应方程式。

(3) 如何检验装置的气密性？

(4) 如何检验 CO_2 气体是否充满？

(5) 为什么(CO_2+瓶+塞)的总质量和(空气+瓶+塞)的总质量要在同一分析天平上称量，而(H_2O+瓶+塞)的总质量可以在台秤上称量？

参考文献

北京师范大学无机化学教研室. 2001. 无机化学实验. 3 版. 北京: 高等教育出版社

中山大学化学系无机化学教研室. 1992. 无机化学实验. 3 版. 北京: 高等教育出版社

实验 5　胆矾结晶水的测定

实验目的

(1) 了解结晶水合物中结晶水含量的测定原理和方法。

(2) 掌握研钵、干燥器等仪器的使用和沙浴、恒量等基本操作。

(3) 进一步熟悉分析天平的使用。

实验原理

许多物质从水溶液中析出晶体时，晶体中常含有一定数目的水分子，这样的水分子称为结晶水。结晶水合物加热到一定温度时，可以脱去一部分或全部结晶水。因此，对于加热能脱去结晶水又不会发生分解的结晶水合物中结晶水的测定，通常是把一定量的结晶水合物置于已灼烧至恒量的坩埚中，加热至较高温度脱水(以不超过被测定物质的分解温度为限)，然后把坩埚移入干燥器中，冷却至室温，再取出用分析天平称量。由结晶水合物经高温加热后的失重值可算出该结晶水合物所含结晶水的质量分数，以及每摩尔该盐所含结晶水的量，从而可确定结晶水合物的化学式。

五水合硫酸铜($CuSO_4 \cdot 5H_2O$)俗称胆矾，为天蓝色晶体，实验证明，$CuSO_4 \cdot 5H_2O$ 中四个水分子与 Cu^{2+} 以配位键结合，第五个水分子以氢键与两个配位水分子和 SO_4^{2-} 结合。图 3.2 为 $CuSO_4 \cdot 5H_2O$ 的结构示意图。

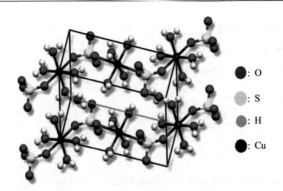

图 3.2 $CuSO_4·5H_2O$ 的结构示意图

加热失水时，先失去 Cu^{2+} 左边的两个水分子，再失去 Cu^{2+} 右边的两个水分子，最后失去以氢键与 SO_4^{2-} 结合的水分子，若继续加热至高于 600℃，白色的 $CuSO_4$ 将分解为 CuO、SO_2、SO_3 和 O_2。因此，$CuSO_4·5H_2O$ 按下列反应逐步脱水(注意：脱水温度与压力、粒度、升温速率等因素有关)：

$$CuSO_4 \cdot 5H_2O \xrightarrow{48℃} CuSO_4 \cdot 3H_2O + 2H_2O$$

$$CuSO_4 \cdot 3H_2O \xrightarrow{99℃} CuSO_4 \cdot H_2O + 2H_2O$$

$$CuSO_4 \cdot H_2O \xrightarrow{218℃} CuSO_4 + H_2O$$

本实验将已知质量的胆矾晶体加热，除去所有结晶水后称量，从而计算出化学式中结晶水的数目。

实验用品

仪器与材料：分析天平、干燥器、坩埚、泥三角、坩埚钳、铁架台、铁圈、沙浴(蒸发皿、沙子)、酒精灯、温度计、滤纸。

试剂：胆矾。

实验步骤

1. 恒量坩埚

将洗净的坩埚及坩埚盖置于泥三角上，小火烘干后，用氧化焰灼烧至红热。冷却至略高于室温，再用干净的坩埚钳将其移入干燥器中，冷却至室温(注意：热坩埚放入干燥器后，一定要在短时间内将干燥器盖子打开一两次，以免内部压力降低，难以打开)。取出坩埚，用分析天平称量。重复加热至脱水温度以上，冷却，称量，直至恒量。

2. 称量胆矾及准备沙浴

在已恒量的坩埚中加入 1.0~1.2 g 研细的胆矾晶体，铺成均匀的一层，再在同一分析天平上准确称量坩埚及胆矾的总质量，减去已恒量坩埚的质量即为胆矾的质量 m_1。将已称量的、内装胆矾晶体的坩埚置于沙浴中，将其 3/4 体积埋入沙内，再在靠近坩埚的沙浴中插入一支温度计(量程 300℃)，其末端应与坩埚底部大致处于同一水平。

3. 胆矾脱水

加热沙浴至约 210℃，然后慢慢升温至 280℃，控制沙浴温度为 260~280℃。当坩埚内粉末由蓝色全部变为灰白色时停止加热（需 15~20 min），用干净的坩埚钳将坩埚移入干燥器内，冷却至室温。将坩埚外壁用滤纸擦干净后，在同一分析天平上称量坩埚和脱水胆矾的总质量，减去已恒量坩埚的质量即为无水硫酸铜的质量 m_2。重复沙浴加热，冷却，称量，直到恒量（本实验要求两次称量之差 ≤ 1 mg）。实验后将无水硫酸铜倒入回收瓶中。

数据记录与处理

胆矾结晶水测定的数据记录与处理见表 3.2。

表 3.2 胆矾结晶水测定的数据记录与处理

空坩埚质量/g			(空坩埚+胆矾质量)/g	(加热后，坩埚+胆矾质量)/g		
第一次称量	第二次称量	平均值		第一次称量	第二次称量	平均值

胆矾的质量 m_1 = _____ g
$CuSO_4$ 的质量 m_2 = _____ g
$CuSO_4$ 的物质的量 $n_2 = m_2 /159.6$ g/mol = _____ mol
结晶水的质量 m_3 = _____ g
结晶水的物质的量 $n_3 = m_3 /18.0$ g/mol = _____ mol
1 mol $CuSO_4$ 的结晶水 = n_3/n_2 = _____
胆矾的化学式_____

思考题

(1) 为什么用沙浴加热并控制温度在 280℃左右？
(2) 为什么要重复灼烧操作？其作用是什么？
(3) 为什么加热后的坩埚冷却至室温才称量？加热后的坩埚为什么要放在干燥器中冷却？

参考文献

北京师范大学无机化学教研室. 2001. 无机化学实验. 3 版. 北京: 高等教育出版社
中山大学化学系无机化学教研室. 1992. 无机化学实验. 3 版. 北京: 高等教育出版社

实验 6 粗盐的提纯与定性检验

实验目的

(1) 学习由粗盐制取试剂级氯化钠的原理和方法。
(2) 掌握称量、量取、溶解、过滤、蒸发、浓缩、结晶、干燥等基本操作。
(3) 了解 SO_4^{2-}、Ca^{2+}、Mg^{2+} 的定性鉴定方法。

实验原理

粗盐中的不溶性杂质（如泥沙等）可通过溶解和过滤的方法除去。粗盐中的可溶性杂质主要是 Ca^{2+}、Mg^{2+}、K^+ 和 SO_4^{2-} 等，选择适当的试剂使它们生成难溶化合物沉淀而被除去。但所选沉淀剂应符合不引进新的杂质或引进的杂质能够在下一步操作中除去的原则。

(1) 除去 SO_4^{2-}：在粗盐中加入过量的 $BaCl_2$ 溶液，过滤，除去难溶物和生成的 $BaSO_4$ 沉淀。

$$Ba^{2+} + SO_4^{2-} = BaSO_4\downarrow$$

(2) 除去 Ca^{2+}、Mg^{2+} 和过量的 Ba^{2+}：在滤液中加入过量的 NaOH 和 Na_2CO_3 溶液，过滤，除去生成的沉淀。

$$Mg^{2+} + 2OH^- = Mg(OH)_2\downarrow$$
$$Ca^{2+} + CO_3^{2-} = CaCO_3\downarrow$$
$$Ba^{2+} + CO_3^{2-} = BaCO_3\downarrow$$

(3) 溶液中过量的混碱可以用盐酸中和除去。

$$2H^+ + CO_3^{2-} = H_2O + CO_2\uparrow$$
$$H^+ + OH^- = H_2O$$

(4) 粗盐中的 K^+ 和上述沉淀剂都不起作用，由于 KCl 的溶解度大于 NaCl 的溶解度，且含量较少，因此在蒸发和浓缩过程中，NaCl 先结晶出来，而 KCl 留在溶液中。

实验用品

仪器与材料：台秤、烧杯、量筒、铁架台、铁圈、玻璃漏斗、布氏漏斗、抽滤瓶、循环水泵、蒸发皿、石棉网、酒精灯、药匙、滤纸、pH 试纸。

试剂：粗盐、$BaCl_2$(1 mol/L)、NaOH(6 mol/L)、Na_2CO_3(饱和)、HCl(6 mol/L)、HAc(6 mol/L)、$(NH_4)_2C_2O_4$(饱和)、镁试剂、NaOH(2 mol/L) 和 Na_2CO_3(饱和，50%)的混合溶液。

实验步骤

1. 粗盐的提纯

(1) 在台秤上称取 10.0 g 粗盐，放入 100 mL 烧杯中，加入 40 mL 水，加热并搅拌使其溶解。至溶液沸腾时，在搅拌下逐滴加入 1 mol/L $BaCl_2$ 溶液(约 2 mL)至沉淀完全。继续加热 5 min，使 $BaSO_4$ 颗粒长大而易于沉淀和过滤。为了检验沉淀是否完全，可将烧杯从石棉网上取下，待沉淀沉降后，倾斜烧杯，沿壁滴加几滴 1 mol/L $BaCl_2$ 溶液检验。用常压过滤或倾析法倒出上清液留存。

(2) 在上述滤液中加入 1 mL 6 mol/L NaOH 溶液和 2 mL 饱和 Na_2CO_3 溶液(或约 4 mL NaOH-Na_2CO_3 混合溶液，pH 约为 11)，加热至沸。为了检验沉淀是否完全，可将烧杯从石棉网上取下，待沉淀沉降后，倾斜烧杯，沿壁滴加几滴 Na_2CO_3 溶液，检查有无沉淀生成。如果不再产生沉淀，用布氏漏斗进行减压过滤，保留滤液。

(3) 在滤液中逐滴加入 6 mol/L HCl 溶液，直至溶液呈微酸性为止(pH 为 2～3)。

(4)将滤液倒入蒸发皿中,用小火加热蒸发,浓缩至稀粥状的稠液为止。切不可将溶液蒸干。

(5)冷却后,减压过滤,尽量将产品抽干。将产品放回蒸发皿中,小火加热干燥,直至不冒水蒸气为止。

(6)将产品冷却至室温,称量。最后把产品放回指定容器中,计算产率。

2. 产品纯度的检验

取粗盐和实验所得的产品各 1 g,分别溶于 5 mL 蒸馏水中。将粗盐溶液过滤,两种澄清溶液分别盛入三支试管中,组成三组,对照检验它们的纯度。

(1)SO_4^{2-} 的检验:在第一组溶液中分别加入 2 滴 6 mol/L HCl,使溶液呈酸性。再加入 3~5 滴 1 mol/L $BaCl_2$ 溶液,如有白色沉淀生成,证明 SO_4^{2-} 存在。记录结果,进行比较。

(2)Ca^{2+} 的检验:在第二组溶液中分别加入 2 滴 6 mol/L HAc,使溶液呈酸性。再加入 3~5 滴饱和 $(NH_4)_2C_2O_4$ 溶液,如有白色沉淀生成,证明 Ca^{2+} 存在。记录结果,进行比较。

(3)Mg^{2+} 的检验:在第三组溶液中分别加入 3~5 滴 6 mol/L NaOH,使溶液呈碱性。再加入 1 滴镁试剂[镁试剂是一种有机染料,在碱性溶液中呈红色或紫色,但被 $Mg(OH)_2$ 沉淀吸附后则呈天蓝色],若有天蓝色沉淀生成,证明 Mg^{2+} 存在。记录结果,进行比较。

数据记录与处理

(1)粗盐的提纯。

产品外观:_____

产品质量:_____g;产率_____%

(2)产品纯度的检验(表 3.3)。

表 3.3 产品纯度检验的实验现象与结论

序号	实验内容	实验现象		结论与反应式
		粗盐	精盐	
(1)	加入 $BaCl_2$ 溶液			
(2)	加入 $(NH_4)_2C_2O_4$ 溶液			
(3)	加入 NaOH 溶液和镁试剂			

注意事项

(1)使用台秤时,注意保持台秤清洁。

(2)食盐溶液浓缩时,切不可蒸干。

(3)普通过滤和减压过滤的装置与操作区别。

(4)pH 试纸的使用方法。

(5)$BaCl_2$、NaOH、Na_2CO_3、HCl 等试剂加入不可过多,除杂顺序不可改变。

思考题

(1) 加入 40 mL 水溶解 10.0 g 食盐的依据是什么？加水过多或过少有什么影响？
(2) 怎样除去实验过程中所加的过量沉淀剂 $BaCl_2$、$NaOH$ 和 Na_2CO_3？
(3) 提纯后的实验溶液浓缩时为什么不能蒸干？
(4) 在检验 SO_4^{2-} 时，为什么要加入盐酸溶液？
(5) 在粗食盐的提纯中，(1)、(2) 两步能否合并过滤？
(6) 为什么在提纯过程中加入 $BaCl_2$ 和 Na_2CO_3 后均要加热至沸？
(7) 哪些因素会造成产品产率过高？

参考文献

北京师范大学无机化学教研室. 2001. 无机化学实验. 3 版. 北京: 高等教育出版社
浙江台州学院医药化工学院. 2011. 基础实验 I (无机化学实验). 杭州: 浙江大学出版社

实验 7　乙酸解离度和解离常数的测定

实验目的

(1) 了解 pH 法测定弱酸解离常数的原理，加深对解离度和解离常数的理解。
(2) 掌握滴定原理、滴定操作及滴定终点的判断。
(3) 学习如何使用酸度计。
(4) 巩固移液管和容量瓶的操作和使用。

实验原理

弱电解质是指在水溶液中仅能部分电离的电解质。当弱电解质在水溶液中达到电离平衡时，测定此时溶液中各物种的浓度，即可求得解离平衡常数 K_a。乙酸 (CH_3COOH，简写作 HAc) 是一元弱酸，在水溶液中存在下列平衡：

$$HAc(aq) \rightleftharpoons H^+(aq) + Ac^-(aq)$$

起始浓度(mol/L)　　　　c　　　　0　　　　0
平衡浓度(mol/L)　　$c(1-\alpha)$　　$c\alpha$　　$c\alpha$

解离平衡常数 K_a 的表达式为

$$K_a = \frac{[H^+][Ac^-]}{[HAc]} = \frac{[H^+]^2}{c-[H^+]} \tag{3.1}$$

电离度

$$\alpha = \frac{[H^+]}{c} \times 100\% \tag{3.2}$$

式中，c 为 HAc 溶液的起始浓度；$[H^+]$、$[Ac^-]$ 和 $[HAc]$ 分别为 H^+、Ac^- 和 HAc 的平衡浓度。

在一定温度下，若 HAc 的总浓度已知(未知浓度样品可以用标准 NaOH 溶液滴定测得)，用 pH 计测定溶液的 pH 并计算出 H^+ 的浓度，就可以借助上述公式计算出 HAc 在该温度下的

K_a 和 α。通过测定一系列不同浓度 HAc 溶液的 K_a 值,其平均值即为在该温度下 HAc 的解离常数。

此外,也可以用半中和法测定 HAc 的解离常数。当用 NaOH 中和 HAc 溶液中酸的 50% 时,体系组成变为等浓度 HAc-NaAc 构成的缓冲溶液。依据缓冲溶液公式(3.3)可知,此时溶液的 pH 等于 pK_a。

$$pH = -\lg[H^+] \tag{3.3}$$

实验用品

仪器与材料:酸度计、鼓风干燥箱、容量瓶、移液管、吸量管、烧杯、温度计、洗耳球、滤纸。

试剂:HAc 标准溶液(约 0.2 mol/L,精确到 4 位有效数字),NaOH 标准溶液(约 0.2 mol/L,精确到 4 位有效数字)。

实验步骤

1. HAc 溶液浓度的测定

以酚酞为指示剂,用已知准确浓度的 NaOH 标准溶液标定 HAc 溶液的准确浓度,结果填入表 3.4。

表 3.4 HAc 溶液浓度的数据记录与处理　　　　室温 ____ ℃

序号		Ⅰ	Ⅱ	Ⅲ
NaOH 溶液的浓度/(mol/L)				
HAc 溶液的用量/mL				
NaOH 溶液的用量/mL				
HAc 溶液的浓度/(mol/L)	测定值			
	平均值			

2. 直接测定 HAc 溶液的 pH

1) 不同浓度 HAc 溶液的配制

用吸量管或移液管分别吸取 2.50 mL、5.00 mL、25.00 mL HAc 标准溶液,移入 50 mL 容量瓶中,用蒸馏水稀释,定容,摇匀。计算出每种 HAc 溶液的浓度。向 4 个干燥的 100 mL 烧杯(编号为 1、2、3、4)中分别倒入上述三种浓度的溶液和 HAc 标准溶液,用于测定 pH。

2) 溶液 pH 的测定

用酸度计按照由稀到浓的次序分别测定上述 HAc 溶液的 pH,填入表 3.5 并计算 K_a 和 α,记录测定时的室温。

表 3.5 HAc 溶液 pH 的数据记录与处理 　　　　　室温 ____ ℃

编号	c_{HAc}/(mol/L)	pH	[H^+]/(mol/L)	α	解离常数 K_a	
					测定值	平均值
1						
2						
3						
4						

注意事项

(1) 酸度计是精密仪器，需正确使用。具体方法及注意事项参见酸度计的使用方法。

(2) 使用标准缓冲溶液校正酸度计时，应避免引入水或不洁物，用完后应回收。

(3) 本实验测定的 K_a 为 1.0×10^{-5}~2.0×10^{-5} 合格（25℃文献值为 1.76×10^{-5}）。

思考题

(1) 依据实验测定和计算结果，说明弱电解质的解离度受浓度和体系离子组成影响的方式。

(2) 为什么按由稀到浓的顺序测定不同浓度 HAc 溶液的 pH？为什么要测量室温？

(3) 对比实验测得值与 HAc 解离常数的标准值，分析二者产生偏差的原因。

(4) 已知 pH 只有 2 位有效数字，则表中[H^+]、K_a 和 α 应保留几位有效数字？

实验 8　生物体中几种必需元素的定性鉴定

实验目的

学习植物体中 Ca、Fe、P 等元素的定性鉴定方法及植物样本的处理方法。

实验原理

生物体是复杂的有机体系，是由多种元素形成的不同类型的化合物构成的。通常认为，生命必需元素包括碳、氢、氧、氮、硫、磷、氯、硅、碘、钠、钾、钙、镁、铁、铜等 20 余种元素。各元素在生物体内的必要性、分布状况和含量随生物的种类、所处地域及器官或组织的不同而差异巨大。例如，硼是某些绿色植物和藻类生长的必需元素，哺乳动物却不需要。钙和磷在动物骨骼中的含量极高，是骨骼中羟基磷灰石的构成元素，同时，磷也是构成核酸的重要元素。作为微量元素的铁则存在于血红蛋白、肌红蛋白、固氮酶、叶绿素等生物大分子中，参与生物体内多种重要代谢活动。上述必需元素或微量元素的缺乏将导致生物体的发育和生物功能受到影响和限制，如缺钙导致痉挛和骨质疏松，缺铁导致人贫血、植物叶片变黄等。

本实验旨在定性检验植物样本（如叶片）中的钙、铁和磷元素。处理植物样本，使其中所含的钙、铁和磷分别转化为 Ca^{2+}、Fe^{3+} 和磷酸盐形式，然后利用特性反应进行鉴定。鉴定反应的机理如下：

$$Ca^{2+} + C_2O_4^{2-} =\!=\!= CaC_2O_4\downarrow\text{（白色）}$$
$$Fe^{3+} + [Fe(CN)_6]^{4-} =\!=\!= Fe_4[Fe(CN)_6]_3\downarrow\text{（蓝色）}$$
$$Fe^{3+} + nSCN^- =\!=\!= [Fe(SCN)_n]^{(3-n)+}\text{（红色）}$$
$$HPO_4^{2-} + 12MoO_4^{2-} + 3NH_4^+ + 23H^+ =\!=\!= (NH_4)_3[P(Mo_{12}O_{40})]\cdot 6H_2O\downarrow\text{（黄色）} + 6H_2O$$

本实验所用的原料处理方法也适用于检验骨头、蛋黄等材料中是否含有钙、铁和磷元素。

实验用品

仪器与材料：瓷坩埚、坩埚钳、泥三角、三脚架、酒精灯、燃烧匙、漏斗、烧杯、试管、镊子、树叶（青黄各半）、棉花、pH 试纸。

试剂：HNO_3（浓、6 mol/L、0.1 mol/L）、氨水（2 mol/L）、KSCN 溶液、$(NH_4)_2MoO_4$ 溶液、$K_4[Fe(CN)_6]$溶液、$(NH_4)_2C_2O_4$ 溶液、红磷、石灰石。

实验步骤

1. 原料的处理

取 6～10 g 新鲜树叶，用镊子或坩埚钳夹持树叶在酒精灯上加热燃烧。将黑色灰烬（碳化物）粉碎后收集在瓷坩埚内，继续加热直至灰化完全（灰白色）。称 0.3～0.5 g 灰粉转移入试管，加入 1 mL 浓 HNO_3 硝化样品，3 min 后再加入 3 mL 水，过滤，用 1 mL 水洗涤滤纸上的沉淀，合并滤液，备用。

2. 定性检验

将滤液分成四等份，前三份分别加入 $(NH_4)_2MoO_4$ 溶液、KSCN 溶液和 $K_4[Fe(CN)_6]$ 溶液。第四份用 2 mol/L $NH_3\cdot H_2O$ 调 pH 至弱碱性，然后加入 $(NH_4)_2C_2O_4$ 溶液。观察和记录现象。

3. 对照试验

(1) 用燃烧匙取少量红磷，加热后在集气瓶内燃烧，将 2 mL 水加入集气瓶并轻摇。待白烟被水吸收后，将水溶液转移到试管并煮沸。再加入 5 滴 6 mol/L HNO_3 溶液和 $(NH_4)_2MoO_4$ 溶液后观察现象。采用玻璃棒摩擦试管内壁的方法促进黄色沉淀 $(NH_4)_3[P(Mo_{12}O_{40})]\cdot 6H_2O$ 的析出。

(2) 向试管内放一小块棉花，加入 1 mL 6 mol/L HNO_3 溶液加热（避免沸腾和硝酸分解），然后加入 3 mL 水，过滤，滤液分成两份，分别加入 KSCN 溶液和 $K_4[Fe(CN)_6]$溶液，观察现象。

(3) 取一块绿豆大小的石灰石，加入 2 mL 0.1 mol/L HNO_3 溶液反应，用 $NH_3\cdot H_2O$ 调 pH 至弱碱性。然后加入 $(NH_4)_2C_2O_4$ 溶液，观察现象。

对比上述对照试验与被测样品的实验现象，判断样品中是否含有钙、铁和磷元素。

注意事项

(1) 树叶灰化处理须完全。

(2) 加热硝化棉花时需要控制加热程度，并避免试管口对人，防止因硝酸沸腾导致沾有热硝酸的棉花从管口喷出伤人。

思考题

部分学生在进行本实验时获得的实验现象不显著,甚至发现个别对照试验的现象不如树叶样品的实验现象显著,分析其中的原因。

实验 9 碘化铅溶度积的测定

实验目的

(1) 了解用离子交换法测定难溶电解质溶度积的原理和方法。

(2) 学习离子交换树脂的一般使用方法,练习酸碱滴定的基本操作。

实验原理

在一定温度下,难溶强电解质(如 PbI_2)在其饱和溶液中存在解离平衡:

$$PbI_2(s) \rightleftharpoons Pb^{2+}(aq) + 2I^-(aq)$$

PbI_2 的溶度积常数表达式为:$K_{sp} = [Pb^{2+}][I^-]^2$,溶度积常数随温度的不同而改变。通过测定在一定温度下 PbI_2 饱和溶液中[Pb^{2+}]或[I^-]([Pb^{2+}]:[I^-] = 1:2),就可以求出该温度下的 K_{sp}。

离子交换树脂是一类含有可交换离子的有机高分子化合物,分为阳离子交换树脂和阴离子交换树脂。前者含有酸性基团,如 R-SO$_3^-$ H$^+$ 含有能与溶液中阳离子交换的 H$^+$。而后者含有碱性基团,如 R-NH$_3^+$ OH$^-$ 含有能与溶液中阴离子交换的 OH$^-$。本实验中采用强酸型阳离子交换树脂,将一定体积的饱和碘化铅溶液通过交换树脂,树脂上的 H$^+$ 即与 Pb^{2+} 进行交换反应。该交换反应可以用下式表示:

$$2R\text{-}H^+ + Pb^{2+} \rightleftharpoons R_2\text{-}Pb^{2+} + 2H^+$$

收集交换后的流出液,并用标准碱溶液滴定其中 H$^+$ 的含量,从而求得 PbI_2 饱和溶液的浓度,计算出溶度积。

已有 Pb^{2+} 交换上去的树脂可用不含 Cl$^-$ 的稀 HNO$_3$ 溶液进行淋洗,使树脂重新转化为酸型(称为树脂的再生)。

实验用品

仪器与材料:层析柱、碱式滴定管、移液管、量筒、小烧杯、锥形瓶、漏斗、漏斗架、铁架台、滴定管夹、玻璃棒、温度计、定量滤纸、pH 试纸。

试剂:强酸型阳离子交换树脂、氢氧化钠标准溶液(约 0.005 mol/L)、硝酸(1 mol/L)、碘化铅饱和溶液、溴百里酚蓝指示剂。

实验步骤

1. 转型和装柱

将阳离子交换树脂在稀酸溶液(1 mol/L)中浸泡 24~48 h,使其完全转变成氢型。用烧杯取适量预活化的阳离子交换树脂，除去浸泡的酸液，在烧杯中用纯水洗涤 3～5 次，将分散在纯水中的树脂注入交换柱，装柱高度约 20 cm。通过调节旋塞控制柱内水量。少量多次加纯水淋洗树脂，直至流出液 pH 与水的 pH 相同，然后关闭旋塞。在以上操作及后续的离子交换过程中，应确保树脂始终浸在溶液中，以避免进入树脂床中的气泡影响离子交换的进行。若树脂床出现气泡，可加入蒸馏水至液面略高出树脂 2～3 cm，用玻璃棒轻轻搅动树脂以赶出气泡。

2. 交换和洗涤

过滤 50～60 mL PbI_2 饱和溶液，用移液管精密吸取 25.00 mL 滤液，放入离子交换柱中。控制交换柱流出液的速度为 30～40 滴/分(过快的流出速度可能导致离子交换不完全)。用干净的锥形瓶盛接流出液。待树脂柱上表面 PbI_2 溶液接近流出时，用纯水分批洗涤交换树脂，直至流出液的 pH 与纯水相同。将流出液与洗涤液全部收集在锥形瓶内。将使用后的树脂回收至指定容器内，经纯水、硝酸、纯水洗涤后即可再次使用。

3. 滴定

向锥形瓶内溶液滴入 2 滴溴百里酚蓝指示剂，用 0.005 mol/L NaOH 标准溶液滴定至终点(溶液由黄色转为亮蓝色)，记录消耗 NaOH 标准溶液的体积，进而求算出 PbI_2 的溶度积。

数据记录与处理

室温_____ ℃
PbI_2 饱和溶液的体积_____ mL
NaOH 标准溶液的浓度_____ mol/L
滴定管内 NaOH 标准溶液的位置：滴定前_____ mL，滴定后_____ mL
NaOH 标准溶液的体积_____ mL
PbI_2 饱和溶液中[Pb^{2+}]_____ mol/L
PbI_2 的 K_{sp}_____

注意事项

(1) 过滤和转移 PbI_2 饱和溶液所用的漏斗、烧杯需干燥。
(2) 本实验测定 K_{sp} 值数量级为 10^{-9}～10^{-8} 合格。

思考题

(1) 为什么要控制离子交换过程中液体的流速不宜太快？
(2) 为什么自始至终要保持液面高于离子交换树脂层？
(3) 对比文献报道值 $K_{sp}(PbI_2)=7.1\times10^{-9}$(298 K)，试分析误差产生原因。
(4) 树脂转型不彻底对实验结果会有什么影响？
(5) 在交换和洗涤过程中，如果流出液有少量损失，对实验结果有什么影响？

实验10　化学反应速率与活化能的测定

实验目的

(1) 了解浓度、温度和催化剂对化学反应速率的影响。
(2) 测定$(NH_4)_2S_2O_8$与KI反应的速率，并计算反应级数、反应速率常数和活化能。
(3) 熟悉量筒、秒表的使用。

实验原理

在水溶液中过二硫酸铵与碘化钾发生如下反应：

$$S_2O_8^{2-} + 3I^- = 2SO_4^{2-} + I_3^- (aq) \tag{3.4}$$

其速率方程可表示为

$$v = k c_{S_2O_8^{2-}}^m c_{I^-}^n$$

式中，v为瞬时速率；$c_{S_2O_8^{2-}}$和c_{I^-}分别为$S_2O_8^{2-}$和I^-的起始浓度；k为反应速率常数；m和n分别为反应物$S_2O_8^{2-}$和I^-的反应级数，$(m+n)$为反应的总级数。

实验中只能测定出一段时间内反应的平均速率：

$$\bar{v} = \frac{-\Delta c_{S_2O_8^{2-}}}{\Delta t}$$

即Δt时间内$S_2O_8^{2-}$的浓度变化。

为了测出在Δt时间内$S_2O_8^{2-}$浓度的改变量，需要在混合$(NH_4)_2S_2O_8$和KI溶液的同时，加入一定体积已知浓度的$Na_2S_2O_3$溶液和淀粉溶液。这样在反应式(3.4)进行的同时还进行另一反应：

$$2S_2O_3^{2-} + I_3^- = S_4O_6^{2-} + 3I^- \tag{3.5}$$

反应(3.5)几乎是瞬间完成，比反应(3.4)快得多。因此，反应(3.4)生成的I_3^-立即与$S_2O_3^{2-}$反应，生成无色$S_4O_6^{2-}$和I^-，观察不到碘与淀粉呈现的特征蓝色。当$S_2O_3^{2-}$消耗殆尽，反应(3.5)不进行，反应(3.4)还在进行，则生成的I_3^-遇淀粉呈蓝色。

从反应开始到溶液出现蓝色这一段时间Δt内，$S_2O_3^{2-}$浓度的改变值为

$$\Delta c = -[c_{S_2O_3^{2-}(后)} - c_{S_2O_3^{2-}(始)}] = c_{S_2O_3^{2-}(始)}$$

再从反应(3.4)和(3.5)对比，则得

$$-\Delta c_{S_2O_8^{2-}} = \frac{c_{S_2O_3^{2-}(始)}}{2}$$

通过改变$S_2O_8^{2-}$和I^-的初始浓度，测定消耗等量的$S_2O_8^{2-}$的物质的量浓度($-\Delta c_{S_2O_8^{2-}}$)所需的不同时间间隔，即可计算反应物不同初始浓度的初速率，确定速率方程和反应速率常数。

实验用品

仪器与材料：恒温水浴，烧杯、量筒、秒表、温度计、玻璃棒或电磁搅拌器、坐标纸。

试剂：$(NH_4)_2S_2O_8$(0.20 mol/L)，KI(0.20 mol/L)，$Na_2S_2O_3$(0.01 mol/L)，KNO_3(0.20 mol/L)，$(NH_4)_2SO_4$(0.20 mol/L)，$Cu(NO_3)_2$(0.02 mol/L)，淀粉溶液(5 g/L)。

实验步骤

1. 浓度对化学反应速率的影响

在室温条件下进行表 3.6 中编号 I 的实验。用量筒分别量取 20.0 mL KI 溶液、8.0 mL $Na_2S_2O_3$ 溶液和 2.0 mL 淀粉溶液，全部加入烧杯中，混合均匀。然后用另一量筒取 20.0 mL $(NH_4)_2S_2O_8$ 溶液，迅速倒入上述混合液中，同时启动秒表，并不断搅动，仔细观察。当溶液刚出现蓝色时，立即停掉秒表，记录反应时间和室温。

用同样方法按照表 3.6 的用量进行编号 II、III、IV、V 的实验。

表 3.6 浓度对化学反应速率的影响　　　　　　室温＿＿＿℃

	实验编号	I	II	III	IV	V
试剂用量/mL	0.20 mol/L $(NH_4)_2S_2O_8$	20.0	10.0	5.0	20.0	20.0
	0.20 mol/L KI	20.0	20.0	20.0	10.0	5.0
	0.01 mol/L $Na_2S_2O_3$	8.0	8.0	8.0	8.0	8.0
	5 g/L 淀粉溶液	2.0	2.0	2.0	2.0	2.0
	0.20 mol/L KNO_3	0	0	0	10.0	15.0
	0.20 mol/L $(NH_4)_2SO_4$	0	10.0	15.0	0	0
混合液中反应的起始浓度 /(mol/L)	$(NH_4)_2S_2O_8$					
	KI					
	$Na_2S_2O_3$					
反应时间 Δt/s						
$S_2O_8^{2-}$ 的浓度变化/(mol/L)						
反应速率 v/[mol/(L·s)]						

2. 温度对化学反应速率的影响

按表 3.7 实验 IV 中的药品用量，将装有 KI、$Na_2S_2O_3$、KNO_3 及淀粉混合溶液的烧杯和装有 $(NH_4)_2S_2O_8$ 溶液的小烧杯同时放入冰水浴中冷却。待它们温度低于室温 10℃时，将 $(NH_4)_2S_2O_8$ 溶液迅速加到混合溶液中，同时计时并不断搅动。当溶液刚出现蓝色时，记录反应时间。此实验编号记为 VI。

用同样方法在热水浴中进行高于室温 10℃的实验。此实验编号记为 VII。

将此两次实验数据 VI、VII 和实验 IV 的数据记入表 3.7 中进行比较。

表 3.7 温度对化学反应速率的影响

实验编号	VI	IV	VII
反应温度 t/℃			
反应时间 Δt/s			
反应速率 v/[mol/(L·s)]			

3. 催化剂对化学反应速率的影响

按表 3.7 实验Ⅳ试剂用量，把 KI、$Na_2S_2O_3$、KNO_3 和淀粉溶液加到 150 mL 烧杯中，再加入 2 滴 $Cu(NO_3)_2$ 溶液，搅匀，然后迅速加入 $(NH_4)_2S_2O_8$ 溶液，搅动、计时。将此实验的反应速率与表 3.7 中实验Ⅳ的反应速率定性地进行比较，可得到什么结论？

数据记录与处理

1. 反应级数和反应速率常数的计算

$$v = k c_{S_2O_8^{2-}}^m c_{I^-}^n$$

两边取对数：

$$\lg v = m \lg c_{S_2O_8^{2-}} + n \lg c_{I^-} + \lg k$$

当 c_{I^-} 不变时（实验Ⅰ、Ⅱ、Ⅲ），以 $\lg v$ 对 $\lg c_{S_2O_8^{2-}}$ 作图得直线，斜率为 m。同理，当 $c_{S_2O_8^{2-}}$ 不变（实验Ⅰ、Ⅳ、Ⅴ）时，以 $\lg v$ 对 $\lg c_{I^-}$ 作图得直线，斜率为 n。此反应级数为 $(m+n)$。利用表 3.8 中实验数据可求出反应速率常数 k。

表 3.8　反应速率常数的计算

实验编号	Ⅰ	Ⅱ	Ⅲ	Ⅳ	Ⅴ
$\lg v$					
$\lg c_{S_2O_8^{2-}}$					
$\lg c_{I^-}$					
m					
n					
反应速率常数 k					

2. 反应活化能的计算

反应速率常数 k 与反应温度 T 一般有以下关系：

$$\lg k = A - \frac{E_a}{2.30RT}$$

式中，E_a 为反应的活化能；R 为摩尔气体常量；T 为热力学温度。测出不同温度下的 k 值，以 $\lg k$ 对 $1/T$ 作图得直线，斜率为 $-\dfrac{E_a}{2.30R}$，可求出反应的活化能 E_a。将数据填入表 3.9。

表 3.9　活化能的计算

实验编号	Ⅵ	Ⅶ	Ⅳ
反应速率常数 k			
$\lg k$			
$1/T$			
反应活化能 E_a			

注意事项

(1) KI、$Na_2S_2O_3$、淀粉、KNO_3 和 $(NH_4)_2SO_4$ 溶液可使用同一个量筒量取，$(NH_4)_2S_2O_8$ 溶液必须单独使用一个量筒。

(2) KI、$Na_2S_2O_3$、淀粉、KNO_3 和 $(NH_4)_2SO_4$ 溶液混合均匀后，将 $(NH_4)_2S_2O_8$ 溶液迅速倒入上述混合溶液中，同时启动秒表并搅拌。

(3) 溶液刚出现蓝色立即按停秒表。

(4) 进行升温或降温的实验时，KI、$Na_2S_2O_3$、淀粉、KNO_3、$(NH_4)_2SO_4$ 混合溶液和 $(NH_4)_2S_2O_8$ 溶液要分别冷却或加热。

(5) 温度计必须分开使用。

(6) 反应时也要保持混合前的温度。

(7) KI 溶液应为无色透明溶液，不宜使用有 I_2 析出的浅黄色溶液。$(NH_4)_2S_2O_8$ 溶液要新配制的，因为时间长了 $(NH_4)_2S_2O_8$ 易分解。如所配制 $(NH_4)_2S_2O_8$ 溶液的 pH 小于 3，说明该试剂已有分解，不适合本实验使用。所用试剂中如混有少量 Cu^{2+}、Fe^{3+} 等杂质，对反应有催化作用，必要时需滴入几滴 0.10 mol/L EDTA 溶液。

(8) 做温度对化学反应速率影响的实验时，如室温低于 10℃，可将温度条件改为室温、高于室温 10℃、高于室温 20℃ 三种情况进行。

(9) 实验指导：

(i) 秒表的使用。

(ii) 加入 $(NH_4)_2S_2O_8$ 溶液和 KI 溶液的顺序及速度。

(iii) 实验时的计时操作。

(iv) 实验后指导学生利用作图法求出反应级数，从而求出反应速率常数和活化能。

(10) 本实验活化能测定值的误差不超过 10%（文献值：51.8 kJ/mol）。

思考题

(1) 以下操作对实验有什么影响？

(i) 量筒没有分开。

(ii) 先加 $(NH_4)_2S_2O_8$ 溶液，最后加 KI 溶液。

(iii) 将 $(NH_4)_2S_2O_8$ 溶液缓慢加入 KI 和其他溶液的混合溶液中。

(2) 为什么在实验 Ⅱ、Ⅲ、Ⅳ 和 Ⅴ 中分别加入 KNO_3 溶液或 $(NH_4)_2SO_4$ 溶液？

(3) 每个实验的计时操作应注意什么？

(4) 如果反应速率用 I^- 或 I_3^- 而不是 $S_2O_8^{2-}$ 的浓度变化表示，反应速率常数 k 是否相同？

(5) 如何测定化学反应的反应级数？用本实验的结果说明。

(6) 用阿伦尼乌斯公式计算反应活化能，与作图得到的数值比较。

(7) 已知 $A(g) \longrightarrow B(l)$ 是一个二级反应，数据如下。试计算反应速率常数 k。

p_A/kPa	40	26.6	19.1	13.3
t/s	0	250	500	1000

实验 11　固体碱熔氧化法制备高锰酸钾

实验目的

(1) 学习碱熔法由二氧化锰制备高锰酸钾的基本原理和操作方法。
(2) 熟悉熔融、浸取等基本操作。
(3) 巩固过滤、结晶和重结晶等基本操作。
(4) 掌握锰的各种氧化态之间相互转化关系。

实验原理

软锰矿的主要成分是二氧化锰，二氧化锰在较强氧化剂(如氯酸钾)存在下与碱共熔时，可被氧化为锰酸钾：

$$3MnO_2 + KClO_3 + 6KOH \xrightarrow{\text{熔融}} 3K_2MnO_4 + KCl + 3H_2O$$

熔块由水浸取后，随着溶液碱性降低，水溶液中的 MnO_4^{2-} 不稳定，发生歧化反应。一般在弱碱性或近中性介质中，歧化反应趋势较小，反应速率也较慢。但在弱酸性介质中，MnO_4^{2-} 易发生歧化反应，生成 MnO_4^- 和 MnO_2。向锰酸钾溶液中通入 CO_2 气体，可发生如下反应：

$$3K_2MnO_4 + 2CO_2 =\!=\!= 2KMnO_4 + MnO_2\downarrow + 2K_2CO_3$$

经减压过滤除去二氧化锰后，将溶液浓缩即可析出暗紫色的针状高锰酸钾晶体。

实验用品

仪器与材料：铁坩埚、启普发生器、坩埚钳、泥三角、布氏漏斗、烘箱、蒸发皿、烧杯、表面皿、滴定分析配套仪器、8 号铁丝。

试剂：二氧化锰、氢氧化钾、氯酸钾、碳酸钙、亚硫酸钠、草酸、工业盐酸、分析纯硫酸。

实验步骤

1. 二氧化锰的熔融氧化

称取 2.5 g 氯酸钾固体和 5.2 g 氢氧化钾固体，放入铁坩埚中，用铁棒将物料混合均匀。将铁坩埚放在泥三角上，用坩埚钳夹紧，小火加热，边加热边用铁棒搅拌。待混合物熔融后，将 3.0 g 二氧化锰固体分多次小心加入铁坩埚中，防止火星外溅。随着熔融物的黏度增大，用力加快搅拌以防结块或黏在坩埚壁上。待反应物干涸后，升高温度，强热 5 min，得到墨绿色锰酸钾熔融物。用铁棒尽量捣碎。

2. 浸取

待盛有熔融物的铁坩埚冷却后，用铁棒尽量将熔块捣碎，并将其倾放于盛有 100 mL 蒸馏水的 250 mL 烧杯中以小火共煮，直到熔融物全部溶解为止，小心用坩埚钳取出坩埚。

3. 锰酸钾的歧化

趁热向浸取液中通二氧化碳气体至锰酸钾全部歧化为止(可用玻璃棒蘸取溶液于滤纸上,如果滤纸上只有紫红色而无绿色痕迹,即表示锰酸钾已歧化完全,pH 为 10~11),然后静置片刻,抽滤。

4. 滤液蒸发结晶

将滤液倒入蒸发皿中,蒸发浓缩至表面开始析出高锰酸钾晶膜为止,自然冷却晶体,然后抽滤,将高锰酸钾晶体抽干。

5. 高锰酸钾晶体的干燥

将晶体转移到已知质量的表面皿中,用玻璃棒将其分开。放入烘箱中(80℃为宜,不能超过240℃)干燥 0.5 h,冷却后称量,计算产率。

6. 纯度分析

实验室备有基准物质草酸和硫酸。设计分析方案,确定所制备的产品中高锰酸钾的含量。

7. 锰各种氧化态之间的相互转化(选做)

利用自制高锰酸钾晶体,如图 3.3 所示设计实验,实现锰的各种氧化态之间的相互转化。写出实验步骤及有关的离子方程式。

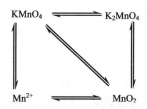

图 3.3 锰各种氧化态之间的相互转化

附注

(1) 参考数据(表 3.10)。

表 3.10 一些化合物溶解度随温度的变化 单位:g/100 g H_2O

化合物	$t/℃$										
	0	10	20	30	40	50	60	70	80	90	100
KCl	27.6	31.0	34.0	37.0	40.0	42.6	45.5	48.3	51.1	54.0	56.7
$K_2CO_3·2H_2O$	51.3	52	52.5	53.2	53.9	54.8	55.9	57.1	58.3	59.6	60.9
$KMnO_4$	2.83	4.4	6.4	9.0	12.56	16.89	22.2	—	—	—	—

(2) 通入 CO_2 过多，溶液的 pH 较低，溶液中会生成大量的 $KHCO_3$。$KHCO_3$ 的溶解度比 K_2CO_3 小得多，溶液浓缩时，$KHCO_3$ 会和 $KMnO_4$ 一起析出。

(3) 启普发生器的安装和调试。

(4) 固体的溶解、过滤和结晶。

思考题

(1) 为什么制备锰酸钾时要用铁坩埚而不用瓷坩埚？

(2) 实验时，为什么用铁棒而不用玻璃棒搅拌？

(3) 在实验步骤 3 中，要用玻璃棒而不用铁棒搅拌溶液，为什么？

(4) 总结启普发生器的构造和使用方法。

(5) 为了使 K_2MnO_4 发生歧化反应，能否用 HCl 代替 CO_2？

(6) 由锰酸钾在酸性介质中歧化的方法得到高锰酸钾的最大转化率是多少？还可采取何种实验方法提高锰酸钾的转化率？

实验 12　硫酸亚铁铵晶体的制备

实验目的

(1) 巩固水浴加热、溶解、结晶、减压过滤等实验操作。

(2) 学会利用溶解度的差异制备硫酸亚铁铵。

(3) 掌握硫酸亚铁及硫酸亚铁铵的性质。

(4) 了解用目视比色法检验产品的纯度。

实验原理

硫酸亚铁铵又称莫尔盐 $[(NH_4)_2SO_4 \cdot FeSO_4 \cdot 6H_2O]$，为浅蓝绿色结晶或粉末，能溶于水，但难溶于乙醇。莫尔盐在空气中不易被氧化，比硫酸亚铁稳定，因此在定量分析中常用来配制亚铁离子的标准溶液。硫酸亚铁铵也是一种重要的化工原料，用途十分广泛。

本实验先将过量的铁屑溶于稀硫酸得到硫酸亚铁：

$$Fe + H_2SO_4 = FeSO_4 + H_2\uparrow$$

等物质的量的硫酸亚铁和硫酸铵作用得到硫酸亚铁铵：

$$FeSO_4 + (NH_4)_2SO_4 + 6H_2O = (NH_4)_2SO_4 \cdot FeSO_4 \cdot 6H_2O$$

从表 3.11 可知，在 0～60℃ 的温度范围内，硫酸亚铁铵在水中的溶解度比组成它的任一组分的溶解度都小。因此，将混合溶液先加热浓缩，然后冷却至室温，就可以从浓的硫酸亚铁和硫酸铵混合溶液中得到结晶的莫尔盐。

表 3.11　三种盐在水中的溶解度　　　　　　　　　　　　单位：g/100 g H_2O

温度/℃	$FeSO_4 \cdot 7H_2O$	$(NH_4)_2SO_4$	$(NH_4)_2SO_4 \cdot FeSO_4 \cdot 6H_2O$
10	40.0	73.0	18.12
20	48.0	75.4	21.2

续表

温度/℃	FeSO$_4$·7H$_2$O	(NH$_4$)$_2$SO$_4$	(NH$_4$)$_2$SO$_4$·FeSO$_4$·6H$_2$O
30	60.0	78.0	24.5
40	73.3	81.0	27.9
50	—	84.5	31.3
70	—	91.9	38.5

实验用品

仪器与材料：台秤、锥形瓶、布氏漏斗、抽滤瓶、比色管、比色架、滤纸。

试剂：铁粉、(NH$_4$)$_2$SO$_4$(s)、(NH$_4$)$_2$SO$_4$·FeSO$_4$·6H$_2$O(s)、H$_2$SO$_4$(3 mol/L，浓)、HCl(3 mol/L)、KSCN(质量分数为0.25)、乙醇(95%)。

实验步骤

1. 硫酸亚铁铵的制备

1) 硫酸亚铁的制备

用台秤称取 2.0 g 铁粉，放入 100 mL 锥形瓶中，加入 10 mL 3 mol/L H$_2$SO$_4$，于通风橱中水浴加热反应(水浴温度 60℃左右)约 30 min，观察锥形瓶中无气泡产生为止。反应过程中适当补加水，以保持原有体积。反应完毕再加入 1 mL 3 mol/L H$_2$SO$_4$。趁热减压过滤，用少量热水洗涤锥形瓶及漏斗上的残渣，抽干。滤液趁热转入蒸发皿中。

2) 硫酸铵饱和溶液的制备

根据溶液中 FeSO$_4$ 的量，按关系式 $n[(NH_4)_2SO_4]:n(FeSO_4)=1:1$ 称取所需的 (NH$_4$)$_2$SO$_4$ 固体，配制成相应温度下 (NH$_4$)$_2$SO$_4$ 饱和溶液。

3) 硫酸亚铁铵的制备

将 (NH$_4$)$_2$SO$_4$ 饱和溶液加到 FeSO$_4$ 溶液中(此时溶液的 pH 应接近 1，如 pH 偏大，可加几滴稀 H$_2$SO$_4$ 调节)，水浴蒸发，浓缩至表面出现晶膜为止。放置缓慢冷却，得硫酸亚铁铵晶体。减压过滤除去母液，用 95% 乙醇洗涤晶体，并尽量抽干。把晶体转移到表面皿上晾干片刻，观察晶体的颜色和形状。称量，计算产率。

2. Fe(Ⅲ)的定量分析

1) Fe(Ⅲ)标准溶液的配制(由预备室配制)

称取 0.5030 g Fe$_2$(SO$_4$)$_3$·9H$_2$O，溶于少量水中，加入 2.5 mL 浓 H$_2$SO$_4$，移入 1000 mL 容量瓶中，用水稀释至刻度。此溶液中 Fe^{3+} 浓度为 0.1000 g/L(0.1000 mg/mL)。

2) 标准色阶的配制

取 0.50 mL Fe(Ⅲ)标准溶液于 25 mL 比色管中，加入 2 mL 3 mol/L HCl 和 1 mL 质量分数为 0.25 的 KSCN 溶液，加不含氧的水稀释至刻度，配制成相当于一级试剂的标准溶液(Fe^{3+}浓度为 0.05 mg/g，即质量分数为 0.005%)。

分别取 1.00 mL 和 2.00 mL Fe(Ⅲ)标准溶液配制成相当于二级和三级试剂的标准溶液（Fe^{3+}浓度分别为 0.10 mg/g 和 0.20 mg/g，即质量分数分别为 0.01%和 0.02%）。

3) 产品级别的确定

称取 1.0 g 产品于 25 mL 比色管中，用 15 mL 不含氧的水溶解。待其全溶后，加入 2 mL 3 mol/L HCl 和 1 mL KSCN（$w = 0.25$）溶液，继续加不含氧的水稀释至刻度，摇匀。与标准色阶比色，确定产品级别。

注意事项

(1) 在水浴加热过程中，应不断振荡锥形瓶，防止结块。
(2) 反应的时间不宜过长。
(3) 硫酸铵饱和溶液制备过程中一定要耐心搅拌。
(4) 硫酸亚铁铵晶体抽滤后要用乙醇洗涤。

思考题

(1) 制备硫酸亚铁时，为什么要使铁粉过量？
(2) 能否将最后产物$(NH_4)_2SO_4 \cdot FeSO_4 \cdot 6H_2O$直接放在蒸发皿内加热干燥？为什么？
(3) 本实验计算硫酸亚铁铵的产率时，应以H_2SO_4的量为准，为什么？
(4) 为什么制备硫酸亚铁铵晶体时，溶液必须呈酸性？蒸发浓缩时是否需要搅拌？
(5) 如何防止亚铁离子被水解和氧化？

实验 13　硝酸钾提纯和溶解度测定

实验目的

(1) 学习硝酸钾溶解度的粗略测定方法，绘制溶解度曲线。
(2) 了解硝酸钾的溶解度与温度的关系。
(3) 利用溶解度曲线对粗硝酸钾进行提纯。
(4) 学习定性检验氯离子。

实验原理

盐类在水中的溶解度是指在一定温度下它们在饱和水溶液中的浓度，一般以每 100 g 水中溶解盐的质量(g)来表示。一般测定硝酸钾的溶解度是将一定量的硝酸钾加入一定量的水中，加热使其完全溶解，然后在不断搅拌下使其冷却至刚有晶体析出，此时溶液浓度就是该温度下的溶解度，以 g/100 g H_2O 表示。

不同物质的溶解度受温度的影响不同：①大多数物质的溶解度受温度的影响很大，如硝酸钾，随着温度的升高，溶解度增大；②一些物质的溶解度随温度变化不大，如氯化钠；③某些物质的溶解度随着温度的升高而降低，如氢氧化钙。因此，可以利用溶解度的差异，通过改变温度对物质进行分离。例如，加热硝酸钾和氯化钠的混合溶液至一定温度，然后冷却降温，利用硝酸钾和氯化钠的溶解度随温度变化的不同对硝酸钾进行提纯。

硝酸钾溶解度的测量方法主要有结晶析出法和溶质质量法。结晶析出法包括升温法和降温法。其中降温法是在一定量的水中，溶入一定量溶质使成不饱和溶液。在使溶液缓缓降温并开始析出晶体(溶液成为饱和状态)的同时测出溶液的温度，即可计算出溶解度。该方法是保持溶质的质量不变，添加溶剂，使其在降温过程中达到饱和状态。本实验装置如图 3.4 所示。

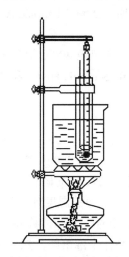

图 3.4　结晶析出法测定硝酸钾溶解度的实验装置

溶质质量法是在一个已知准确质量的蒸发皿中，称量一定温度下的硝酸钾饱和溶液的质量，然后加热饱和溶液，蒸发掉全部水分，再称量蒸发皿的质量，可计算出该温度下硝酸钾的溶解度。

实验用品

仪器与材料：台秤、布氏漏斗、抽滤瓶、烧杯、大试管、温度计、蒸发皿、三脚架、泥三角、玻璃棒、酒精灯、滤纸。

试剂：KNO_3(分析纯，s)、粗 KNO_3(混有 NaCl，s)、HNO_3(5 mol/L)、$AgNO_3$(0.1 mol/L)。

实验步骤

1. 硝酸钾的提纯

(1) 溶解蒸发。

称取 10.0 g 粗 KNO_3 放入一支硬质试管中，加入 35 mL 水，将试管置于甘油浴中加热。待盐全部溶解后，继续加热蒸发使溶液为原有体积的 2/3，趁热过滤，晶体是_____。再用热漏斗过滤，滤液盛于小烧杯中自然冷却，析出晶体是_____。减压过滤，尽量抽干，将析出的 KNO_3 晶体水浴烤干。

(2) 粗产品的重结晶。

保留 0.1~0.2 g 粗产品供纯度检验用，剩余的粗产品按粗产品:水 = 2:1(质量比)的比例溶于蒸馏水中。加热、搅拌待晶体全部溶解后停止加热。待溶液冷却至室温，抽滤，水浴烘干，得到纯度较高的 KNO_3 晶体。

(3) 纯度检验。

2. 定性检验硝酸钾的纯度

分别将 0.1 g 粗产品和一次重结晶的产品放入 2 支小试管中，加入 2 mL 蒸馏水溶解，各加入 1 滴 5 mol/L HNO_3 酸化，再各加入 2 滴 0.1 mol/L $AgNO_3$ 溶液。根据试剂级别标准检验试样中总氯量。

称取 1.0 g 样品（精确至 0.01 g），加热至 400℃使其分解，于 700℃灼烧 15 min。冷却后溶于蒸馏水中稀释至 25 mL，加 2 mL 5 mol/L HNO_3 和 0.1 mol/L $AgNO_3$ 溶液，摇匀，放置 10 min。所呈浊度不得大于标准。

3. 升温法测定硝酸钾的溶解度

(1) 用台秤分别称量 3.5 g、1.5 g、1.5 g、2.0 g、2.5 g 硝酸钾。
(2) 在大试管中加入 10 mL 蒸馏水，加入 3.5 g 硝酸钾，水浴加热，边加热边搅拌至完全溶解。
(3) 从水浴中拿出试管，插入一支干净的温度计，用玻璃棒轻轻搅拌并摩擦管壁。同时观察温度计的读数，当开始有晶体析出时，立即读数并记录。
(4) 把试管再放入水浴中加热使晶体全部溶解，然后重复上述(3)的操作，再测定开始析出晶体的温度。
(5) 向试管中再加入 1.5 g 硝酸钾[试管中共有硝酸钾 3.5 + 1.5 = 5.0(g)]，然后重复上述(3)、(4)的操作。
(6) 重复(5)的操作，依次测得加入 1.5 g、2.0 g、2.5 g（试管中依次共有硝酸钾 6.5 g、8.5 g、11.0 g）开始析出晶体的温度（温度计不必洗涤）。
(7) 根据所得数据，以温度为横坐标、溶解度为纵坐标绘制溶解曲线。从图上可以清楚地得到溶解度和温度的关系。

数据记录与处理

升温法测定硝酸钾溶解度的数据记录与处理见表 3.12。

表 3.12　升温法测定硝酸钾溶解度的数据记录与处理

编号		1	2	3	4	5
KNO_3 晶体的质量/g						
水的质量/g						
溶液中开始析出晶体时的温度/℃	t_1					
	t_2					
	平均					
KNO_3 晶体在水中的溶解度/(g/100 g H_2O)						

注意事项

(1) 试管应干燥、洁净。

(2) 用纸槽把固体送入试管底部。
(3) 使用温度计小心缓慢地轻轻搅拌。
(4) 当室温不够低时，可把试管浸入冷水中冷却降温。在降温过程中，用玻璃棒轻轻搅拌溶液并摩擦管壁，防止溶液出现过饱和。
(5) 通过温度计读数，需把握刚刚开始析出晶体的时刻，以免增大误差。

思考题

(1) 测定溶解度时，硝酸钾的量及水的体积是否需要准确？测定装置选用什么玻璃器皿较为合适？
(2) 测定溶解度时，水的蒸发对实验有何影响？应采取什么措施？
(3) 溶解和结晶过程是否需要搅拌？
(4) 纯化粗硝酸钾应采用什么操作步骤，如溶解、蒸发、结晶等？

实验 14 氧化还原反应及平衡

实验目的

(1) 掌握电极电势、介质的酸碱性以及氧化还原电对中物质的浓度与氧化还原反应的方向、产物、速率之间的关系。
(2) 学会装配原电池，了解化学电池电动势。

实验原理

化学反应可分为两大类：一类是在反应过程中，反应物之间没有电子的转移，如中和反应、沉淀反应、配位反应等；另一类是在反应过程中，反应物之间发生了电子的转移，这一类就是氧化还原反应。

氧化还原反应的本质是电子的转移，即电子的得失或偏移。氧化剂在反应中得到电子，氧化数降低；还原剂在反应中失去电子，氧化数升高。物质得失电子能力的大小，或者说氧化还原能力的强弱，可以用同一物质的氧化态和还原态所组成的电对的电极电势相对高低衡量。电对的电极电势值(φ)越小，电对的还原态物质还原能力越强；电对的电极电势值(φ)越大，电对的氧化态物质氧化能力越强。因此，根据电极电势的大小，可判断一个氧化还原反应进行的方向。

当氧化态物质或还原态物质的浓度改变时，会影响电极电势的数值，进而影响氧化还原反应的速率和产物。特别是当有沉淀剂或配位剂存在，能够大大减少溶液中某一离子的浓度时，甚至可以改变反应的方向。当有 H^+ 或 OH^- 参加电极反应时，介质的 pH 会对电极电势产生影响，甚至影响氧化还原反应的方向和产物。

实验用品

仪器与材料：试管、离心试管、伏特计(或酸度计)、表面皿、U 形管、电极(锌片、铜片)、回形针、红色石蕊试纸(或酚酞试纸)、导线、砂纸、滤纸。

试剂：HAc(6 mol/L)、H_2SO_4(1 mol/L)、NaOH(6 mol/L)、$NH_3 \cdot H_2O$(浓)、$ZnSO_4$(1 mol/L)、

$CuSO_4$(0.01 mol/L、1 mol/L)、KI(0.1 mol/L)、KBr(0.1 mol/L)、$KMnO_4$(0.01 mol/L)、Na_2SO_3(0.1 mol/L)、$FeCl_3$(0.1 mol/L)、$Fe_2(SO_4)_3$(0.1 mol/L)、$FeSO_4$(0.1 mol/L)、H_2O_2(3%)、KIO_3(0.1 mol/L)、溴水、碘水(0.1 mol/L)、氯水(饱和)、KCl(饱和)、CCl_4、酚酞指示剂、淀粉溶液(0.4%)、琼脂、氟化铵。

实验步骤

1. 装配原电池及浓度对电极电势的影响

1) 装配原电池

在一个小烧杯中加入约 20 mL 1 mol/L $ZnSO_4$ 溶液，在其中插入锌片(砂纸打磨)。在另一个小烧杯中加入约 20 mL 1 mol/L $CuSO_4$ 溶液，在其中插入铜片(砂纸打磨)。用盐桥将两个烧杯相连，组成原电池。用导线将锌片和铜片分别与伏特计(或酸度计)的负极和正极相连接，测量两电极之间的电势差(图 3.5)。

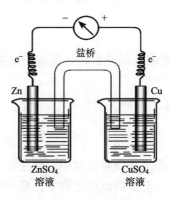

图 3.5 Cu-Zn 原电池

2) 浓度对电极电势的影响

在上述原电池中的 $ZnSO_4$ 溶液中注入浓氨水至生成的沉淀恰好溶解，生成无色溶液：

$$Zn^{2+} + 4NH_3 \rightleftharpoons [Zn(NH_3)_4]^{2+}$$

测量电势差，观察有何变化。

在 $CuSO_4$ 溶液中注入浓氨水至生成的沉淀恰好溶解，生成深蓝色溶液：

$$Cu^{2+} + 4NH_3 \rightleftharpoons [Cu(NH_3)_4]^{2+}$$

测量电势差，观察又有何变化。利用能斯特方程解释实验现象。

3) 设计浓差电池

自行设计并测定下列浓差电池电动势，比较实验值与计算值。

$(-)$ Cu | $CuSO_4$(0.01 mol/L) ‖ $CuSO_4$(1 mol/L) | Cu $(+)$

2. 氧化还原反应和电极电势

(1) 在试管中加入 0.5 mL 0.1 mol/L KI 溶液，然后加入 2 滴 0.1 mol/L $FeCl_3$ 溶液，摇匀后

加入 0.5 mL CCl_4，充分振荡，观察 CCl_4 层颜色变化。

(2) 用 KBr 溶液代替 KI 溶液进行同样实验，观察现象。

(3) 在分别加入 3 滴碘水和溴水的试管中加入约 0.5 mL 0.1 mol/L $FeSO_4$ 溶液，摇匀后加入 0.5 mL CCl_4，充分振荡，观察 CCl_4 层颜色变化。

根据以上实验结果，定性比较 Br_2/Br^-、I_2/I^- 和 Fe^{3+}/Fe^{2+} 的电极电势。

3. 酸度和浓度对氧化还原反应的影响

1) 酸度的影响

在 3 支均加入 0.5 mL 0.1 mol/L Na_2SO_3 溶液的试管中分别加入 0.5 mL 1 mol/L H_2SO_4 溶液、0.5 mL 蒸馏水和 0.5 mL 6 mol/L NaOH 溶液，混匀后，再各加入 2 滴 0.01 mol/L $KMnO_4$ 溶液，观察变化，写出反应方程式。

在试管中加入 10 滴 0.1 mol/L KI 溶液和 2 滴 0.1 mol/L KIO_3 溶液，再加入几滴淀粉溶液，混合后观察溶液颜色的变化。然后加入 2~3 滴 1 mol/L H_2SO_4 溶液酸化该混合溶液，观察有什么变化。最后加入 2~3 滴 6 mol/L NaOH 使混合溶液显碱性，观察又有什么变化。写出反应方程式。

根据以上实验结果，说明酸度对氧化还原反应的影响。

2) 浓度的影响

在盛有 H_2O、CCl_4 及 0.1 mol/L $Fe_2(SO_4)_3$ 各 0.5 mL 的试管中加入 0.5 mL 0.1 mol/L KI 溶液，振荡后观察 CCl_4 层的颜色。

在盛有 CCl_4、1 mol/L $FeSO_4$ 及 0.1 mol/L $Fe_2(SO_4)_3$ 各 0.5 mL 的试管中加入 0.5 mL 0.1 mol/L KI 溶液，振荡后观察 CCl_4 层的颜色。比较与上一实验中 CCl_4 层颜色的区别。

在第一步实验的试管中加入少许 NH_4F 固体，振荡，观察 CCl_4 层的颜色。

根据以上实验结果，说明浓度对氧化还原反应的影响。

3) 酸度对氧化还原反应速率的影响

在两支各盛有 0.5 mL 0.1 mol/L KBr 溶液的试管中分别加入 0.5 mL 1 mol/L H_2SO_4 和 6 mol/L HAc 溶液，再各加入 2 滴 0.01 mol/L $KMnO_4$ 溶液，观察两支试管中紫红色褪去的速度。写出反应方程式。

4. 氧化数居中的物质的氧化还原性

(1) 在试管中加入 0.5 mL 0.1 mol/L KI 溶液和 2 滴 1 mol/L H_2SO_4 溶液，再加入 2 滴 3% H_2O_2，观察溶液颜色的变化。

(2) 在试管中加入 2 滴 0.01 mol/L $KMnO_4$ 溶液，再加入 3 滴 1 mol/L H_2SO_4 溶液，摇匀后加入 2 滴 3% H_2O_2，观察溶液颜色的变化。

注意事项

(1) 伏特计的使用。

(2) 盐桥的使用。

思考题

(1) 酸度对 Cl_2/Cl^-、Br_2/Br^-、I_2/I^-、Fe^{3+}/Fe^{2+}、Cu^{2+}/Cu、Zn^{2+}/Zn 电对的电极电势有无影响?为什么?

(2) $KMnO_4$ 溶液的氧化性在哪种介质中最强?为什么?

(3) 为什么 H_2O_2 既具有氧化性,又具有还原性?试从电极电势加以说明。

第4章 元素无机化学实验

实验 15 s 区元素（碱金属、碱土金属）

实验目的

(1) 比较碱金属和碱土金属的活泼性。
(2) 比较碱土金属氢氧化物及其盐类的溶解度。
(3) 比较锂盐和镁盐的相似性。
(4) 了解焰色反应的操作并熟悉使用金属钾、钠的安全措施。

实验用品

仪器与材料：蒸发皿、试管、离心机、烧杯、镊子、砂纸、镍丝、滤纸、点滴板、钴玻璃片。

试剂：钾、钠、镁、钙、Na_2CO_3(0.1 mol/L、1 mol/L)、LiCl(1 mol/L)、NaCl(1 mol/L)、NaF(1 mol/L)、Na_2HPO_4(1 mol/L)、KCl(1 mol/L)、$CaCl_2$(1 mol/L)、$SrCl_2$(1 mol/L)、$BaCl_2$(1 mol/L)、K_2CrO_4(1 mol/L)、$MgCl_2$(0.5 mol/L、1 mol/L)、Na_2SO_4(1 mol/L)、$NaHCO_3$(1 mol/L)、K[Sb(OH)$_6$](饱和)、$NaHC_4H_4O_6$(饱和)、$KMnO_4$ (0.01 mol/L)、NH_4Cl(饱和)、$(NH_4)_2C_2O_4$(饱和)、$(NH_4)_2CO_3$(0.5 mol/L)、Na_3PO_4(0.5 mol/L)、$(NH_4)_2SO_4$(饱和)、NaOH(2 mol/L，新制)、氨水(1 mol/L、2 mol/L，新制)、H_2SO_4(1 mol/L)、HCl (2 mol/L、6 mol/L)、HAc(2 mol/L)、HNO_3(浓)。

实验步骤

1. 碱金属、碱土金属的活泼性

(1) 取一小块金属钠，用滤纸吸干表面的煤油，立即放在蒸发皿中加热。金属钠开始燃烧时即停止加热。观察现象，写出反应式。产物冷却后，用玻璃棒轻轻捣碎产物，转移入试管中，加入少量水使其溶解，冷却，观察有无气体放出。用 1 mol/L H_2SO_4 溶液酸化溶液，然后加一滴 0.01 mol/L $KMnO_4$ 溶液，观察现象，写出反应式。

(2) 取一小段金属镁条，用砂纸擦去表面氧化物，点燃，观察现象，写出反应式。

(3) 与水的作用。

取一小块(绿豆大小)金属钠及钾，用滤纸吸干其表面煤油，把它们分别放入两个盛有水的烧杯中。为了安全起见，立即用倒置漏斗覆盖在烧杯口处。观察现象，检验溶液的酸碱性，写出反应式。

取两小段镁条，用砂纸擦去表面氧化物，分别投入盛有冷水和热水的两支试管中。对比反应现象，写出反应式。

取一小块金属钙置于试管中，加入少量水。观察现象，检验溶液的酸碱性，写出反应式。

2. 碱土金属氢氧化物的溶解性

以 1 mol/L $MgCl_2$、$CaCl_2$、$BaCl_2$ 和新配制的 2 mol/L NaOH 溶液及氨水作试剂,设计系列实验,说明碱土金属氢氧化物溶解度的大小顺序。

3. 碱金属及碱土金属的难溶盐

1)碱金属微溶盐

取少量 1 mol/L LiCl 溶液分别与 1 mol/L NaF 溶液、Na_2HPO_4 溶液反应,观察现象(必要时可微热试管观察),写出反应式。

在装有少量 1 mol/L NaCl 溶液的试管中加入少量 $K[Sb(OH)_6]$ 饱和溶液,放置数分钟。如无晶体析出,可用玻璃棒摩擦试管内壁。观察晶形沉淀 $Na[Sb(OH)_6]$ 的析出。

在装有少量 1 mol/L KCl 溶液的试管中加入 1 mL 酒石酸氢钠($NaHC_4H_4O_6$)饱和溶液,观察难溶盐 $KHC_4H_4C_6$ 晶体的析出。

2)碱土金属难溶盐

(1)碳酸盐:分别取少量 1 mol/L $MgCl_2$ 溶液、$CaCl_2$ 溶液、$BaCl_2$ 溶液与 1 mol/L Na_2CO_3 溶液反应,生成的沉淀经离心分离后分别与 2 mol/L HAc 溶液及 HCl 溶液反应,观察沉淀是否溶解。

另分别取少量 1 mol/L $MgCl_2$ 溶液、$CaCl_2$ 溶液、$BaCl_2$ 溶液,加入 1~2 滴 NH_4Cl 饱和溶液、2 滴 1 mol/L 氨水和 2 滴 0.5 mol/L $(NH_4)_2CO_3$ 溶液,观察沉淀是否生成。写出反应式,并解释实验现象。

(2)草酸盐:分别向 1 mol/L $MgCl_2$ 溶液、$CaCl_2$ 溶液、$BaCl_2$ 溶液中滴加 $(NH_4)_2C_2O_4$ 饱和溶液,生成的沉淀经离心分离后再分别与 2 mol/L HAc 溶液及 HCl 溶液反应,观察现象,写出反应式。

(3)铬酸盐:分别向 1 mol/L $MgCl_2$ 溶液、$CaCl_2$ 溶液、$BaCl_2$ 溶液中滴加 1 mol/L K_2CrO_4 溶液,观察沉淀是否生成。沉淀经离心分离后再分别与 2 mol/L HAc 溶液及 HCl 溶液反应,观察现象,写出反应式。

(4)硫酸盐:分别向 1 mol/L $MgCl_2$ 溶液、$CaCl_2$ 溶液、$BaCl_2$ 溶液中滴加 1 mol/L Na_2SO_4 溶液,观察沉淀是否生成。沉淀经离心分离后再试验其在 $(NH_4)_2SO_4$ 饱和溶液及浓 HNO_3 中的溶解性,解释现象,写出反应式。

4. 锂盐和镁盐的相似性

(1)分别向 1 mol/L LiCl 溶液、$MgCl_2$ 溶液中滴加 1 mol/L NaF 溶液,观察现象,写出反应式。

(2)试验 1 mol/L LiCl 溶液与 0.1 mol/L Na_2CO_3 溶液反应及 0.5 mol/L $MgCl_2$ 溶液与 1 mol/L $NaHCO_3$ 溶液反应。观察现象,写出反应式。

(3)向 1 mol/L LiCl 溶液和 0.5 mol/L $MgCl_2$ 溶液中分别滴加 0.5 mol/L Na_3PO_4 溶液。观察现象,写出反应式。

由以上实验总结锂、镁盐的相似性并给予解释。

5. 焰色反应

取一根镶有镍丝的玻璃棒,镍丝尖端弯成小环状。蘸取 6 mol/L HCl 溶液,在氧化焰中灼烧至近于无色,重复该过程两三次以达到洁净镍丝的目的。用洁净的镍丝分别蘸取 1 mol/L LiCl 溶液、NaCl 溶液、KCl 溶液、$CaCl_2$ 溶液、$BaCl_2$ 溶液、$SrCl_2$ 溶液在氧化焰中灼烧。观察火焰的颜色并记录结果。注意:对于钾离子的焰色,应通过钴玻璃片滤光后观察。

思考题

(1) 如果实验中发生镁燃烧的事故,是否可以用水或二氧化碳灭火器灭火?为什么?
(2) 如何分离 Ca^{2+}、Mg^{2+}?为什么 $Mg(OH)_2$ 与 $MgCO_3$ 都可溶于 NH_4Cl 饱和溶液中?

实验 16　p 区非金属元素(一)(卤素、氧、硫)

实验目的

(1) 掌握卤素的氧化性和卤素离子的还原性。
(2) 掌握次卤酸盐及卤酸盐的氧化性。
(3) 了解卤素的歧化反应。
(4) 了解某些金属卤化物的性质。
(5) 掌握硫化氢、硫代硫酸盐的还原性,过硫酸盐的强氧化性。

实验用品

仪器与材料:试管、碘化钾-淀粉试纸、pH 试纸。

试剂:$KClO_3(s)$、$Na_2S_2O_3(s)$、$K_2S_2O_8(s)$、KBr(0.1 mol/L)、NaF(0.1 mol/L)、NaCl(0.1 mol/L)、KI (0.01 mol/L、0.1 mol/L)、K_2CrO_4(0.1 mol/L)、$K_2Cr_2O_7$(0.1 mol/L)、$KMnO_4$ (0.1 mol/L)、H_2O_2 (3%)、$MnSO_4$(0.002 mol/L、0.1 mol/L)、$AgNO_3$(0.1 mol/L)、氯水、碘水、NaOH (2 mol/L、40%)、KOH (2 mol/L)、H_2SO_4(1 mol/L、3 mol/L、6 mol/L)、HCl (2 mol/L)、H_2S 水溶液(饱和)、HNO_3(2 mol/L)、氨水 (2 mol/L)、四氯化碳、乙醚、乙醇、品红溶液、淀粉溶液。

实验步骤

1. 氯水对溴、碘离子混合溶液的氧化顺序

在试管中加入 10 滴 0.1 mol/L KBr 溶液、2 滴 0.01 mol/L KI 溶液和 1 滴管 CCl_4,然后逐滴加入氯水,仔细观察 CCl_4 层颜色的变化,写出反应式。

2. 氯的含氧酸盐氧化性

(1) 用滴管吸取 2 mol/L KOH 溶液逐滴加入 4 mL 氯水中,至溶液呈弱碱性(用 pH 试纸检验),将溶液分装在四支试管中。第一支试管滴加 2 mol/L HCl 溶液,选择合适的试纸检验气体产物。第二支试管滴加品红溶液,第三支试管滴加 3～4 滴 0.1 mol/L KI 溶液及 1 滴淀粉溶

液，第四支试管滴加 0.1 mol/L $MnSO_4$ 溶液。写出反应式。

(2) 取绿豆大小的 $KClO_3$ 晶体，用 1～2 mL 水溶解后，加入 1 滴管 CCl_4 及 3～4 滴 0.1 mol/L KI 溶液，摇动试管，观察水相及有机相的变化。再加入 2～3 滴 6 mol/L H_2SO_4 溶液酸化，又有什么变化？写出反应式。

3. 卤化物的溶解度比较

分别向盛有 10 滴 0.1 mol/L NaF 溶液、NaCl 溶液、KBr 溶液、KI 溶液的试管中滴加 0.1 mol/L $AgNO_3$ 溶液，制得的卤化银沉淀经离心分离后分别与 2 mol/L HNO_3 溶液、2 mol/L 氨水及 0.5 mol/L $Na_2S_2O_3$ 溶液反应，观察沉淀是否溶解。写出反应式，解释氟化物与其他卤化物溶解度的差异，总结变化规律。

4. 卤化银的感光性

将制得的 AgCl 沉淀均匀地涂在滤纸上，滤纸上放一把钥匙。光照约 10 min 后移开钥匙，可清楚地看到钥匙的轮廓。

5. 过氧化氢的鉴定和性质

1) 过氧化氢的鉴定

取 10 滴 3% H_2O_2 溶液，加入 10 滴乙醚，并加入 3～4 滴 1 mol/L H_2SO_4 溶液酸化，再加入 2～3 滴 0.1 mol/L K_2CrO_4 溶液，振荡试管，观察水层和乙醚层颜色的变化，写出反应式。

2) 酸性

在试管中加入 10 滴 40% NaOH 溶液、10 滴 3% H_2O_2 溶液及 10 滴乙醇，振荡试管，观察现象，写出反应式。

3) 介质酸碱性对 H_2O_2 氧化还原性的影响

在 10 滴 3% H_2O_2 溶液中加入 2～3 滴 2 mol/L NaOH 溶液，再加入 5～6 滴 0.1 mol/L $MnSO_4$ 溶液，观察现象，写出反应式。将溶液静置后倾去清液，向沉淀中加入 2～3 滴 3 mol/L H_2SO_4 溶液，然后滴加 3% H_2O_2 溶液，观察又有什么变化，写出反应式并给予解释。

6. 硫代硫酸盐的性质

取黄豆大小的 $Na_2S_2O_3 \cdot 5H_2O$ 晶体溶于约 3 mL 水中，均分到四支试管中，进行以下实验：
(1) 在第一支试管中滴加 2 mol/L HCl 溶液，观察现象，写出反应式。
(2) 在第二支试管中滴加碘水，观察现象，写出反应式。
(3) 在第三支试管中滴加氯水，设法证实反应后溶液中有 SO_4^{2-} 存在，写出反应式。
(4) 将第四支试管中的 $Na_2S_2O_3$ 溶液逐滴加入装有 4 滴 0.1 mol/L $AgNO_3$ 溶液的试管中，仔细观察现象，写出反应式。

7. 过二硫酸钾的氧化性

在有 2 滴 0.002 mol/L $MnSO_4$ 溶液的试管中加入约 5 mL 1 mol/L H_2SO_4 溶液和黄豆大小的 $K_2S_2O_8$ 固体，混匀后均分到两支试管中。其中一支试管再加入 2 滴 0.1 mol/L $AgNO_3$ 溶液，一起水浴加热。观察溶液颜色的变化，比较现象并解释原因，写出反应式。

8. 硫化氢的还原性

(1) 在装有 1 滴 0.1 mol/L $KMnO_4$ 溶液的试管中加入 2 滴 1 mol/L H_2SO_4 溶液酸化后,再滴加 H_2S 饱和溶液,观察现象,写出反应式。

(2) 在装有 1 滴 0.1 mol/L $K_2Cr_2O_7$ 溶液的试管中加入 2 滴 1 mol/L H_2SO_4 溶液酸化后,再滴加 H_2S 饱和溶液,观察现象,写出反应式。

思考题

(1) 为什么实验室经常用固体过二硫酸盐而不预先配成溶液?

(2) 设计实验说明 $NaClO$ 和 $KClO_3$ 氧化性的强弱。

(3) 根据实验结果,比较①$S_2O_8^{2-}$ 与 MnO_4^- 的氧化性;②$S_2O_3^{2-}$ 与 I^- 的还原性。

实验 17 p 区非金属元素(二)(氮、磷、碳、硅、硼)

实验目的

(1) 掌握氨和铵盐、硝酸和硝酸盐的主要性质。

(2) 掌握磷酸盐的主要性质。

(3) 掌握亚硝酸及其盐的性质。

(4) 掌握活性炭的吸附作用,以及二氧化碳、碳酸盐和碳酸氢盐在水溶液中相互转化的条件。

(5) 掌握硅酸盐和硼酸盐的性质。

实验用品

仪器与材料:坩埚、表面皿、试管、烧杯、温度计、pH 试纸。

试剂:$H_3BO_3(s)$、$CaCl_2 \cdot 6H_2O(s)$、$CuSO_4 \cdot 5H_2O(s)$、$Co(NO_3)_2 \cdot 6H_2O(s)$、$NiSO_4 \cdot 7H_2O(s)$、$MnSO_4(s)$、$ZnSO_4 \cdot 7H_2O(s)$、$FeCl_3 \cdot 6H_2O(s)$、活性炭、靛蓝溶液、乙醇、$NaNO_2$(0.1 mol/L、饱和)、$KMnO_4$(0.01 mol/L)、$Na_3PO_4$(0.1 mol/L)、$Na_2HPO_4$(0.1 mol/L)、$NaH_2PO_4$(0.1 mol/L)、$Na_4P_2O_7$(0.1 mol/L)、$NaPO_3$(0.1 mol/L)、$CaCl_2$(0.1 mol/L)、$NaHCO_3$(0.5 mol/L)、$Pb(NO_3)_2$(0.001 mol/L、0.1 mol/L)、$FeCl_3$(0.2 mol/L)、$Na_2SiO_3$(20%)、$K_2CrO_4$(0.1 mol/L)、$Na_2CO_3$(1 mol/L)、$MgCl_2$(0.1 mol/L)、$CuSO_4$(0.1 mol/L)、$AgNO_3$(0.1 mol/L)、$KI$(0.1 mol/L)、$NaOH$(6 mol/L)、$H_2SO_4$(1 mol/L、3 mol/L、浓)、氨水(2 mol/L、浓)、HCl(2 mol/L、浓)、甘油。

实验步骤

1. 氨的加合作用

在坩埚内滴入 4~5 滴浓氨水,再把一个内壁用浓盐酸湿润过的烧杯罩在坩埚上,观察现象,写出反应式。

2. 铵盐的检出(气室法)

取 4~5 滴铵盐溶液置于一表面皿中心,另一表面皿中心贴有一小条湿润的 pH 试纸。然后在铵盐溶液中滴加 6 mol/L NaOH 溶液至呈碱性,将贴有 pH 试纸的表面皿盖在铵盐的表面皿上形成"气室"。将气室置于水浴上微热,观察 pH 试纸颜色的变化。

3. 亚硝酸的生成与分解

分别取 $NaNO_2$ 饱和溶液和 3 mol/L H_2SO_4 溶液各 1 mL 放置在两支试管中,用冰水冷却 2 min 后混合均匀,观察现象。溶液放置一段时间后又有什么变化?写出反应式。

4. 亚硝酸的氧化还原性

1) 亚硝酸的氧化性

取 4 滴 0.1 mol/L KI 溶液,加 2 滴 1 mol/L H_2SO_4 溶液酸化后,再滴加 0.1 mol/L $NaNO_2$ 溶液,观察现象及产物的颜色。微热试管,溶液又有什么变化?写出反应式。

2) 亚硝酸的还原性

取 1 滴 0.01 mol/L $KMnO_4$ 溶液,加 2 滴 1 mol/L H_2SO_4 溶液酸化后,再滴加 0.1 mol/L $NaNO_2$ 溶液,观察现象,写出反应式。

5. 磷酸盐的性质和溶解度

1) 磷酸盐的性质

用 pH 试纸测定正磷酸盐、焦磷酸盐、偏磷酸盐水溶液的 pH。

用 pH 试纸测定浓度同为 0.1 mol/L 的 Na_3PO_4 溶液、Na_2HPO_4 溶液、NaH_2PO_4 溶液的 pH。分别向三支试管中加入 0.5 mL 0.1 mol/L Na_3PO_4 溶液、Na_2HPO_4 溶液、NaH_2PO_4 溶液,然后分别滴加适量的 0.1 mol/L $AgNO_3$ 溶液,观察是否有沉淀生成。反应溶液的 pH 又有何变化?试给予解释。

2) 磷酸盐的溶解度

分别向浓度同为 0.1 mol/L 的 Na_3PO_4、Na_2HPO_4 和 NaH_2PO_4 溶液中加入 0.1 mol/L $CaCl_2$ 溶液,观察有无沉淀生成。再加入 2 mol/L 氨水后又有何变化?继续加入 2 mol/L HCl 溶液后又有什么变化?试给予解释,并写出反应式。

6. 活性炭的吸附作用

(1) 在溶液中对有色物质的吸附:往 2 mL 靛蓝溶液中加入一小勺活性炭,振荡试管,然后过滤除去活性炭,观察溶液的颜色变化。

(2) 对无机离子的吸附作用:往 0.001 mol/L $Pb(NO_3)_2$ 溶液中加入几滴 0.1 mol/L K_2CrO_4 溶液,观察黄色 $PbCrO_4$ 沉淀的生成。再往另一支试管中加入约 2 mL 0.001 mol/L $Pb(NO_3)_2$ 溶液及一小勺活性炭,振荡试管。过滤除去活性炭后向清液加几滴 0.1 mol/L K_2CrO_4 溶液,观察现象并加以解释。

7. 一些金属离子与碳酸盐的反应

分别向盛有 0.2 mol/L $FeCl_3$ 溶液和 0.1 mol/L $MgCl_2$、$Pb(NO_3)_2$、$CuSO_4$ 溶液的试管中滴加 1 mol/L Na_2CO_3 溶液，观察现象。再分别向四支盛有以上溶液的试管中滴加 0.5 mol/L $NaHCO_3$ 溶液，观察现象，通过计算初步确定反应物，并分别写出反应式。

8. 硼的性质

(1) 取少量硼酸晶体(绿豆大小)溶于约 2 mL 水中(为方便溶解，可微热)。冷却至室温后测其 pH。再向硼酸溶液中加入 4~5 滴甘油，测 pH。写出反应式，并作解释。

(2) 硼酸的鉴定反应：取少量硼酸晶体(绿豆大小)放在蒸发皿中，加入 0.5 mL 乙醇和几滴浓硫酸，混合后点燃，观察火焰的颜色，并完成反应式。

9. 难溶性硅酸盐的生成——"水中花园"

在一个 50 mL 烧杯中加入约 30 mL Na_2SiO_3 溶液(20%)，然后把 $CaCl_2$、$CuSO_4$、$Co(NO_3)_2$、$NiSO_4$、$MnSO_4$、$ZnSO_4$、$FeCl_3$ 固体各一小粒投入烧杯中，并使各固体之间保持一定间隔，记住其位置。放置约 1 h 后观察现象。

思考题

(1) 在化学反应中，为什么一般不用硝酸和盐酸作酸化试剂？

(2) 硼酸为弱酸，为什么硼酸溶液加甘油后酸性会增强？

(3) 实验室中为什么可以用磨口玻璃仪器储存酸液而不能用来储存碱液？为什么盛过硅酸钠溶液的容器在实验后必须立即洗净？

(4) 如何区别碳酸钠、硅酸钠和硼酸钠？

(5) 是否能用二氧化碳灭火器扑灭金属镁的火焰吗？为什么？

实验 18 p 区金属元素(铝、锡、铅、锑、铋)

实验目的

(1) 掌握铝的化学性质。

(2) 掌握锡、铅、锑、铋的氢氧化物酸碱性及其不同氧化态的氧化还原性。

实验用品

仪器与材料：离心机。

试剂：$Al(s)$、$PbO_2(s)$、$K[Sb(OH)_6](s)$、$NaBiO_3(s)$、$(NH_4)_2SO_4$(饱和)、$Al_2(SO_4)_3$(1 mol/L)、$FeCl_3$(0.1 mol/L)、$NaNO_3$(0.5 mol/L)、$SnCl_2$(0.1 mol/L)、$HgCl_2$(0.1 mol/L)、$Bi(NO_3)_3$(0.1 mol/L)、$MnSO_4$(0.1 mol/L)、$KMnO_4$(0.1 mol/L)、$SbCl_3$(0.1 mol/L)、$BiCl_3$(0.1 mol/L)、$AsCl_3$(0.1 mol/L)、KI(0.1 mol/L)、K_2CrO_4(0.1 mol/L)、$Pb(NO_3)_2$(0.1 mol/L)、$NaHCO_3$(饱和)、$NaAc$(饱和)、$BaCl_2$(0.1 mol/L)、$NaOH$(2 mol/L、6 mol/L、40%)、H_2SO_4

（1 mol/L、2 mol/L、3 mol/L）、HCl（2 mol/L、浓）、HNO_3（6 mol/L）、HAc（6 mol/L）、CCl_4。

实验步骤

  1. 单质铝的性质

  将铝片用砂纸打磨除去表面氧化膜后分别试验其与：①热水；②冷水；③2 mol/L HCl 溶液；④2 mol/L NaOH 溶液；⑤0.5 mol/L $NaNO_3$ 溶液在足量的 40% NaOH 溶液中的反应，并证实⑤反应产物中 NH_3 的生成。写出反应式，并由实验简单总结铝的性质。

  2. 铝盐成矾作用

  取 1 mL 1 mol/L $Al_2(SO_4)_3$ 溶液加入 1 mL $(NH_4)_2SO_4$ 饱和溶液，稍稍静置，观察现象。如溶液仍是澄清透明，可稍摩擦试管壁。观察现象，写出反应式。

  3. Sn(Ⅱ)的还原性

  (1) 在 0.1 mol/L $FeCl_3$ 溶液中滴加 0.1 mol/L $SnCl_2$ 溶液，观察现象，写出反应式。试加 1 滴 KSCN 溶液检验溶液中是否还存在 Fe^{3+}。

  (2) 在 0.1 mol/L $HgCl_2$ 溶液中滴加 0.1 mol/L $SnCl_2$ 溶液，观察现象，写出反应式。

  (3) 在 0.1 mol/L $SnCl_2$ 溶液中滴加两滴 0.1 mol/L $Bi(NO_3)_3$ 溶液，观察现象，写出反应式。通过以上实验比较 Sn(Ⅱ)与 Fe(Ⅱ)、Sn(Ⅱ)与 Hg(Ⅰ)还原性的强弱。

  4. Pb(Ⅳ)的氧化性

  (1) 在试管中放入少量 PbO_2(s)，然后滴加浓盐酸溶液，观察现象，写出反应式。

  (2) 在有少量 PbO_2(s)的试管中加入 3 mol/L H_2SO_4 溶液酸化，再加入 1 滴 0.1 mol/L $MnSO_4$ 溶液，水浴加热，观察现象，写出反应式。

  5. Sb(Ⅲ)和 Bi(Ⅲ)的还原性

  (1) 在 5 mL 40% KOH 溶液中加入 2～3 滴 0.1 mol/L $KMnO_4$ 溶液，制备 K_2MnO_4 溶液后把溶液分为两份，分别加入 0.1 mol/L $SbCl_3$ 溶液和 $BiCl_3$ 溶液，观察现象，写出反应式。

  (2) 在两支试管中制备$[Ag(NH_3)_2]^+$溶液后，分别加入少量 Na_3AsO_3 溶液（自制）、Na_3SbO_3 溶液（自制）和 0.1 mol/L $Bi(NO_3)_3$ 溶液，微热试管，观察现象，写出反应式。

  (3) 在两支试管中分别加入 0.1 mol/L $AsCl_3$ 和 $SbCl_3$ 溶液，再加入 $NaHCO_3$ 饱和溶液至溶液呈弱酸性。滴加碘水，观察现象，写出反应式。

  (4) 取少量 0.1 mol/L $Bi(NO_3)_3$ 溶液滴加 6 mol/L NaOH 溶液至白色沉淀生成后，加入氯水（或溴水），观察现象，写出反应式。

  通过以上实验说明 Sb(Ⅲ)和 Bi(Ⅲ)的还原性。

  6. Sb(Ⅴ)和 Bi(Ⅴ)的氧化性

  (1) 在两支试管中各加入少量 $K[Sb(OH)_6]$(s)、$NaBiO_3$(s)及少量的水，用稀酸酸化溶液（用什么酸酸化？）。再加入少量 0.1 mol/L KI 溶液及四氯化碳，观察现象，写出反应式。

(2) 在两支试管中分别加入两滴 0.1 mol/L $MnSO_4$ 溶液，用 2 mol/L H_2SO_4 溶液酸化后分别加入少量 $K[Sb(OH)_6](s)$ 和 $NaBiO_3(s)$，观察现象，写出反应式。

通过以上实验说明 Sb(V) 和 Bi(V) 的氧化性。

7. 难溶物

1) 卤化物

在少量水中加入数滴 0.1 mol/L $Pb(NO_3)_2$ 溶液，再滴加几滴 2 mol/L HCl 溶液，有什么现象？加热后又有什么变化？再把溶液冷却又有什么现象？试给予解释。

在 0.1 mol/L $Pb(NO_3)_2$ 溶液中滴加浓盐酸，有什么现象？取少量白色沉淀，继续滴加浓盐酸，又有什么现象？用水稀释后又有什么变化？写出反应式。

取数滴 0.1 mol/L $Pb(NO_3)_2$ 溶液，用少量水稀释后再加入 1~2 滴 0.1 mol/L KI 溶液，观察现象，试验沉淀在热水中的溶解情况。

2) 铅的含氧酸盐

(1) 铬酸盐：取少量 0.1 mol/L $Pb(NO_3)_2$ 溶液，再滴加几滴 0.1 mol/L K_2CrO_4 溶液，观察 $PbCrO_4$ 沉淀的生成。分别考察沉淀在 6 mol/L HNO_3 溶液、6 mol/L NaOH 溶液、6 mol/L HAc 溶液和 NaAc 饱和溶液中的溶解情况，写出反应式。再用 0.1 mol/L $BaCl_2$ 溶液代替 $Pb(NO_3)_2$ 溶液重复以上实验，观察现象有何异同，写出反应式。

(2) 硫酸盐：观察 0.1 mol/L $Pb(NO_3)_2$ 溶液与 1 mol/L H_2SO_4 溶液反应生成沉淀的颜色，再分别试验沉淀在 2 mol/L NaOH 溶液及 NaAc 饱和溶液中的反应，写出反应式。再用 0.1 mol/L $BaCl_2$ 溶液代替 $Pb(NO_3)_2$ 溶液重复以上实验，观察现象，写出反应式。

思考题

(1) 结合实验说明锡、铅氧化还原性变化不同的原因。

(2) 设计实验方案对 $SbCl_3$ 和 $Bi(NO_3)_3$ 混合溶液进行分离和鉴定。

实验 19　d 区元素化合物的性质（一）

实验目的

(1) 掌握 d 区元素某些氢氧化物的酸碱性。

(2) 掌握 d 区元素某些化合物可变价态的氧化还原性。

(3) 了解 d 区元素某些金属离子的水解性。

实验用品

试剂：锌粒（或锌粉）、$NaBiO_3(s)$、$Na_2SO_3(s)$、$MnO_2(s)$、HCl（2 mol/L、6 mol/L、浓）、H_2SO_4（2 mol/L）、NaOH（2 mol/L、6 mol/L、40%）、$TiOSO_4$（0.1 mol/L）、$Cr_2(SO_4)_3$（0.1 mol/L）、$MnSO_4$（0.1 mol/L）、$FeCl_3$（0.1 mol/L、1.0 mol/L）、$(NH_4)_2Fe(SO_4)_2$（0.1 mol/L）、$CoCl_2$（0.1 mol/L）、$NiSO_4$（0.1 mol/L）、$KMnO_4$（0.01 mol/L）、$(NH_4)_2MoO_4$（饱和）、Na_2WO_4（饱和）、NH_4VO_3（饱和）、H_2O_2（3%）、碘化钾-淀粉试纸、溴水。

实验步骤

1. 氢氧化物的酸碱性

分别向少量 0.1 mol/L $TiOSO_4$ 溶液、$Cr_2(SO_4)_3$ 溶液、$MnSO_4$ 溶液、$(NH_4)_2Fe(SO_4)_2$ 溶液、$FeCl_3$ 溶液、$CoCl_2$ 溶液和 $NiSO_4$ 溶液中滴加适量 2 mol/L NaOH 溶液,观察沉淀的产生。将上述所得沉淀均分为两份,分别加入过量的 2 mol/L NaOH 溶液和 HCl 溶液,观察沉淀是否溶解。将可溶解于稀碱的溶液加热煮沸,观察现象,写出反应式。

2. 某些化合物的氧化还原性

1) 铁(Ⅱ)、钴(Ⅱ)、镍(Ⅱ)的还原性

分别在 0.1 mol/L $(NH_4)_2Fe(SO_4)_2$ 溶液、$CoCl_2$ 溶液、$NiSO_4$ 溶液中滴加几滴溴水,观察现象,写出反应式。

分别在 0.1 mol/L $(NH_4)_2Fe(SO_4)_2$ 溶液、$CoCl_2$ 溶液、$NiSO_4$ 溶液中加入 6 mol/L NaOH 溶液,观察现象。将沉淀放置一段时间后观察有何变化。再将 Co(Ⅱ)、Ni(Ⅱ)生成的沉淀各分成两份,分别加 3% H_2O_2 溶液和溴水,它们各有何变化?写出反应式。

根据实验结果比较 Fe(Ⅱ)、Co(Ⅱ)、Ni(Ⅱ)还原性的差异。

2) 铁(Ⅲ)、钴(Ⅲ)、镍(Ⅲ)氧化性

制取 $Fe(OH)_3$、$CoO(OH)$、$NiO(OH)$ 沉淀,分别加入浓盐酸,观察现象,并用碘化钾-淀粉试纸检验所放出的气体,写出反应式。

根据实验结果比较 Fe(Ⅲ)、Co(Ⅲ)、Ni(Ⅲ)氧化性的差异。

3) 锰化合物的氧化还原性

(1) 锰(Ⅱ)的还原性:分别试验 0.1 mol/L $MnSO_4$ 溶液在碱性介质中与空气、溴水的作用以及在酸性介质中与固体 $NaBiO_3$ 的作用,观察现象,写出反应式。

(2) 锰(Ⅳ)的氧化还原性:在少许 MnO_2 固体中加入浓盐酸,微热并检验有无氯气产生,写出反应式。

取少量固体 MnO_2,加入数滴 40% NaOH 溶液和少量 0.01 mol/L $KMnO_4$ 溶液,微热片刻,观察现象,写出反应式。

(3) 锰(Ⅶ)的氧化性:分别试验 Na_2SO_3 溶液在酸性、中性和碱性介质中与 $KMnO_4$ 的反应,写出反应式。

4) 铬、钼、钨的氧化还原性

(1) 不同氧化态铬的氧化还原性:利用 0.1 mol/L $Cr_2(SO_4)_3$ 溶液、3% H_2O_2 溶液、2 mol/L NaOH 溶液、2 mol/L H_2SO_4 溶液等试剂设计系列试管实验,说明在不同介质下铬的不同氧化态的氧化还原性和它们之间相互转化条件。写出反应式。

(2) 钼(Ⅵ)和钨(Ⅵ)的氧化性:取少量 $(NH_4)_2MoO_4$ 饱和溶液用 6 mol/L 盐酸溶液酸化后,加一粒锌粒(或锌粉),振荡,观察溶液颜色有什么变化。放置一段时间后(在进一步反应过程中可补加几滴盐酸),观察又有何变化。写出反应式。

用 Na_2WO_4 饱和溶液代替 $(NH_4)_2MoO_4$ 饱和溶液重复上述实验,观察现象,写出反应式。

5) 钛、钒的氧化还原性

(1) 钛(Ⅳ)和钛(Ⅲ)的氧化还原性：在 0.1 mol/L $TiOSO_4$ 溶液中加入一粒锌粒(或锌粉)，观察现象。反应一段时间后，将溶液分装于两支试管中，分别试验它们在空气中与少量 $CuCl_2$ 溶液的反应，观察现象，写出反应式。

(2) 钒的常见氧化态的水合离子颜色及其氧化还原性：取 NH_4VO_3 饱和溶液，用 6 mol/L 盐酸溶液酸化后加入少量锌粉，放置片刻，仔细观察溶液颜色的变化。分别试验溶液与不同量 $KMnO_4$ 溶液的反应使 V^{2+} 氧化为 V^{3+}、VO^{2+}、VO_2^+，观察它们在溶液中的颜色，写出反应式。

3. 金属离子的水解

1) 铁(Ⅲ)盐的水解

将少量蒸馏水加热煮沸后，加入数滴 1.0 mol/L $FeCl_3$ 溶液，煮沸片刻，观察现象，写出反应式。

2) 铬(Ⅲ)盐的水解

向 0.1 mol/L $Cr_2(SO_4)_3$ 溶液中滴加 Na_2CO_3 溶液，观察现象，写出反应式，并解释实验结果。

3) 钛(Ⅳ)盐的水解

取 1~2 滴 0.1 mol/L $TiOSO_4$ 溶液，加入适量蒸馏水，加热煮沸，观察现象，写出反应式。

4. 小设计

设计实验方案，将含有 Cr^{3+}、Fe^{3+}、Mn^{2+} 的混合溶液分离检出。

附注

(1) $TiOSO_4$ 溶液的制备：在 2 mL $TiCl_4$ 溶液中加入 30 mL 6 mol/L H_2SO_4 溶液，用水稀释至 200 mL。

(2) $Fe(OH)_2$ 的制备：取 2 mL 蒸馏水置于试管中，加入 2 滴 2 mol/L H_2SO_4 溶液，煮沸以赶尽其中空气，然后在其中溶解少许 $(NH_4)_2Fe(SO_4)_2$ 晶体。在另一支试管中小心煮沸 2 mol/L NaOH 溶液，冷却后，用滴管吸取 NaOH 溶液插入 $(NH_4)_2Fe(SO_4)_2$ 溶液中，然后慢慢放出(整个操作过程都要避免将空气带入溶液中)，便可观察到近乎白色的 $Fe(OH)_2$ 沉淀生成。

思考题

(1) 如何把 Fe^{3+}、Al^{3+}、Cr^{3+} 从混合溶液中分离？

(2) 如何实现 Cr^{3+} 与 CrO_4^{2-}、MnO_2 与 Mn^{2+}、MnO_2 与 MnO_4^{2-}、MnO_2 与 MnO_4^-、MnO_4^{2-} 与 MnO_4^-、MnO_4^- 与 Mn^{2+} 等价态之间的相互转化？

(3) 钛和钒各有几种常见氧化态？指出它们在水溶液中的状态和颜色。

实验 20 d 区元素化合物的性质(二)

实验目的

(1) 观察和掌握 d 区某些水合离子的颜色。

(2) 了解 d 区元素某些金属离子的配合物及形成配合物后对其性质的影响。

(3) 了解 d 区元素某些配合物在鉴定金属离子中的应用。

实验用品

试剂：NaF(s)、$Na_2C_2O_4$(s)、EDTA(s)、HCl(2 mol/L)、H_2SO_4(6 mol/L)、NaOH(2 mol/L、6 mol/L)、氨水(2 mol/L、6 mol/L)、$Cr(NO_3)_3$(1 mol/L)、$TiOSO_4$(0.1 mol/L)、NH_4VO_3(0.1 mol/L、饱和)、$Cr_2(SO_4)_3$(0.1 mol/L)、$MnSO_4$(0.1 mol/L)、$FeCl_3$(0.1 mol/L)、$(NH_4)_2Fe(SO_4)_2$(0.1 mol/L)、$CoCl_2$(0.1 mol/L)、$NiSO_4$(0.1 mol/L)、KI(0.1 mol/L)、$AgNO_3$(0.1 mol/L)、KSCN(饱和)、H_2O_2(3%)、乙二胺(1%)、丁二酮肟(1%)、乙醚、戊醇、丙酮、四氯化碳。

实验步骤

1. 观察和熟悉下列水合离子的颜色

1) 水合阳离子

$[Ti(H_2O)_6]^{3+}$、$[Cr(H_2O)_6]^{3+}$、$[Mn(H_2O)_6]^{2+}$、$[Fe(H_2O)_6]^{2+}$、$[Co(H_2O)_6]^{2+}$、$[Ni(H_2O)_6]^{2+}$。

2) 阴离子

CrO_4^{2-}、$Cr_2O_7^{2-}$、MnO_4^{2-}、MnO_4^-、MoO_4^{2-}、WO_4^{2-}、VO_3^-。

以表格形式写出实验结果。

2. 某些金属元素离子的颜色变化

1) Cr^{3+} 的水合异构现象

取少量 1 mol/L $Cr(NO_3)_3$ 溶液加热，观察加热前后溶液颜色的变化。

$$[Cr(H_2O)_6](NO_3)_3 \underset{冷}{\overset{热}{\rightleftharpoons}} [Cr(H_2O)_5(NO_3)](NO_3)_2 + H_2O$$

2) 观察不同配体的 Co(II)配合物的颜色

向 KSCN 饱和溶液滴加 0.1 mol/L $CoCl_2$ 溶液至呈蓝紫色，将此溶液分装三支试管，在其中两支试管中分别加入蒸馏水和丙酮，对比三支试管溶液颜色差异，并作解释。

$$[Co(NCS)_4]^{2-} + 6H_2O \underset{}{\overset{丙酮}{\rightleftharpoons}} [Co(H_2O)_6]^{2+} + 4NCS^-$$

3. 某些金属离子配合物

1) 氨合物

分别向少量 0.1 mol/L $Cr_2(SO_4)_3$ 溶液、$MnSO_4$ 溶液、$FeCl_3$ 溶液、$(NH_4)_2Fe(SO_4)_2$ 溶液、$CoCl_2$ 溶液和 $NiSO_4$ 溶液中滴加 6 mol/L 氨水，观察现象，写出反应式，并总结上述金属离子形成氨合物的能力。

2) 配合物的形成对氧化还原性的影响

在 0.1 mol/L KI 和 CCl_4 混合溶液中加入 0.1 mol/L $FeCl_3$ 溶液，观察现象。若上述试液在加入 $FeCl_3$ 之前先加入少量固体 NaF，现象有什么不同？作出解释并写出反应式。

在室温下，比较 0.1 mol/L $(NH_4)_2Fe(SO_4)_2$ 溶液在有 EDTA 和没有 EDTA 存在下与

0.1 mol/L $AgNO_3$ 溶液的反应，并给予解释。

3）配合物稳定性与配位体的关系

在 0.1 mol/L $Cr_2(SO_4)_3$ 溶液中加入少量固体 $Na_2C_2O_4$，振荡，观察溶液颜色的变化，再逐滴加入 2 mol/L NaOH 溶液，观察有无沉淀生成，并作解释，写出反应式。

在 0.1 mol/L $FeCl_3$ 溶液中加入少量 KSCN 饱和溶液，观察现象。然后加入少量固体 $Na_2C_2O_4$，观察溶液颜色变化，并作解释，写出反应式。

在 0.1 mol/L $NiSO_4$ 溶液中加入过量 2 mol/L 氨水，观察现象。然后逐滴加入 1% 乙二胺溶液，再观察现象。

4. 配合物应用：金属离子的鉴定

1）铁的鉴定

根据所学知识进行铁(Ⅱ)和铁(Ⅲ)的鉴定。

2）钴(Ⅱ)的鉴定

在 0.1 mol/L $CoCl_2$ 溶液中加入戊醇(或丙醇)后，再滴加 KSCN 饱和溶液，观察现象，写出反应式。

3）镍(Ⅱ)的鉴定

在 0.1 mol/L $NiSO_4$ 溶液中加入 2 mol/L 氨水至溶液呈弱碱性，再加入 1 滴 1% 丁二酮肟溶液，观察现象。反应式为

$$Ni^{2+} + 2 \begin{array}{c} CH_3-C=NOH \\ | \\ CH_3-C=NOH \end{array} + 2NH_3 = Ni\begin{pmatrix} CH_3-C=NOH \\ | \\ CH_3-C=NO \end{pmatrix}_2 \downarrow + 2NH_4^+$$

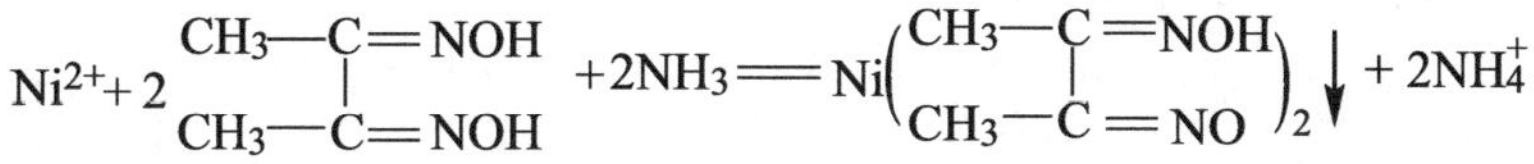

4）铬(Ⅲ)的鉴定

在 0.1 mol/L $Cr_2(SO_4)_3$ 溶液中滴加过量 6 mol/L NaOH 溶液，再加入 3% H_2O_2 溶液，观察现象。用稀 H_2SO_4 酸化，再加入少量乙醚(或丙醇)，继续滴加 3% H_2O_2，观察现象，写出反应式。

5）钛(Ⅳ)的鉴定

在少量 0.1 mol/L $TiOSO_4$ 溶液中滴加 3% H_2O_2 溶液，观察现象。再加入少量 6 mol/L 氨水，又有什么现象？反应式为

$$TiO^{2+} + H_2O_2 = [TiO(H_2O_2)]^{2+}(橙红色)$$

$$[TiO(H_2O_2)]^{2+} + NH_3 \cdot H_2O = H_2Ti(O_2)O_2 \downarrow (黄色) + NH_4^+ + H^+$$

6）钒(Ⅴ)的鉴定

取少量 NH_4VO_3 饱和溶液用盐酸酸化，再加入几滴 3% H_2O_2 溶液，观察现象。反应式为

$$NH_4VO_3 + H_2O_2 + 4HCl = [V(O_2)]Cl_3 + NH_4Cl + 3H_2O$$

5. 小设计

已知混合溶液中含有 Fe^{3+}、Co^{3+}、Ni^{2+} 三种离子，设计一方案分别检出。

思考题

(1) 为什么 d 区元素水合离子具有颜色？

(2) 利用 KI 定量测定 Cu^{2+} 时，杂质 Fe^{3+} 的存在会产生干扰，如何排除干扰？

(3) 根据氧化还原电对的电极电势，常温下 Fe^{3+} 难以将 Ag^+ 还原为单质银，如何应用配合物性质，用 Fe^{3+} 回收银盐溶液中的银？

实验 21　ds 区金属（铜、银、锌、镉、汞）

实验目的

(1) 了解铜、银、锌、镉、汞氧化物或氢氧化物的酸碱性及其硫化物的溶解性。

(2) 掌握 Cu(Ⅰ) 和 Cu(Ⅱ) 重要化合物的性质及相互转化。

(3) 试验并熟悉铜、银、锌、镉、汞的配位能力以及 Hg^{2+} 和 Hg_2^{2+} 的转化。

实验用品

仪器与材料：试管、烧杯、离心机、离心试管、pH 试纸、玻璃棒。

试剂：碘化钾、碎铜屑、HCl(2 mol/L、浓)、H_2SO_4(2 mol/L)、HNO_3(2 mol/L、浓)、NaOH(2 mol/L、6 mol/L、40%)、氨水(2 mol/L、浓)、$CuSO_4$(0.2 mol/L)、$ZnSO_4$(0.2 mol/L)、$CdSO_4$(0.2 mol/L)、$CuCl_2$(0.5 mol/L)、$Hg(NO_3)_2$(0.2 mol/L)、$SnCl_2$(0.2 mol/L)、$AgNO_3$(0.1 mol/L)、Na_2S(1 mol/L)、KI(0.2 mol/L)、KSCN(0.1 mol/L)、$Na_2S_2O_3$(0.5 mol/L)、NaCl(0.2 mol/L)、金属汞、葡萄糖(10%)。

实验内容

1. 铜、银、锌、镉、汞氢氧化物或氧化物的生成和性质

1) 铜、锌、镉氢氧化物的生成和性质

向三支分别盛有 0.5 mL 0.2 mol/L $CuSO_4$ 溶液、$ZnSO_4$ 溶液、$CdSO_4$ 溶液的离心试管中滴加新配制的 2 mol/L NaOH 溶液，观察溶液颜色及沉淀的生成。

将沉淀离心分离、洗涤，然后将每种沉淀分成两份：一份加入 2 mol/L H_2SO_4 溶液，另一份继续滴加 6 mol/L NaOH 溶液。观察现象，写出反应式。

2) 银、汞氧化物的生成和性质

(1) 氧化银的生成和性质：取 0.5 mL 0.1 mol/L $AgNO_3$ 溶液，滴加新配制的 2 mol/L NaOH 溶液，观察 Ag_2O（为什么不是 AgOH）的颜色和状态。将沉淀离心分离、洗涤，分成两份：一份加入 2 mol/L HNO_3 溶液，另一份加入 2 mol/L 氨水。观察现象，写出反应式。

(2) 氧化汞的生成和性质：取 0.5 mL 0.2 mol/L $Hg(NO_3)_2$ 溶液，滴加新配制的 2 mol/L NaOH 溶液，观察溶液颜色和沉淀的状态。将沉淀离心分离、洗涤，分成两份：一份加入 2 mol/L HNO_3 溶液，另一份加入 40% NaOH 溶液。观察现象，写出有关反应式。

2. 铜、银、锌、镉、汞硫化物的生成和性质

向五支分别盛有 0.5 mL 0.2 mol/L $CuSO_4$ 溶液、0.1 mol/L $AgNO_3$ 溶液、0.2 mol/L $ZnSO_4$ 溶液、0.2 mol/L $CdSO_4$ 溶液、0.2 mol/L $Hg(NO_3)_2$ 溶液的离心试管中滴加 1 mol/L Na_2S 溶液。观察沉淀的生成和颜色。

将沉淀离心分离、洗涤，然后将每种沉淀分成四份：分别加入 2 mol/L 盐酸、浓盐酸、浓硝酸和王水（自配），分别水浴加热。观察沉淀溶解情况。

根据实验现象并查阅有关数据，填写表 4.1，对铜、银、锌、镉、汞硫化物的溶解情况得出结论，并写出反应式。

表 4.1 所制备硫化物的物理性质及相关数据

性质 硫化物	颜色	溶解度				K_{sp}
		2 mol/L 盐酸	浓盐酸	浓硝酸	王水	
CuS						
Ag_2S						
ZnS						
CdS						
HgS						

3. 铜、银、锌、汞的配合物

1) 氨合物的生成

向四支分别盛有 0.5 mL 0.2 mol/L $CuSO_4$ 溶液、0.2 mol/L $ZnSO_4$ 溶液、0.2 mol/L $Hg(NO_3)_2$ 溶液、0.1 mol/L $AgNO_3$ 溶液的试管中滴加 2 mol/L 氨水，观察沉淀的生成。继续加入过量的 2 mol/L 氨水，又有什么现象发生？写出反应式。

比较 Cu^{2+}、Ag^+、Zn^{2+}、Hg^{2+} 与氨水反应有什么不同。

2) 汞配合物的生成和应用

向 0.5 mL 0.2 mol/L $Hg(NO_3)_2$ 溶液中滴加 0.2 mol/L KI 溶液，观察沉淀的生成和颜色。再向该沉淀中加入少量碘化钾固体（直至沉淀刚好溶解为止，不要过量），溶液显什么颜色？写出反应式。

在所得的溶液中加入几滴 40% NaOH 溶液，再与氨水反应，观察沉淀的颜色。

向 5 滴 0.2 mol/L $Hg(NO_3)_2$ 溶液中逐滴加入 0.1 mol/L KSCN 溶液，最初生成白色 $Hg(SCN)_2$ 沉淀，继续滴加 KSCN 溶液，沉淀溶解生成无色配离子 $[Hg(SCN)_4]^{2-}$。再在该溶液中加几滴 0.2 mol/L $ZnSO_4$ 溶液，观察白色 $Zn[Hg(SCN)_4]$ 沉淀的生成（该反应可定性检验 Zn^{2+}），必要时用玻璃棒摩擦试管壁。

4. 铜、银、汞的氧化还原性

1) 氧化亚铜的生成和性质

取 0.5 mL 0.2 mol/L $CuSO_4$ 溶液，滴加过量的 6 mol/L NaOH 溶液，使起初生成的蓝色沉淀溶解成深蓝色溶液。然后在溶液中加入 1 mL 10%葡萄糖溶液，混匀后微热，有黄色沉淀产生进而变成红色沉淀。写出反应式。

将沉淀离心分离、洗涤、然后沉淀分成两份：一份沉淀与 1 mL 2 mol/L H_2SO_4 溶液作用，静置一段时间，注意沉淀变化，然后加热至沸，观察现象；另一份沉淀中加入 1 mL 浓氨水，振荡后，静置一段时间，观察溶液颜色。放置一段时间后，溶液为什么会变成深蓝色？

2)氧化亚铜的生成和性质

取 10 mL 0.5 mol/L $CuCl_2$ 溶液,加入 3 mL 浓盐酸和少量碎铜屑,加热沸腾至其中液体呈深棕色(绿色完全消失),继续加热,直至溶液近无色,取几滴上述溶液加入 10 mL 蒸馏水中,如有白色沉淀产生,则迅速把全部溶液倒入 100 mL 蒸馏水中,将白色沉淀洗涤至无蓝色为止。

取少许沉淀分成两份:一份与 3 mL 浓氨水作用,观察有何变化;另一份与 3 mL 浓盐酸作用,观察又有何变化。写出反应式。

3)碘化亚铜的生成和性质

在盛有 0.5 mL 0.2 mol/L $CuSO_4$ 溶液的试管中,边滴加 0.2 mol/L KI 溶液边振荡,溶液变为棕黄色(CuI 为白色沉淀,I_2 溶于 KI 呈黄色)。再滴加适量 0.5 mol/L $Na_2S_2O_3$ 溶液,以除去反应中生成的碘。观察产物的颜色和状态,写出反应式。

4)汞(Ⅱ)与汞(Ⅰ)的相互转化

(1) Hg^{2+} 的氧化性:在 5 滴 0.2 mol/L $Hg(NO_3)_2$ 溶液中逐滴加入 0.2 mol/L $SnCl_2$ 溶液(由适量至过量),观察现象,写出反应式。

(2) Hg^{2+} 转化为 Hg_2^{2+} 和 Hg_2^{2+} 的歧化分解:在 0.5 mL 0.2 mol/L $Hg(NO_3)_2$ 溶液中滴入 1 滴金属汞,充分振荡。用滴管把清液转入两支试管中(余下的汞要回收),在一支试管中加入 0.2 mol/L NaCl 溶液,另一支试管中滴入 2 mol/L 氨水。观察现象,写出反应式。

思考题

(1)在白色氯化亚铜沉淀中加入浓氨水或浓盐酸后形成什么颜色的溶液?放置一段时间后会变成蓝色溶液,为什么?

(2)上述实验中深棕色溶液是什么物质?将近无色溶液倒入蒸馏水中发生了什么反应?

(3)加入硫代硫酸钠是为了与溶液中产生的碘作用,而便于观察碘化亚铜白色沉淀的颜色;但若硫代硫酸钠过量,则看不到白色沉淀,为什么?

(4)使用汞时应注意什么?为什么汞要用水封存?

(5)用平衡原理预测在 $Hg_2(NO_3)_2$ 溶液中通入 H_2S 气体后,生成的沉淀物为何物,并加以解释。

(6)制备氯化亚铜时,能否用氯化铜和碎铜屑在用盐酸酸化呈微弱的酸性条件下反应?为什么?若用浓氯化钠溶液代替盐酸,此反应能否进行?为什么?

(7)根据钠、钾、钙、镁、铝、锡、铅、铜、银、锌、镉、汞的标准电极电势,推测这些金属的活动顺序。

(8)当 SO_2 通入 $CuSO_4$ 饱和溶液和 NaCl 饱和溶液的混合液时,将发生什么反应?能看到什么现象?试说明之,写出相应的反应式。

(9)选用什么试剂来溶解下列沉淀?

氢氧化铜,硫化铜,溴化铜,碘化银。

(10)现有三瓶已失标签的硝酸汞溶液、硝酸亚汞溶液和硝酸银溶液。至少用两种方法鉴别之。

(11)试用实验证明:黄铜的组成是铜和锌(其他组成可不考虑)。

实验22 未知阳离子的分离与鉴定

实验目的

(1)熟悉常见阳离子的性质。

(2) 掌握常见阳离子的分离、鉴定的原理及方法。
(3) 掌握试剂的取用、水浴加热、离心分离和沉淀的洗涤等基本操作。

实验原理

离子的分离与鉴定是以各离子对试剂的不同反应为依据的,这种反应常伴随有特殊的现象,如沉淀的生成或溶解、特殊颜色的出现、气体的产生等。各离子对试剂作用的相似性和差异性都构成离子分离与检出方法的基础,也就是说,离子的基本性质是分离与检出的基础。

离子的分离和检出只有在一定条件下才能进行。一定的条件主要是指溶液的酸度、反应物浓度、反应温度、促进或妨碍此反应的物质是否存在等。为使反应向期望的方向进行,必须选择适当的反应条件。因此,除了要熟悉离子的有关性质外,还要学会运用离子平衡(酸碱平衡、沉淀溶解平衡、氧化还原平衡和配位解离平衡)的规律控制反应条件,这对进一步了解离子分离条件和检出条件的选择将有很大帮助。

常见阳离子与常见试剂的反应:

(1) 与 HCl 反应

$$\left. \begin{array}{l} Ag^+ \\ Hg^{2+} \\ Pb^{2+} \end{array} \right\} \xrightarrow{HCl} \left\{ \begin{array}{l} AgCl \downarrow 白色,溶于氨水 \\ Hg_2Cl_2 \downarrow 白色,溶于浓HNO_3和H_2SO_4 \\ PbCl_2 \downarrow 白色,溶于热水、NH_4Ac和NaOH \end{array} \right.$$

(2) 与 H_2SO_4 反应

$$\left. \begin{array}{l} Ba^{2+} \\ Sr^{2+} \\ Ca^{2+} \\ Pb^{2+} \\ Ag^+ \end{array} \right\} \xrightarrow{H_2SO_4} \left\{ \begin{array}{l} BaSO_4 \downarrow 白色,难溶于酸 \\ SrSO_4 \downarrow 白色,溶于煮沸的酸 \\ CaSO_4 \downarrow 白色,溶解度较大,当Ca^{2+}浓度很大时才析出沉淀 \\ PbSO_4 \downarrow 白色,溶于NH_4Ac、NaOH和浓H_2SO_4,不溶于稀H_2SO_4 \\ Ag_2SO_4 \downarrow 白色,在浓溶液中产生沉淀,溶于热水 \end{array} \right.$$

(3) 与 NaOH 反应

$$\left. \begin{array}{l} Al^{3+} \\ Zn^{2+} \\ Pb^{2+} \\ Sb^{3+} \\ Sn^{2+} \end{array} \right\} \xrightarrow{过量NaOH} \left\{ \begin{array}{l} AlO_2^- 或 [Al(OH)_4]^- \\ ZnO_2^{2-} 或 [Zn(OH)_4]^{2-} \\ PbO_2^{2-} 或 [Pb(OH)_4]^{2-} \\ SbO_2^- 或 [Sb(OH)_4]^- \\ SnO_2^{2-} 或 [Sn(OH)_4]^{2-} \end{array} \right. \qquad Cu^{2+} \xrightarrow[\triangle]{浓NaOH} [Cu(OH)_4]^{2-}$$

(4) 与 NH_3 反应

$$\left. \begin{array}{l} Ag^+ \\ Cu^{2+} \\ Cd^{2+} \\ Zn^{2+} \end{array} \right\} \xrightarrow{过量NH_3} \left\{ \begin{array}{l} [Ag(NH_3)_2]^+ \\ [Cu(NH_3)_4]^{2+} 深蓝色 \\ [Cd(NH_3)_4]^{2+} \\ [Zn(NH_3)_4]^{2+} \end{array} \right.$$

(5) 与 $(NH_4)_2CO_3$ 反应

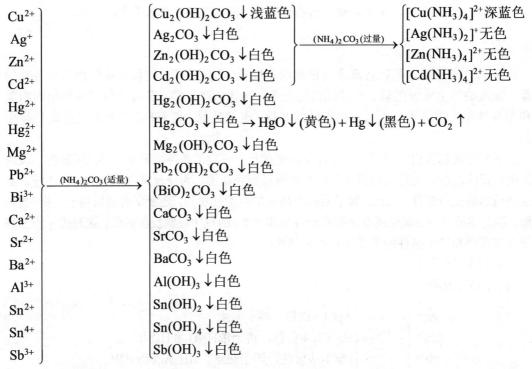

(6) 与 H_2S 或 $(NH_4)_2S$ 反应

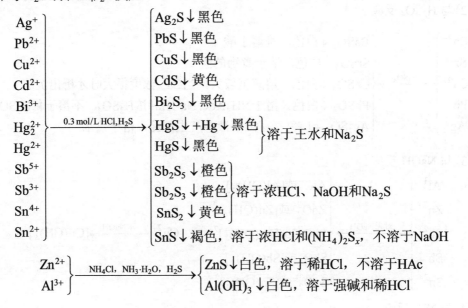

实验用品

仪器与材料：试管、烧杯、离心机、离心试管、酒精灯。

试剂：$MgCl_2$(0.5 mol/L)、$CaCl_2$(0.5 mol/L)、$BaCl_2$(0.5 mol/L)、$(NH_4)_2CO_3$(1 mol/L)、NH_3-NH_4Cl 缓冲溶液、HAc(2 mol/L、3 mol/L、6 mol/L)、K_2CrO_4(0.5 mol/L)、氨水(2 mol/L、

6 mol/L)、$(NH_4)_2C_2O_4$(0.5 mol/L)、$(NH_4)_2SO_4$(1 mol/L)、$(NH_4)_2HPO_4$(0.5 mol/L)、NaOH (6 mol/L)、镁试剂、铝试剂、$AgNO_3$(0.1 mol/L)、$Ba(NO_3)_2$(0.5 mol/L)、$Cd(NO_3)_2$(0.5 mol/L)、$Al(NO_3)_3$(0.5 mol/L)、$NaNO_3$(0.5 mol/L)、HCl(6 mol/L)、HNO_3(2 mol/L、6 mol/L)、NaAc (2 mol/L)、Na_2S(0.5 mol/L)、H_2SO_4(3 mol/L、6 mol/L)、Na_2CO_3(饱和)、$KSbC_4H_4O_6$(饱和)、$Fe(NO_3)_3$(0.5 mol/L)、H_2O_2(6%)、KSCN(1 mol/L)、$FeCl_3$(0.5 mol/L)、$CoCl_2$(0.5 mol/L)、$Ni(NO_3)_2$(0.5 mol/L)、$MnCl_2$(0.5 mol/L)、$CrCl_3$(0.5 mol/L)、$ZnCl_2$(0.5 mol/L)、NH_4F (0.5 mol/L)、NH_4SCN(饱和)、$Pb(Ac)_2$(0.5 mol/L)、$(NH_4)_2Hg(SCN)_4$(饱和)、$NaBiO_3$(s)、戊醇、丁二酮肟(1%)。

实验步骤

1. 分离检出试液中可能含有的 Mg^{2+}、Ca^{2+}、Ba^{2+}

1) Ca^{2+}、Ba^{2+}与Mg^{2+}的分离

取 2 mL 试液于离心试管中，加 1 mL NH_3-NH_4Cl 缓冲溶液，将离心试管置于约 60℃热水中加热，在搅拌下加入 1 mol/L $(NH_4)_2CO_3$ 溶液至沉淀完全。继续加热几分钟。然后离心分离，将清液移至另一支离心试管中，按下文第 4)步中操作处理，沉淀供第 2)步用。

2) Ba^{2+}与Ca^{2+}的分离和检出

将所得 $CaCO_3$、$BaCO_3$ 沉淀用少量水洗涤，离心分离，弃去洗涤液，加入 3 mol/L HAc，不断搅拌并水浴加热。待沉淀溶完后，滴入 0.5 mol/L K_2CrO_4 溶液至 Ba^{2+}沉淀完全。离心分离，清液留待检出 Ca^{2+}。

3) Ca^{2+}的检出

往清液中加 1 滴 6 mol/L 氨水和几滴 0.5 mol/L $(NH_4)_2C_2O_4$ 溶液，加热，产生白色沉淀，表示有 Ca^{2+}。为了消除 CrO_4^{2-}的黄色对观察 CaC_2O_4 颜色的干扰，可离心分离，弃去黄色溶液，加少量水洗涤沉淀，再离心分离弃去洗涤液，然后观察。

4) Mg^{2+}的检出

(1) 残余 Ba^{2+}、Ca^{2+}的除去：往第 1)步所得清液中加入 0.5 mol/L $(NH_4)_2C_2O_4$ 和 1 mol/L $(NH_4)_2SO_4$ 各 1 滴，加热几分钟，如果溶液浑浊，离心分离，并弃去沉淀，清液用来检出 Mg^{2+}。

(2) Mg^{2+}的检出：取 1 mL 清液于试管中，再加 0.5 mL 6 mol/L 氨水和 0.5 mL 0.5 mol/L $(NH_4)_2HPO_4$ 溶液，用玻璃棒摩擦试管壁，产生白色沉淀，表示有 Mg^{2+}。

另取 2 滴 1)的清液，滴在点滴板上，再加 2 滴 6 mol/L NaOH 溶液和 1 滴镁试剂，产生蓝色沉淀，表示有 Mg^{2+}。

2. Ag^+、Cd^{2+}、Al^{3+}、Ba^{2+}、Na^+混合离子的分离和鉴定

混合离子由相应的硝酸盐溶液配制。取 2 滴 Ag^+试液和 Cd^{2+}、Al^{3+}、Ba^{2+}、Na^+试液各 5 滴加到离心试管中，混合均匀后，参照以下步骤进行分离鉴定。

1) Ag^+的分离和鉴定

在混合试液中加 1 滴 6 mol/L 盐酸，搅拌，在沉淀生成时再滴加 1 滴 6 mol/L 盐酸至沉淀完全，搅拌片刻，离心分离，把清液转移到另一支离心试管中，按下文第 2)步中操作处理。

沉淀用 1 滴 6 mol/L 盐酸和 10 滴蒸馏水洗涤，离心分离，洗涤液并入上面的清液中。在沉淀上加入 2～3 滴 6 mol/L 氨水，搅拌使其溶解，在所得清液中加入 1～2 滴 6 mol/L HNO_3 溶液酸化，有白色沉淀析出，说明有 Ag^+ 存在。

2) Al^{3+} 的分离和鉴定

在第 1) 步的清液中滴加 6 mol/L 氨水至碱性，搅拌片刻，离心分离，把清液转移到另一支离心试管中，按下文第 3) 步中操作处理。在沉淀中加入 2 mol/L HAc 和 2 mol/L NaAc 各 2 滴，再加入 2 滴铝试剂，搅拌后微热，产生红色沉淀，说明有 Al^{3+} 存在。

3) Ba^{2+} 的分离和鉴定

在第 2) 步的清液中滴加 6 mol/L H_2SO_4 溶液至产生白色沉淀，再过量 2 滴，搅拌片刻，离心分离，把清液转移到另一支试管中，按下文第 4) 步中操作处理。沉淀用热蒸馏水 10 滴洗涤，离心分离，清液并入上面的清液中。在沉淀中加入 3～4 滴 Na_2CO_3 饱和溶液，搅拌片刻，再加入 2 mol/L HAc 溶液和 2 mol/L NaAc 溶液各 3 滴，继续搅拌片刻，然后加入 2～4 滴 0.5 mol/L K_2CrO_4 溶液，产生黄色沉淀，说明有 Ba^{2+} 存在。

4) Cd^{2+}、Na^+ 的分离和鉴定

取少量第 3) 步中的清液于一支试管中，加入 2～3 滴 0.5 mol/L Na_2S 溶液，产生亮黄色沉淀，说明有 Cd^{2+} 存在。另取少量第 3) 步中的清液于另一支试管中，加入几滴酒石酸锑钾饱和溶液，产生白色结晶状沉淀，说明有 Na^+ 存在。

3. Fe^{3+}、Co^{2+}、Ni^{2+}、Mn^{2+}、Cr^{3+}、Zn^{2+} 混合离子的分离和鉴定

取 3 mL 上述混合试剂，加到离心试管中，参照以下步骤进行分离和鉴定。

1) Fe^{3+}、Co^{2+}、Ni^{2+}、Mn^{2+}、Cr^{3+}、Zn^{2+} 的分离

在溶液中加入 H_2O_2 和足量 NaOH，可以把本组离子分为两组，沉淀是 FeO(OH)、CoO(OH)、$Ni(OH)_2$、$MnO(OH)_2$，而 Cr^{3+}、Zn^{2+} 分别成为 CrO_4^{2-}、$Zn(OH)_4^{2-}$ 留在溶液中。这种分组方法常称为"碱过氧化氢法"。

在溶液中加入 5～6 滴 6 mol/L NaOH 溶液至呈强碱性(pH>12)后，再多加 2～3 滴 NaOH 溶液，然后逐滴加入 6% H_2O_2 溶液，并用玻璃棒搅拌，待沉淀转为棕黑色即停止加 H_2O_2。继续搅拌 2～3 min，水浴加热，使胶状沉淀凝聚和过量的 H_2O_2 分解，加热至不再有气泡产生为止。离心分离，把清液移至另一支离心试管中，记为清液 1，留待下面步骤 7)、8) 处理。用热水洗涤沉淀一次，离心分离，弃去洗涤液。

2) 沉淀的溶解

向第 1) 步所得沉淀上加几滴 3 mol/L H_2SO_4 溶液和 2 滴 6% H_2O_2 溶液。搅拌后，将离心试管置于水浴中加热至沉淀全部溶解，同时使多余的 H_2O_2 分解。待溶液冷至室温，进行 Fe^{3+}、Co^{2+}、Ni^{2+}、Mn^{2+} 的检出。

3) Fe^{3+} 的检出

取 1 滴第 2) 步的溶液加到点滴板凹穴中，加 1 滴 1 mol/L KSCN 溶液。溶液出现血红色，加入 0.5 mol/L NH_4F 溶液，血红色褪去，表示有 Fe^{3+}。

4) Co^{2+} 的检出

在试管中加 2 滴第 2) 步的溶液和少量 0.5 mol/L NH_4F 溶液，再加入少量戊醇，最后加入 NH_4SCN 饱和溶液，戊醇层呈蓝色(或蓝绿色)，表示有 Co^{2+}。利用 F^- 与 Fe^{3+} 形成无色配离子

掩蔽 Fe^{3+}，消除 SCN^- 鉴定 Co^{2+} 时 Fe^{3+} 的干扰。

5) Ni^{2+} 的检出

在离心试管中加 2 滴第 2) 步的溶液，并加几滴 2 mol/L $NH_3 \cdot H_2O$ 至呈弱碱性（此时析出的沉淀为何物？氨水加得过多有何缺点？），离心分离，往上层清液中加 1～2 滴 1%丁二酮肟，产生桃红色沉淀，表示有 Ni^{2+}。

6) Mn^{2+} 的检出

取 1 滴第 2) 步的溶液，加入少量 2 mol/L HNO_3 溶液及 $NaBiO_3$ 固体，搅拌后静置，溶液变紫红色，表示有 Mn^{2+}。如果第 2) 步溶液中有多余的 H_2O_2，此时它将与 $NaBiO_3$ 发生氧化还原反应，消耗少量 $NaBiO_3$。

7) Cr^{3+} 的检出

用 6 mol/L HAc 溶液酸化步骤 1) 中余下的清液 1，留一半溶液用于检出 Zn^{2+}，往其余溶液中加几滴 0.5 mol/L $Pb(Ac)_2$ 溶液，产生黄色沉淀，表示有 CrO_4^{2-}，即原溶液中有 Cr^{3+}。

如果清液 1 中有多余的 H_2O_2，在酸性介质中会与 CrO_4^{2-} 反应，减少 CrO_4^{2-} 的量，使检出的灵敏度降低。

8) Zn^{2+} 的检出

往步骤 7) 留下的溶液中加入等体积的 $(NH_4)_2Hg(SCN)_4$ 饱和溶液，振荡试管，生成白色 $ZnHg(SCN)_4$ 沉淀，表示有 Zn^{2+}。如果现象不明显，用玻璃棒摩擦试管壁，以破坏过饱和溶液。

写出各步的反应式。

思考题

(1) 在 Ca^{2+}、Ba^{2+} 混合液中，为什么可以在 Ca^{2+} 存在下用 K_2CrO_4 检出 Ba^{2+}？

(2) 可能存在 Mg^{2+}、Ca^{2+}、Ba^{2+} 的试液与不含 CO_3^{2-} 的 NaOH 溶液反应，没有沉淀出现，是否能否定 Mg^{2+} 的存在？如果产生白色沉淀，是否能肯定 Mg^{2+} 的存在，为什么？

(3) 分离 Fe^{3+}、Co^{2+}、Ni^{2+}、Mn^{2+}、Cr^{3+}、Zn^{2+} 时，为什么加过量的 NaOH，同时还加 H_2O_2？如果碱加得过多，或多余的 H_2O_2 没有分解完，有何缺点？

(4) 为了使 $FeO(OH)$、$CoO(OH)$、$Ni(OH)_2$、$MnO(OH)_2$ 等沉淀溶解，除了加 H_2SO_4 外，还要加 H_2O_2，为什么？

(5) 检出 CrO_4^{2-}、Zn^{2+} 时，为什么要先用 HAc 酸化溶液？

实验 23　未知阴离子的分离与鉴定

实验目的

(1) 熟悉常见阴离子性质的初步检验步骤。

(2) 设计非金属阴离子混合溶液分离与鉴定的方法。

实验原理

形成阴离子的元素大部分是非金属元素，如 S^{2-}、Cl^-、NO_3^- 和 SO_4^{2-} 等。有一些金属元素也可以以复杂阴离子的形式存在，如 VO_3^-、CrO_4^{2-}、$Al(OH)_4^-$ 等，但是一般都在阳离子分析中鉴定。常见的重要阴离子有 Cl^-、Br^-、I^-、S^{2-}、SO_3^{2-}、$S_2O_3^{2-}$、SO_4^{2-}、NO_3^-、NO_2^-、PO_4^{3-}、CO_3^{2-} 共 11 种,这里主要介绍它们的分离与鉴定的一般方法。

许多阴离子只在碱性溶液中存在或共存,一旦溶液被酸化,它们就会分解或相互间发生反应。酸性条件下,NO_2^-、SO_3^{2-}、$S_2O_3^{2-}$、S^{2-}、CO_3^{2-} 易分解;氧化性离子 NO_3^-、NO_2^-、SO_3^{2-} 可与还原性离子 I^-、SO_3^{2-}、$S_2O_3^{2-}$、S^{2-} 发生氧化还原反应。还有些离子易被空气氧化,如 NO_2^-、SO_3^{2-}、S^{2-} 易被空气氧化成 NO_3^-、SO_4^{2-} 和 S 等。分析不当也容易造成错误。

由于阴离子间的相互干扰较少,实际上许多离子共存的机会也较少,因此大多数阴离子分析一般都采用分别分析的方法,如 S^{2-}、SO_3^{2-}、$S_2O_3^{2-}$;只有少数相互有干扰的离子才采用系统分析法,如 Cl^-、Br^-、I^-。常见阴离子的鉴定反应列于表 4.2。

表 4.2 常见阴离子的主要鉴定反应

离子	试剂	鉴定反应	介质条件	主要干扰离子
Cl^-	$AgNO_3$	$Cl^- + Ag^+ =\!=\!= AgCl\downarrow$(白色) AgCl 溶于过量氨水或 $(NH_4)_2CO_3$ 中,用 HNO_3 酸化,沉淀重新析出	酸性介质	
Br^-	Cl_2 水 CCl_4(或苯)	$2Br^- + Cl_2 =\!=\!= Br_2 + 2Cl^-$ 析出的 Br_2 溶于 CCl_4(或苯)溶剂中,呈橙黄(或橙红)色	中性或酸性介质	
I^-	Cl_2 水 CCl_4(或苯)	$2I^- + Cl_2 =\!=\!= I_2 + 2Cl^-$ 析出的 I_2 溶于 CCl_4(或苯)溶剂中,呈紫红色	中性或酸性介质	
SO_4^{2-}	$BaCl_2$	$SO_4^{2-} + Ba^{2+} =\!=\!= BaSO_4\downarrow$(白色)	酸性介质	
SO_3^{2-}	稀 HCl	$SO_3^{2-} + 2H^+ =\!=\!= SO_2\uparrow + H_2O$ SO_2 的检验: (1) SO_2 可使稀 $KMnO_4$ 还原而褪色 (2) SO_2 可将 I_2 还原为 I^-,使淀粉-I_2 溶液褪色 (3) SO_2 可使品红溶液褪色	酸性介质	$S_2O_3^{2-}$、S^{2-} 存在干扰鉴定
	$Na_2[Fe(CN)_5NO]$ $ZnSO_4$ $K_4[Fe(CN)_6]$	生成红色沉淀	中性介质	S^{2-} 与 $Na_2[Fe(CN)_5NO]$ 生成紫红色配合物,干扰 SO_3^{2-} 鉴定
$S_2O_3^{2-}$	稀 HCl	$S_2O_3^{2-} + 2H^+ =\!=\!= SO_2\uparrow + S + H_2O$ 反应中因有硫析出,而使溶液变浑浊	酸性介质	SO_3^{2-}、S^{2-} 存在干扰鉴定
	$AgNO_3$	$S_2O_3^{2-} + 2Ag^+ =\!=\!= Ag_2S_2O_3\downarrow$(白色) $Ag_2S_2O_3$ 沉淀不稳定,生成后立即发生水解,且这种水解常伴随着显著的颜色变化:由白→黄→棕,最后变为黑色物质 $Ag_2S_2O_3 + H_2O =\!=\!= Ag_2S\downarrow$(黑色)$+ 2H^+ + SO_4^{2-}$	中性介质	S^{2-} 存在干扰鉴定
S^{2-}	稀 HCl	$2H^+ + S^{2-} =\!=\!= H_2S$ H_2S 气体的检验: (1) 根据 H_2S 气体的特殊气味 (2) H_2S 气体可使蘸有 $Pb(NO_3)_2$ 或 $Pb(Ac)_2$ 的试纸变黑	酸性介质	SO_3^{2-}、$S_2O_3^{2-}$ 存在干扰鉴定
	$Na_2[Fe(CN)_5NO]$	$S^{2-} + [Fe(CN)_5NO]^{2-} =\!=\!= [Fe(CN)_5NOS]^{4-}$(紫红色)	碱性介质	

续表

离子	试剂	鉴定反应	介质条件	主要干扰离子
NO_2^-	对氨基苯磺酸和 α-萘胺	$NO_2^- + H_2N-C_{10}H_7 + H_2N-C_6H_4-SO_3H + H^+ \rightarrow$ $H_2N-C_{10}H_6-N=N-C_6H_4-SO_3H + 2H_2O$	中性或乙酸介质	MnO_4^- 等强氧化剂存在干扰鉴定
NO_3^-	$FeSO_4$、浓 H_2SO_4	$NO_3^- + 3Fe^{2+} + 4H^+ = 3Fe^{3+} + NO + 2H_2O$ $Fe^{2+} + NO = [Fe(NO)]^{2+}$(棕色) 在混合液与浓硫酸分层处形成棕色环	酸性介质	NO_2^- 有同样的反应,干扰鉴定
CO_3^{2-}	稀 HCl	$CO_3^{2-} + 2H^+ = CO_2\uparrow + H_2O$ $CO_2 + 2OH^- + Ba^{2+} = BaCO_3\downarrow + H_2O$(白色)	酸性介质	
PO_4^{3-}	$AgNO_3$	$PO_4^{3-} + 3Ag^+ = Ag_3PO_4\downarrow$	中性或弱酸性介质	CrO_4^{2-}、AsO_4^{3-}、AsO_3^{3-}、I^-、$S_2O_3^{2-}$ 等离子能与 Ag^+ 生成有色沉淀,干扰鉴定
PO_4^{3-}	$(NH_4)_2MoO_4$ HNO_3	$PO_4^{3-} + 3NH_4^+ + 12MoO_4^{2-} + 24H^+ =$ $(NH_4)_3PO_4 \cdot 12MoO_3 \cdot 6H_2O\downarrow + 6H_2O$	HNO_3 介质	(1) SO_3^{2-}、$S_2O_3^{2-}$、S^{2-}、I^-、Sn^{2+} 等还原性物质存在时易将 $(NH_4)_2MoO_4$ 还原为低价钼的化合物钼蓝,而使溶液呈深蓝色,严重干扰 PO_4^{3-} 的检出 (2) SiO_3^{2-}、AsO_4^{3-} 与钼酸铵试剂也能形成相似的黄色沉淀,干扰鉴定 (3) 大量 Cl^- 存在时,可与 $Mo(VI)$ 形成配位化合物而降低反应的灵敏度
SiO_3^{2-}	饱和 NH_4Cl	$SiO_3^{2-} + 2NH_4^+ + 2H_2O \xrightarrow{\triangle} H_2SiO_3\downarrow$(白色胶状沉淀) $+ 2NH_3\uparrow + H_2O$	碱性介质	
F^-	浓 H_2SO_4	$CaF_2 + H_2SO_4 \xrightarrow{\triangle} 2HF\uparrow + CaSO_4$ 放出的 HF 与硅酸盐或 SiO_2 作用,则生成 SiF_4 气体,当 SiF_4 与水作用时,立即分解并转化为不溶性的硅酸沉淀而使水变浑 $Na_2SiO_3 \cdot CaSiO_3 \cdot 4SiO_2$(玻璃) $+ 28HF = 4SiF_4\uparrow +$ $Na_2SiF_6 + CaSiF_6 + 14H_2O$ $SiF_4 + 4H_2O = H_4SiO_4\downarrow + 4HF$	酸性介质	

对未知样品的分析,并不是用这 11 种阴离子的鉴定反应逐一检出,而应预先做一些初步

实验，以消除某些离子存在的可能性，简化分析步骤。预备实验大致有：

(1) 与稀硫酸作用：在试样中加稀硫酸并加热。产生气泡，表示可能含有 CO_3^{2-}、S^{2-}、SO_3^{2-}、$S_2O_3^{2-}$、NO_2^-。如试样是溶液，所含离子的浓度又不高时，就不一定观察到明显的气泡。

(2) 与 $BaCl_2$ 溶液作用：在中性或弱碱性试液中滴加 $BaCl_2$ 溶液，生成白色沉淀，表示可能存在 SO_4^{2-}、CO_3^{2-}、SO_3^{2-}、PO_4^{3-}、$S_2O_3^{2-}$；若没有沉淀生成，表示 SO_4^{2-}、CO_3^{2-}、SO_3^{2-}、PO_4^{3-} 不存在，$S_2O_3^{2-}$ 则不能肯定其是否存在，因为 $S_2O_3^{2-}$ 浓度大时（大于 4.5 g/L）才生成沉淀。

(3) 与 $AgNO_3$ 和 HNO_3 作用：若有沉淀，表示可能有 S^{2-}、$S_2O_3^{2-}$、Cl^-、Br^-、I^-。由沉淀的颜色可做出进一步判断。

(4) 还原性阴离子的检验：强还原性阴离子 S^{2-}、SO_3^{2-}、$S_2O_3^{2-}$ 可以被碘氧化，因此根据加入碘-淀粉溶液后是否褪色，可判断这些阴离子是否存在。若用强氧化剂 $KMnO_4$ 溶液试验，Cl^-、Br^-、I^-、NO_2^- 也可与之反应。

(5) 氧化性阴离子的检验：在酸化的试液中加 KI 溶液和 CCl_4，若振荡后 CCl_4 层显紫色，则有氧化性阴离子。在我们讨论的阴离子中，只有 NO_2^- 有此反应。

在鉴定某些离子相互干扰的情况下，也需要采取一些适当的分离步骤。例如，Cl^-、Br^-、I^- 共存时，可按下述方法进行分离鉴定：

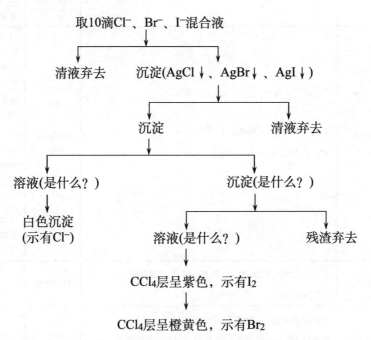

如果溶液中同时存在 S^{2-}、SO_3^{2-} 和 $S_2O_3^{2-}$，需先除去 S^{2-}，因为 S^{2-} 的存在干扰其他两个离子的鉴定。除去的方法是向混合液中加入 $PbCO_3$ 固体，使 $PbCO_3$ 转化为溶解度更小的 PbS 沉淀，离心分离后，再分别鉴定清液中的 SO_3^{2-}、$S_2O_3^{2-}$。

NO_2^- 的存在也干扰 NO_3^- 的鉴定。用棕色环法鉴定 NO_3^- 时需先将 NO_2^- 除去。方法是向混合液中加入 NH_4Cl 饱和溶液并加热，反应如下：

$$NH_4^+ + NO_2^- = N_2\uparrow + 2H_2O$$

实验用品

仪器与材料：试管、离心试管、离心机、点滴板、酒精灯、$Pb(Ac)_2$试纸。

试剂：$PbCO_3(s)$、$FeSO_4(s)$、H_2SO_4(2 mol/L、浓)、HNO_3(2 mol/L、6 mol/L、浓)、HCl(6 mol/L)、HAc(6 mol/L)、氨水(6 mol/L)、Na_2S(0.1 mol/L)、Na_2SO_3(0.1 mol/L)、$Na_2S_2O_3$(0.1 mol/L)、Na_3PO_4(0.1 mol/L)、NaCl(0.1 mol/L)、KBr(0.1 mol/L)、KI(0.1 mol/L)、KNO_3(0.1 mol/L)、$NaNO_2$(0.1 mol/L)、Na_2SO_4(0.1 mol/L)、$AgNO_3$(0.1 mol/L)、$BaCl_2$(1 mol/L)、$ZnSO_4$(饱和)、氯水、$(NH_4)_2MoO_4$(饱和)、$Na_2[Fe(CN)_5NO]$(1%)、$K_4[Fe(CN)_6]$(0.1 mol/L)、CCl_4、$(NH_4)_2CO_3$(12%)、对氨基苯磺酸、α-萘胺。

实验步骤

1. 特定阴离子的鉴定

1) S^{2-}的鉴定

(1) 在点滴板上滴入 0.1 mol/L Na_2S溶液，然后滴入 1% $Na_2[Fe(CN)_5NO]$溶液，出现紫红色，表示有 S^{2-}存在。

(2) 在试管中加入 0.5 mL 0.1 mol/L Na_2S溶液，再加入 0.5 mL 6 mol/L HCl。将湿润的 $Pb(Ac)_2$试纸悬于试管口，微热，试纸变黑，表示有 S^{2-}存在。

2) SO_3^{2-}的鉴定

在点滴板上滴加 $ZnSO_4$饱和溶液，然后加入 1 滴 0.1 mol/L $K_4[Fe(CN)_6]$溶液和 1% $Na_2[Fe(CN)_5NO]$溶液，用氨水调至中性，再滴加 0.1 mol/L Na_2SO_3溶液，生成红色沉淀，表示有 SO_3^{2-}存在。

3) $S_2O_3^{2-}$的鉴定

在点滴板上滴加一滴 0.1 mol/L $Na_2S_2O_3$溶液，然后加入 2 滴 0.1 mol/L $AgNO_3$溶液，沉淀颜色由白→黄→棕→黑，表示有 $S_2O_3^{2-}$存在。

4) SO_4^{2-}的鉴定

在离心试管中加 3~4 滴 0.1 mol/L Na_2SO_4溶液，加入 1 滴 1 mol/L $BaCl_2$溶液，离心分离，在沉淀中加入数滴 6 mol/L HCl，沉淀不溶解则表示有 SO_4^{2-}存在。

5) PO_4^{3-}的鉴定

取少量 0.1 mol/L Na_3PO_4溶液，加入 10 滴浓硝酸和 20 滴钼酸铵饱和溶液，微热至 40~50℃，即可观察到黄色沉淀生成。

6) Cl^-的鉴定

取 2 滴 0.1 mol/L NaCl 溶液，加入 1 滴 2 mol/L HNO_3和 2 滴 0.1 mol/L $AgNO_3$溶液，观察沉淀的颜色。离心分离后，弃去清液，向沉淀中加入数滴 6 mol/L 氨水，沉淀溶解，然后加入 6 mol/L HNO_3酸化，又有白色沉淀析出，此法可鉴定 Cl^-。

7) Br^-的鉴定

取 2 滴 0.1 mol/L KBr 溶液，加入 1 滴 2 mol/L H_2SO_4和 5~6 滴 CCl_4，然后逐滴加入新配制的氯水，边加边摇，若 CCl_4层出现棕色至黄色，表示有 Br^-存在。

8) I^-的鉴定

同 Br^- 的鉴定。若 CCl_4 层出现紫色，表示有 I^- 存在。若加入过量氯水，紫色又褪去，这是因为碘生成 IO_3^- 而重新返回水相。

9) NO_3^- 的鉴定：棕色环实验

取 1 mL 0.1 mol/L KNO_3，加入 1～2 小粒 $FeSO_4$ 晶体，振荡，溶解后，沿试管壁滴加 5～10 滴浓硫酸，观察浓 H_2SO_4 和溶液两个液面交界处有无棕色环出现。

10) NO_2^- 的鉴定

取 1 滴 0.1 mol/L $NaNO_2$ 溶液，滴加 6 mol/L HAc 酸化，再加对氨基苯磺酸和 α-萘胺溶液各 1 滴，溶液立即显红色，表示有 NO_2^- 存在。

该反应适用于检验少量的 NO_2^-。若 NO_2^- 浓度过大，溶液的粉红色很快褪去，生成黄色溶液或褐色沉淀。

11) CO_3^{2-} 的鉴定

一般用 $Ba(OH)_2$ 气体瓶检出 CO_3^{2-}，即 CO_2 气体可以使 $Ba(OH)_2$ 水溶液变浑浊。

2. 已知阴离子混合液的分离与鉴定

(1) 在试管中加入 0.1 mol/L NaCl、KBr、KI 溶液各 2 滴，按照实验原理中给出的方法，对 Cl^-、Br^-、I^- 进行分离并鉴定。

(2) 在试管中加入 0.1 mol/L Na_2S、Na_2SO_3、$Na_2S_2O_3$ 溶液各 3 滴，按照实验原理中给出的方法，先将 S^{2-} 分离出去，再对各离子分别进行鉴定。

3. 未知阴离子混合液分析

领取一份含有三种阴离子的未知溶液（其中可能含有 NO_3^-、NO_2^-、S^{2-}、Cl^-、SO_4^{2-}、$S_2O_3^{2-}$）。按照实验原理中给出的方法进行初步检验后，再自行设计分离和鉴定实验方案，进行实验。

根据初步实验结果，判断哪些阴离子可能存在，说明原因并写出反应式。然后画出分离鉴定示意图，得出结论。

思考题

(1) 选用一种试剂区别以下五种溶液：NaCl、$NaNO_2$、Na_2S、$Na_2S_2O_3$、Na_2HPO_4。

(2) 某碱性无色阴离子混合液，被 H_2SO_4 酸化后变浑浊，混合液中可能含有哪些阴离子？

(3) NO_3^- 与 NO_2^- 共存时，其鉴定反应如何进行？

第5章 综合与设计实验

实验24 一种钴(Ⅲ)配合物的制备与组成测定

实验目的

(1) 掌握制备金属配合物常用的方法——水溶液中的取代反应和氧化还原反应。
(2) 对配合物组成进行初步推断。
(3) 学习使用电导率仪测定配合物组成的原理和方法。

实验原理

1. 合成

借助水溶液取代反应制备金属配合物,其实质是利用一种适当的配体在水溶液中取代金属盐中水合配离子的配位水分子。氧化还原反应是将其他氧化态的金属配合物在配体存在下经过适当的氧化或还原过程制备目标价态的金属配合物。

根据标准电极电势可知,在酸性溶液中,二价钴盐比三价钴盐稳定;相反,在生成配合物后,三价钴又比二价钴稳定。因此,常采用空气或过氧化氢氧化钴(Ⅱ)配合物来制备钴(Ⅲ)配合物。常见的钴(Ⅲ)配合物有:$[Co(NH_3)_6]^{3+}$(橙色)、$[Co(NH_3)_5H_2O]^{3+}$(粉红色)、$[Co(NH_3)_5Cl]^{2+}$(紫红色)、$[Co(NH_3)_4CO_3]^+$(紫红色)、$[Co(NH_3)_3(NO_2)_3]$(黄色)、$[Co(CN)_6]^{3-}$(紫色)、$[Co(NO_2)_6]^{3+}$(黄色)等。

2. 组成分析

用化学方法分析确定某配合物的组成,首先确定配合物的外界,然后将配离子破坏后确定其内界组成。配离子的稳定性受很多因素影响,如加热和改变溶液的酸碱性。本实验先用定性、半定量以及估量的分析方法初步推断配合物的化学式,然后借助电导率仪测定确定浓度的配合物溶液的电导率,通过与已知电解质溶液的电导率对比来确定该配合物化学式中含有几个离子,进一步确定该化学式。

游离的 Co(Ⅱ)与硫氰化钾作用生成蓝色配合物$[Co(NCS)_4]^{2-}$,以此判断 Co(Ⅱ)的存在。因为$[Co(NCS)_4]^{2-}$在水中解离度大,故常用硫氰化钾浓溶液或粉末来制备,并加入戊醇和乙醚以使配合物稳定。

游离的 NH_4^+ 可用奈斯勒试剂($K_2[HgI_4]$ + KOH)检验。

电解质溶液的导电性可以用电导(G)表示:

$$G = \gamma/K$$

式中,γ 为电导率,常用单位为 S/cm;K 为电导池常数,单位为 cm^{-1}。测定已知电导率的电解质溶液的电导,然后根据上式即可求得电导池常数 K。一般采用 KCl 溶液作为标准电导溶液。

实验用品

仪器与材料：电子台秤、烧杯、锥形瓶、量筒、漏斗、铁架台、酒精灯、试管、滴管、试管夹、铁圈、温度计、电导率仪、pH 试纸、滤纸、药匙、石棉网。

试剂：氯化铵、六水合氯化钴、硫氰化钾、浓氨水、浓硝酸、浓盐酸、6 mol/L HCl 溶液、30% H_2O_2、2 mol/L $AgNO_3$ 溶液、0.5 mol/L $SnCl_2$ 溶液、奈斯勒试剂、乙醚、戊醇。

实验步骤

1. 钴(Ⅲ)配合物的制备

在 100 mL 锥形瓶中加入 1.0 g 氯化铵和 6 mL 浓氨水，振荡锥形瓶使固体溶解。将 2.0 g 氯化钴粉末分 4~5 次加入并不断摇动，加完后继续摇动使溶液呈棕褐色稀浆。在摇动下滴加 3~5 mL 30% H_2O_2，加完后继续摇动 10 min 使固体完全溶解。当溶液停止起泡时，缓慢加入 6 mL 浓盐酸并摇动，转入不超过 85℃ 的水浴中继续摇动加热 10~15 min。在室温下待反应液完全冷却后减压过滤。先用 5 mL 冷水分 2~3 次洗涤沉淀，再用 5 mL 6 mol/L 盐酸洗涤。所得固体产物在 105℃ 下烘干并称量。

2. 配合物组成的初步推断

(1)将 0.2 g 产物加 30 mL 蒸馏水在小烧杯中搅拌溶解，测溶液 pH 以检验其酸碱性。

(2)取 15 mL 上述溶液，慢慢滴加 2 mol/L $AgNO_3$ 溶液直至不再有沉淀生成(如何判断沉淀完全？)。过滤除去沉淀，向滤液中加入 1~2 mL 浓硝酸并搅拌，再往溶液中滴加 $AgNO_3$ 溶液，观察是否有沉淀生成。若有，收集沉淀并与前次沉淀比较沉淀量的多少。

(3)取 2 mL 产物溶液于试管中，加 5 滴 0.5 mol/L $SnCl_2$ 溶液(为什么？)，振荡后再加入少量 KSCN 粉末，振荡后依次加入戊醇和乙醚各 1 mL，振荡后观察有机层的颜色(为什么？)。

(4)取 2 mL 产物溶液于试管中，再加 2 滴奈斯勒试剂，观察现象。

(5)将(1)中剩余的溶液加热，使其完全变成棕黑色(判断方法？)，将其冷却后用 pH 试纸检验溶液的酸碱性，然后用双层滤纸过滤。取所得滤液重复实验(3)和(4)，观察现象与原来的有什么不同。依据以上结果初步推测化合物的化学式。

(6)按照初步推断的化学式配制 20~30 mL 浓度为 0.01 mol/L 的该配合物溶液，用电导率仪测量其电导率。稀释 10 倍后再次测量溶液的电导率，与表 5.1 对比以确定其化学式中所含离子数。

表 5.1 几种电解质的电导率

电解质	类型(离子数)	电导率/S	
		0.01 mol/L	0.001 mol/L
KCl	1-1 型(2)	1230	133
$BaCl_2$	1-2 型(3)	2150	250
$K_3[Fe(CN)_6]$	1-3 型(4)	3400	420

思考题

(1) 将氯化钴加入氯化铵和浓氨水的混合液中，可发生什么反应？生成什么配合物？氯化铵除了参与反应，还起什么作用？

(2) 氯化钴为什么要分数次加入？

(3) 实验中加过氧化氢起什么作用？如不用过氧化氢，还可以用哪些物质，用这些物质有什么缺点？实验中加入浓盐酸的作用是什么？

(4) 要使本实验制备产物的产率高，哪些步骤是比较关键的？为什么？

实验 25　硫酸四氨合铜(Ⅱ)大晶体的制备及组成分析

实验目的

(1) 掌握硫酸四氨合铜(Ⅱ)的制备及其组成的测定方法。

(2) 掌握蒸发、结晶、减压过滤等基本操作。

(3) 掌握蒸馏法测定氨的技术。

实验原理

硫酸四氨合铜($[Cu(NH_3)_4]SO_4$)常用作杀虫剂、媒染剂，在碱性镀铜中也常作为电镀液的主要成分，在工业上用途广泛。硫酸四氨合铜为中度稳定的蓝色晶体，常温下在空气中易与水和二氧化碳反应，生成铜的碱式盐，使晶体变成绿色的粉末。

1. 制备

$CuSO_4$ 与浓氨水作用，生成浅蓝色碱式硫酸铜沉淀。氨过量时，最终生成深蓝色硫酸四氨合铜溶液。

$$2CuSO_4 + 2NH_3 \cdot H_2O == Cu_2(OH)_2SO_4 \downarrow (浅蓝色) + (NH_4)_2SO_4$$

$$Cu_2(OH)_2SO_4 + 8NH_3 == 2[Cu(NH_3)_4]^{2+} + SO_4^{2-} + 2OH^-$$

由于硫酸四氨合铜在受热时易分解失氨，其晶体的制备不宜选用蒸发、浓缩等常规方法。利用硫酸四氨合铜在乙醇中的溶解度远小于在水中的溶解度的性质，向其水溶液中加入乙醇可析出深蓝色晶体。

2. 组成分析

1) NH_3 含量的测定

$$Cu(NH_3)_4SO_4 + 2NaOH == CuO \downarrow + 4NH_3 \uparrow + Na_2SO_4 + H_2O$$

$$NH_3 + HCl(过量) == NH_4Cl$$

$$HCl(剩余量) + NaOH == NaCl + H_2O$$

2) SO_4^{2-} 含量的测定

$$SO_4^{2-} + Ba^{2+} == BaSO_4 \downarrow$$

实验用品

仪器与材料：烧杯、锥形瓶、表面皿、分液漏斗、长颈漏斗、酒精灯、布氏漏斗、抽滤瓶、水泵、橡皮塞、洗瓶、药匙。

试剂：$CuSO_4 \cdot 5H_2O$ 晶体、浓氨水、95%乙醇、0.5 mol/L HCl 标准溶液、0.5 mol/L NaOH 标准溶液、2 mol/L NaOH 溶液、6 mol/L HCl 溶液、0.1 mol/L $BaCl_2$ 溶液、0.1%甲基红溶液、0.1 mol/L $AgNO_3$ 溶液。

实验步骤

1. 硫酸四氨合铜（Ⅱ）的制备

称取 5.0 g $CuSO_4 \cdot 5H_2O$ 溶于 7 mL 水中，然后慢慢加入 10 mL 浓氨水并搅拌，沉淀逐渐消失，溶液呈透明的蓝色。沿烧杯壁慢慢滴加 18 mL 95%乙醇，然后盖上表面皿静置。待溶液析出晶体后减压过滤，用 5 mL 体积比 1∶2 的 95%乙醇和浓氨水的混合液洗涤固体产物 2 次。在空气中自然风干 $[Cu(NH_3)_4]SO_4 \cdot H_2O$ 晶体，称量后保存待用。

2. 组成分析

1) NH_3 含量的测定

在锥形瓶中准确加入 30 mL HCl 标准溶液（约 0.5 mol/L），放入冰水浴中冷却。称取 0.30 g 样品放入带支管的 250 mL 烧瓶中，用 80 mL 水溶解。借助橡皮塞将盛有 10 mL 2 mol/L NaOH 溶液的分液漏斗装配在锥形瓶上，并使漏斗下端插入样品溶液液面下 2～3 cm。烧瓶的支管引出导气管通入冰水浴中的 HCl 标准溶液。大火加热样品溶液，同时打开漏斗旋塞让 NaOH 溶液加入样品溶液。当反应液接近沸腾时改用小火，保持微沸状态，蒸馏反应持续 1 h 左右即可将样品所含的氨全部蒸出。蒸馏完毕后，取出插入 HCl 溶液中的导管，用蒸馏水冲洗导管内外，洗涤液收集在氨吸收瓶中，从冰水浴中取出吸收瓶，加 2 滴 0.1%甲基红溶液，用 NaOH 标准溶液（0.5 mol/L）滴定剩余的 HCl 溶液。

NH_3 含量依下式计算：

$$w_{NH_3}/\% = (c_1V_1 - c_2V_2) \times 17.04 / (m_s \times 1000) \times 100$$

式中，c_1 和 V_1 分别为 HCl 标准溶液的浓度和体积；c_2 和 V_2 分别为 NaOH 标准溶液的浓度和体积；m_s 为样品质量。

2) SO_4^{2-} 含量的测定

称取约 0.65 g（含硫量约 90 mg）试样置于 400 mL 烧杯中，加 25 mL 蒸馏水使其溶解，用水稀释至 200 mL。

(1) 沉淀的制备：在上述溶液中加入 2 mL 6 mol/L HCl 溶液，盖上表面皿，加热至近沸。取 30～35 mL 0.1 mol/L $BaCl_2$ 溶液于小烧杯中加热至近沸，然后用滴管将热 $BaCl_2$ 溶液逐滴加入样品溶液中并不断搅拌。当 $BaCl_2$ 溶液即将加完时，向静置后的上清液中加入 1～2 滴 $BaCl_2$ 溶液，观察是否有白色浑浊出现，以检验是否已沉淀完全。盖上表面皿，继续加热陈化约 10 min，然后冷却至室温。

(2) 沉淀的过滤和洗涤：将上清液用倾注法倒入长颈漏斗，用洁净烧杯收集滤液（检查有无沉淀穿滤现象。若有，应重新换滤纸)。用少量蒸馏水洗涤沉淀 3~4 次，然后将沉淀小心地转移至滤纸上。用洗瓶吹洗烧杯内壁，洗涤液并入漏斗中，并用撕下的滤纸角擦拭玻璃棒和烧杯内壁，将滤纸角放入漏斗中。再用少量蒸馏水洗涤滤纸上的沉淀，直至滤液检测不出 Cl^-。

(3) 沉淀的干燥和灼烧：取下滤纸，将沉淀包好，置于已恒量的坩埚中，先用小火烘干炭化，再用大火灼烧至滤纸灰化。然后将坩埚转入马弗炉中，在 800~850℃ 灼烧 30 min。取出坩埚，待红热退去，置于干燥器中，冷却后称量。再重复灼烧 20 min，冷却，取出，称量，直至恒量。根据沉淀($BaSO_4$)的质量计算试样中 SO_4^{2-} 的百分含量。

思考题

(1) 为什么溶液需加热近沸，但不应煮沸？
(2) 为什么 $BaSO_4$ 沉淀的灼烧温度需要控制在 800~850℃？

实验 26 葡萄糖酸锌的制备与质量分析

实验目的

(1) 了解锌元素的生物意义和葡萄糖酸锌的制备方法。
(2) 巩固并综合运用蒸发、浓缩、过滤、重结晶、滴定等实验操作。
(3) 了解葡萄糖酸锌的质量分析方法。

实验原理

锌是人体必需的微量元素之一，成人体内锌含量为 2.0~2.5 g，是脑中含量最多的微量元素。目前已知锌存在于人体 70 种以上的酶系(200 多种酶)中，如呼吸酶、乳酸脱氢酶、碳酸脱氢酶、超氧化物歧化酶、碱性磷酸酶、DNA 和 RNA 聚合酶等，对维持机体的正常生理功能起着重要作用。锌与核酸、蛋白质的合成，与碳水化合物、维生素 A 的代谢以及胰腺、性腺和垂体的活动都有关系，能促进皮肤、骨骼和性器官的正常发育，维持消化和代谢活动。缺锌会导致生长缓慢、皮肤炎、味觉与嗅觉异常、视力衰退、性腺不成熟、胃肠道疾病、免疫功能减退等。补充锌可降低糖尿病与心脏病致病率，增强创伤组织的再生能力，改善食欲和消化机能，对抗感冒，强化免疫系统等。但体内锌过量时可抑制铁的利用，易导致顽固性贫血等疾病。

葡萄糖酸锌是一种目前被广泛采用的补锌药物，具有见效快、吸收率高、副作用小等优点。相同锌剂量的葡萄糖酸锌的生物利用度约为硫酸锌的 1.6 倍。葡萄糖酸锌可以由葡萄糖酸或葡萄糖酸钙直接与氧化锌或锌盐反应制取。本实验采用等物质的量的葡萄糖酸钙与硫酸锌直接反应制取葡萄糖酸锌。

$$[CH_2OH(CHOH)_4COO]_2Ca + ZnSO_4 \Longrightarrow [CH_2OH(CHOH)_4COO]_2Zn + CaSO_4 \downarrow$$

利用硫酸钙的低溶解度，可以过滤除去副产物 $CaSO_4$ 沉淀，滤液经浓缩后加入乙醇可析出无色或白色葡萄糖酸锌结晶。产物无味，易溶于水，极难溶于乙醇。

采用配位滴定法测定产品中目标化合物的含量，用 EDTA 标准溶液在 NH_3-NH_4Cl 弱碱性缓冲溶液中滴定锌离子，据此计算产物中葡萄糖酸锌的含量。

实验用品

仪器与材料：抽滤瓶、布氏漏斗、水泵、酸式滴定管、酒精灯、移液管、蒸发皿、容量瓶、量筒、烧杯、锥形瓶、温度计、电子台秤、分析天平。

试剂：葡萄糖酸钙、$ZnSO_4 \cdot 7H_2O$、95%乙醇、25% $BaCl_2$ 溶液、2 mol/L HCl 溶液、硫酸钾标准溶液、0.05 mol/L EDTA 标准溶液、NH_3-NH_4Cl 缓冲溶液(pH=10.0)、铬黑 T 指示剂。

实验步骤

1. 葡萄糖酸锌的制备

量取 12 mL 蒸馏水于 100 mL 烧杯中，加入 3.3 g $ZnSO_4 \cdot 7H_2O$ 搅拌使其溶解，于恒温水浴加热至 85~90℃，用 12 mL 蒸馏水溶解 4.5 g 葡萄糖酸钙得到白色糊状液体。将其缓慢加入 $ZnSO_4$ 溶液(用时 10~15 min)并不断搅拌，加完后保持恒温反应 20 min。

用双层滤纸趁热抽滤反应液，用 2~3 mL 水洗涤烧杯和沉淀，滤液转移至蒸发皿中并在沸水浴中浓缩至黏稠状(体积约为 10 mL，如浓缩液有沉淀，需过滤掉)。冷却滤液接近室温后转移至冰水浴，缓慢加入 20 mL 95%乙醇并不断搅拌，此时有大量白色胶状葡萄糖酸锌析出。静置后用倾析法除去上清液。再次向胶状沉淀中加入 20 mL 95%乙醇(可分两批加入)，在冰水浴中用玻璃棒持续搅拌胶状沉淀使其逐渐转变硬，并挤压使其分散成白色粉末状颗粒(尽量使颗粒小、松散且不黏)。抽滤并用 3~5 mL 乙醇洗涤产物一次。用滤纸进一步吸干产品表面残余的溶剂，称量，备用。

2. 硫酸盐的检查

称取 0.5 g 葡萄糖酸锌，用 20 mL 水溶解后加入 2 mL 2 mol/L 盐酸溶液，转移至 25 mL 比色管中即得供试溶液。另取 2.5 mL 硫酸钾标准溶液，加 17 mL 水和 2 mL 2 mol/L 盐酸溶液，转移至 25 mL 比色管中即得对照溶液(0.05%)。分别向供试溶液和对照溶液中加入 2 mL 25%氯化钡溶液，用水稀释至 25 mL，充分混匀，放置 10 min 后，在黑色背景下从比色管上方向下观察和比较浑浊程度。

3. 锌含量的测定

用分析天平准确称取 0.7 g 产品，加 50 mL 水完全溶解，然后加 5 mL NH_3-NH_4Cl 缓冲液(pH = 10.0)与数滴铬黑 T 指示剂(或约 15 小粒铬黑 T 固体)。然后用 0.05 mol/L EDTA 标准溶液滴定，溶液自紫红色刚好转变为蓝色为终点，记录所用 EDTA 标准溶液的体积。平行测定三份，按下式计算样品中锌的含量：

$$w_{Zn}/\% = \frac{c_{EDTA} \times V_{EDTA} \times 65}{m \times 1000} \times 100$$

式中，c_{EDTA} 和 V_{EDTA} 分别为滴定剂的浓度和消耗的体积；m 为样品的质量(g)。

数据记录与处理

(1)硫酸盐检查：①现象描述；②检查结论。

(2) 葡萄糖酸锌中锌含量的测定(表 5.2)。

表 5.2 葡萄糖酸锌中锌含量的测定

实验编号	1	2	3
m_1(称量瓶+葡萄糖酸锌)/g			
m_2(称量瓶+剩余葡萄糖酸锌)/g			
m(葡萄糖酸锌)/g			
$V_{始}$(EDTA)/mL			
$V_{终}$(EDTA)/mL			
ΔV(EDTA)/mL			
w_{Zn}/%			
w_{Zn}/%(平均)			

注意事项

硫酸钾标准溶液的配制：称取 0.181 g 硫酸钾置于 1 L 容量瓶中，加适量水使其溶解并稀释至刻度，摇匀，即得硫酸钾标准溶液(1 mL 相当于 100 μg SO_4^{2-})。

思考题

(1) 如果选用葡萄糖酸为原料，以下四种含锌化合物应选择哪种？为什么？
①ZnO；②$ZnCl_2$；③$ZnCO_3$；④$Zn(CH_3COO)_2$。
(2) 若葡萄糖酸锌含量的测定结果不符合规定，可能由哪些原因引起？
(3) 在沉淀与结晶葡萄糖酸锌时都加入 95%乙醇，其作用是什么？
(4) 为什么葡萄糖酸锌的制备必须在热水浴中进行？

实验 27　纳米二氧化钛的制备和光催化性能

实验目的

(1) 了解纳米 TiO_2 的制备和光催化性能原理。
(2) 掌握控制水解法制备 TiO_2 纳米光催化剂、光催化降解有机废水的方法。
(3) 学习使用磁力搅拌器、马弗炉、离心机、超声波清洗机、光化学反应装置和紫外-可见分光光度计等。

实验原理

1. 半导体光催化技术和二氧化钛光催化剂

环境污染制约社会的可持续发展，威胁人们的健康。医药中间体、有机染料等有机污染物通常具有较稳定的化学结构、较高的生物毒性且容易发生迁移，仅利用微生物分解难以有效去除，而吸附、萃取等物理方法无法从根本上达到无害化处理。半导体纳米粒子的光催化性能可以将有机物氧化降解为小分子和 CO_2、H_2O 等无毒物质，因而半导体光催化剂的制备及在有机废水治理上的应用受到广泛关注和深入研究。

二氧化钛(TiO_2，俗称钛白粉)是一种重要的化工产品。纳米 TiO_2 具有比表面积大、光催化活性好、耐光腐蚀、无环境毒性等优良性能，是迄今为止研究最深入、最接近商业化应用的半导体光催化剂。TiO_2 有三种晶相结构，分别是锐钛矿相、金红石相和板钛矿相(亚稳相)，其应用研究主要围绕锐钛矿相和金红石相，二者均属于四方晶系，一般认为金红石相(禁带宽度约 3.0 eV)为热力学稳定相，锐钛矿相为动力学稳定相，比金红石相具有更大的禁带宽度(3.2 eV)和更强的催化活性。无定形水合二氧化钛一般在约 300℃开始转化为锐钛矿结构 TiO_2，约 550℃时开始向金红石型 TiO_2 转变；进一步升高温度，金红石型的比例增大，高于 700℃时锐钛矿结构则几乎全转化为金红石型。

纳米 TiO_2 的光催化原理简述如下：半导体充满电子的价带和空的导带之间存在一个禁带。当辐照光的光子能量大于禁带宽度的能量(锐钛矿：$\lambda_{max} \leqslant 387$ nm；金红石相：$\lambda_{max} \leqslant 410$ nm)时，价带电子被激发后跃迁到导带，形成的电子(e^-)和空穴(h^+)会分别向 TiO_2 粒子的表面迁移。在纳米 TiO_2 的水悬浮液中，h^+ 被溶液中的 H_2O、OH^- 俘获，生成反应活性高、氧化能力极强的羟基自由基($\cdot OH$)，e^- 被吸附在 TiO_2 表面的 O_2 俘获，生成超氧阴离子自由基($\cdot O_2^-$)，进一步反应可产生 $\cdot OOH$ 等氧化性自由基。这些氧化性自由基能将有机分子氧化并逐渐降解，最终矿化产物为 CO_2、H_2O 等无机物。因此，光催化反应的实质就是在纳米半导体中把吸收的光能转化为化学能、实现有机物降解的过程。光催化过程中氧化性物种生成的反应如下：

电子-空穴对的生成

$$TiO_2 + h\nu \longrightarrow TiO_2 + h^+ + e^-$$

电子-空穴对的复合

$$h^+ + e^- \longrightarrow 复合 + 能量$$

羟基自由基的生成

$$H_2O + h^+ \longrightarrow \cdot OH + H^+$$

$$OH^- + h^+ \longrightarrow \cdot OH$$

氧化性物种的相互转变

$$O_2 + e^- \longrightarrow \cdot O_2^-$$

$$H_2O + \cdot O_2^- \longrightarrow \cdot OOH + OH^-$$

$$2\cdot OOH \longrightarrow H_2O_2 + O_2$$

$$\cdot OOH + H_2O + e^- \longrightarrow H_2O_2 + OH^-$$

$$H_2O_2 + e^- \longrightarrow \cdot OH + OH^-$$

2. 纳米 TiO_2 的制备方法和性能影响因素

纳米 TiO_2 的制备通常包括固相法(如粉碎法)、气相法(如化学气相沉积法)和液相法(溶胶-凝胶法、控制水解法等)。控制水解法也是通过将钛酸酯或易水解的 Ti(Ⅳ)盐在水或醇/水溶液中缓慢水解获得水合二氧化钛，经高温煅烧后获得高结晶度的纳米 TiO_2。控制水解法的反应时间短、操作简便且经济性好。钛酸酯的水解和缩合反应如下：

$$—Ti—OR + H_2O \longrightarrow —Ti—OH + ROH$$

$$—Ti—OH + RO—Ti— \longrightarrow —Ti—O—Ti— + ROH$$

然而，纳米 TiO_2 的光催化反应需要借助紫外光实现，而到达地面的日光中紫外光所占比例极少(不足 5%)。人造光源的使用提高了光催化反应的成本，不利于商业化应用。采用金属或非金属元素掺杂、半导体复合、染料敏化和贵金属复合等策略可以制备可见光响应的 TiO_2 光催化剂或其复合材料，为利用清洁的太阳能光催化降解有机污染物提供了可能性。

当掺杂适量浓度的金属离子(如 Fe^{3+})时，TiO_2 晶体内部的电荷缺陷密度增大，同时 Fe^{3+} 形成的杂质能级处于 TiO_2 禁带之中，使得催化剂的吸收带从紫外区红移到可见光区(400 nm 以上)且展宽，能更有效地利用可见光。通过调节掺杂浓度，获得适度的缺陷密度，将有助于提高 TiO_2 的光催化活性。影响光催化反应效率的因素包括催化剂(尺寸、晶体结构、形貌、比表面积、催化剂表面羟基数量等)、光反应条件(辐照光波长、强度)和反应体系(初始污染物浓度、pH、外加催化剂或助剂、气氛等)。

实验用品

仪器与材料：磁力搅拌器、鼓风干燥箱、马弗炉、高速离心机、超声波清洗机、光化学反应器、紫外-可见分光光度计、循环水泵、三口烧瓶、瓷坩埚、量筒、烧杯、恒压漏斗、布氏漏斗、抽滤瓶。

试剂：钛酸四丁酯、九水合硝酸铁、无水乙醇、冰醋酸、0.1 mol/L 硝酸、罗丹明 B。

实验步骤

1. 纳米 TiO_2 的制备

将装有 15 mL 蒸馏水的 250 mL 三口烧瓶安装在磁力搅拌器上。用干燥的量筒取 3 mL 钛酸四丁酯，溶解在 30 mL 无水乙醇中(所用烧杯也需干燥)，将钛酸四丁酯溶液用漏斗缓慢地滴入三口烧瓶中并快速搅拌(12~15 min 滴完)。继续搅拌 10 min 后滴加 0.5 mL 冰醋酸并继续搅拌 10 min。将所得白色沉淀减压过滤，用 8 mL 无水乙醇洗涤一次，将滤饼晾晒至乙醇挥发完全后放入鼓风干燥箱，80℃下烘干。所得白色固体研磨成粉后于 500℃下煅烧 2 h(升温速率：10℃/min)，得到纳米 TiO_2。

Fe 掺杂纳米 TiO_2 的制备：改用 15 mL 硝酸铁溶液代替 15 mL 蒸馏水，其他过程同上。硝酸铁溶液配制方法为：用 2 mL 0.1 mol/L 硝酸溶解一定量的九水合硝酸铁(用量可以参考表 5.3)，溶液滴入装有 13 mL 蒸馏水的 250 mL 三口烧瓶中并搅拌均匀。

表 5.3 制备 Fe 掺杂纳米 TiO_2 的九水合硝酸铁用量

Fe^{3+}掺杂百分数/%	0.2	0.5	1.0	1.5
九水合硝酸铁的质量/mg	6	15	30	45

2. 纳米 TiO_2 的光催化性能

将 50 mg 自制纳米 TiO_2 和 33 mL 水加入 100 mL 烧杯，超声分散 10 min，加入 17 mL 30 mg/L 罗丹明 B 溶液搅拌均匀，得到的反应液中纳米 TiO_2 和罗丹明 B 的质量浓度分别为 1.0 mg/mL 和 10 mg/L。将烧杯置于磁力搅拌器上避光搅拌 20 min 后取样 5 mL，然后开启光化学反应器的电源，依据实验需要对反应液进行紫外光或可见光辐照，每隔 15 min 关闭光化学反应器，取出 5 mL 反应液，然后再次开启反应器进行反应，总反应时间为 1 h。将每次取出的反应液立即离心分离，吸出 3 mL 上层清液，借助紫外-可见分光光度计测试其在 554 nm 处的吸光度 A。在实验浓度范围内，罗丹明 B 溶液的浓度和吸光度成正比(遵循朗伯-比尔定律)。计算罗丹明 B 的脱色率(D，即罗丹明 B 溶液吸光度的下降百分数)，用以衡量罗丹明 B 的降解程度。绘制脱色率-时间(D-t)曲线，用以评价 TiO_2 纳米光催化剂催化降解罗丹明 B 的性能。

$$D = (A_0 - A)/A_0 \times 100\% = (c_0 - c)/c_0 \times 100\%$$

式中，A_0 和 A、c_0 和 c 分别表示为 0 和 t min 时溶液的吸光度、对应的浓度。

分别对比纯纳米 TiO_2 和不同 Fe 掺杂浓度的纳米 TiO_2 在紫外和可见光下的光催化性能，分析 Fe 掺杂对改善催化剂活性的作用。

注意事项

(1) 取用和盛放钛酸四丁酯溶液的量筒等仪器在使用后需尽快先用 5～8 mL 乙醇荡洗，再用水洗涤。

(2) 减压过滤二氧化钛水合物沉淀时最好使用双层滤纸以防止滤纸破裂。

(3) 光化学反应装置的使用应严格遵守仪器的操作流程，操作时需佩戴防紫外护目镜、手套等防护装备，取样期间需要关闭光化学反应装置。

(4) 本实验制备的纯 TiO_2 和 Fe 掺杂 TiO_2 的 XRD 谱图可以与锐钛矿相(JCPDS 89-4921)和金红石相(JCPDS 89-4920)的 TiO_2 标准图卡进行比对。

(5) 罗丹明 B 的最大吸收峰在 554 nm。

思考题

(1) 为什么取用和盛放钛酸四丁酯或其溶液的仪器需要严格干燥？

(2) 将钛酸四丁酯溶于无水乙醇的目的是什么？制备 Fe 掺杂纳米 TiO_2 时硝酸的作用是什么？

(3) 如何设计实验评判自制纳米 TiO_2 催化剂的性能？

(提示：①P25 是一种商品纳米 TiO_2 光催化剂，被普遍作为参照剂用于对比自制光催化剂的性能；②有机染料罗丹明 B 在紫外光或可见光下会发生不同程度的降解)

参考文献

卞国庆, 纪顺俊. 2007. 综合化学实验. 苏州: 苏州大学出版社
田玉美. 2008. 新大学化学实验. 北京: 科学出版社
徐如人, 庞文琴. 2006. 无机合成与制备化学. 北京: 高等教育出版社
张霞. 2009. 无机化学实验. 北京: 冶金工业出版社

实验 28　七水合硫酸锌的制备

实验目的

(1) 学习通过锌灰提取制备 $ZnSO_4·7H_2O$ 的方法。
(2) 培养学生综合应用无机化学基础理论和基本实验技能解决问题的能力。

实验原理

锌灰是锌冶炼中的工业废渣，中小企业每年都以数百吨计，其中含锌(以 ZnO 计)30%～40%、硅(以 SiO_2 计)约 30%，以及 Fe、Al、Cd、Cu、Ni 等金属元素。传统的矿渣填埋既污染环境、又造成资源浪费。鉴于锌灰中 Zn、Fe、Al、Cd、Cu、Ni 等元素具有重要的回收利用价值，在分析原料成分的基础上通过矿渣提取、杂质元素的分离和纯化、产物结晶等路线制备 $ZnSO_4·7H_2O$ 具有重要的经济价值。

对于一种以 $Zn_5(OH)_8Cl_2$ 为主要矿物成分、Fe 和 Si 含量较高的锌灰，制备 $ZnSO_4·7H_2O$ 的流程简述如下：利用硫酸浸取锌灰获得含锌的浸取液，加入氨水调高浸取液的 pH，促使 Al^{3+}、Fe^{3+} 和可溶性硅物种 Si(IV) 水解、聚合并沉淀析出(Fe^{2+} 利用过氧化氢氧化后水解)。滤去杂质沉淀后，利用锌粉还原滤液中微量 Cd^{2+}、Cu^{2+} 和 Ni^{2+} 杂质离子，得到含有 Zn^{2+}、NH_4^+、SO_4^{2-} 和 Cl^- 的溶液。用碳酸铵沉淀 Zn^{2+}，洗去沉淀中的杂质离子 NH_4^+ 和 Cl^-，用稀硫酸溶解沉淀得到硫酸锌溶液。利用乙醇中 $ZnCl_2$ 的溶解度远大于 $ZnSO_4$ 的特点，将乙醇加入硫酸锌溶液使 $ZnSO_4·7H_2O$ 析出，并进一步降低产物中残余 Cl^- 的含量。

实验用品

仪器与材料：分析天平、台秤、滴定管、吸量管、锥形瓶、容量瓶、量筒、烧杯、滴瓶、铁架台、酒精灯、循环水泵。

试剂：锌灰、0.05 mol/L EDTA 标准溶液、碳酸铵、锌粉、30%过氧化氢、无水乙醇、3 mol/L 硫酸、6 mol/L 氨水、铬黑 T 指示剂、NH_3-NH_4Cl 缓冲溶液(pH = 10.0)。

实验步骤

1. 七水合硫酸锌的制备

称取 10 g 锌灰放入 250 mL 烧杯中，加入 20 mL 蒸馏水和 10 mL 3 mol/L H_2SO_4，置于温度为 60～70℃的水浴中加热浸取并持续搅拌(依据液体体积减少程度可酌情补水少量)。30 min 后再次加入等量的蒸馏水和稀硫酸，继续加热搅拌反应 30 min，以使样品中锌被进一步浸出。

停止加热，待反应液冷却至室温后抽滤，用 3～5 mL 水洗涤残渣 1～2 次。测定滤液的 pH，用 6 mol/L 氨水调节滤液 pH 至 4.0 左右(精密 pH 试纸)。搅拌 10 min 后滴入 H_2O_2 溶液(30%，强腐蚀性)，并用氨水维持溶液 pH 4.0 左右。反应 20 min 以确保溶液中的亚铁离子被完全氧化并水解生成 $Fe(OH)_3$ 沉淀。继续向烧杯中滴加氨水至 pH = 5.2～5.4，反应 15 min 后减压过滤，向滤液中加入 0.1 g 锌粉，搅拌 5 min 后再次滤去过量的锌粉，得到滤液。

配制 15～20 mL 碳酸铵饱和溶液，在搅拌下将其缓慢滴入滤液至不再有白色沉淀产生。减压过滤并用水反复洗涤沉淀至刚滴出的滤液中检测不到铵离子为止。将洗净的沉淀收集到烧杯中，缓慢滴入 3 mol/L H_2SO_4 使沉淀刚好溶解完全。再加入适量的无水乙醇(10～15 mL)搅拌 5 min 后静置。待析出固体后过滤得到 $ZnSO_4·7H_2O$ 粗产物。

借助重结晶可以进一步提高产品的纯度。加入适量的蒸馏水溶解粗产物，蒸发浓缩后使产物结晶析出，冷却，过滤后取出样品，在较低温度下风干，称量并计算产率。

2. 产品的分析检验

用分析天平准确称取约 0.7 g 样品放入锥形瓶，加 50 mL 水使样品完全溶解，然后加入 5 mL NH_3-NH_4Cl 缓冲溶液(pH = 10.0)和数滴铬黑 T 指示剂，然后用 0.05 mol/L EDTA 标准溶液滴定至溶液自紫红色刚好转变为蓝色为止，记录 EDTA 标准溶液的用量。平行测定三份，计算样品中 Zn 的平均含量，折算出硫酸锌的纯度。

思考题

(1)加入氨水调节溶液 pH 的用量有多少？在溶液 pH = 4.0 阶段和 5.2～5.4 阶段分别是为了除去什么？

(2)在 Zn 粉置换反应后，为什么先采用碳酸盐沉淀、然后用硫酸溶解沉淀、最后加乙醇析出产物的路线，而没有采用直接蒸发浓缩滤液后结晶制备产物的方法？

实验 29　二水合二草酸根合铜(Ⅱ)酸钾的制备与组成分析

实验目的

(1)进一步掌握溶解、沉淀、抽滤、蒸发、浓缩等基本操作。

(2)了解二水合二草酸根合铜(Ⅱ)酸钾晶体的制备和组成测定方法。

实验原理

二水合二草酸根合铜(Ⅱ)酸钾的制备方法很多，可以由硫酸铜与草酸钾直接混合制备，也可以由氢氧化铜或氧化铜与草酸氢钾反应制备。本实验由氧化铜与草酸氢钾反应制备二水合二草酸根合铜(Ⅱ)酸钾。$CuSO_4$ 在碱性条件下生成 $Cu(OH)_2$ 沉淀，加热沉淀则转化为易过滤的 CuO。一定量的 $H_2C_2O_4$ 溶于水后加入 K_2CO_3 得到 KHC_2O_4 和 $K_2C_2O_4$ 混合溶液，该混合溶液与 CuO 作用生成二草酸根合铜(Ⅱ)酸钾，经水浴蒸发、浓缩，冷却后得到蓝色 $K_2[Cu(C_2O_4)_2]·2H_2O$ 晶体。涉及的反应有：

$$CuSO_4 + 2NaOH = Cu(OH)_2\downarrow + Na_2SO_4$$

$$Cu(OH)_2 = CuO + H_2O$$

$$2H_2C_2O_4 + K_2CO_3 = 2KHC_2O_4 + CO_2 + H_2O$$

$$2KHC_2O_4 + CuO = K_2[Cu(C_2O_4)_2] + H_2O$$

称取一定量试样在氨水中溶解、定容。将溶液分成两份：取一份用 H_2SO_4 中和，并在硫酸溶液中用 $KMnO_4$ 滴定试样中的 $C_2O_4^{2-}$；另取一份在 HCl 溶液中加入 PAR 指示剂，在 pH = 6.5～7.5 的条件下加热近沸，趁热用 EDTA 滴定至绿色为终点，以测定晶体中的 Cu^{2+}。通过消耗的 KMnO₄ 和 EDTA 的体积及其浓度计算 $C_2O_4^{2-}$ 及 Cu^{2+} 的含量,并确定 $C_2O_4^{2-}$ 及 Cu^{2+} 组成比,进而推算出产物的实验式。

PAR[4-(2-吡啶基偶氮)间苯二酚]指示剂属于吡啶基偶氮化合物,结构式为

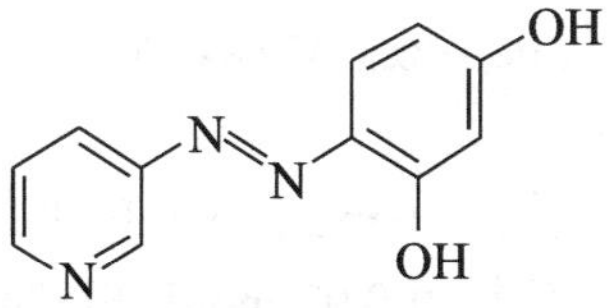

与 PAN 相比,PAR 分子中较多的亲水基团使其自身及金属螯合物的水溶性强,在 pH = 5~7 对 Cu^{2+} 的滴定有更明显的终点。

实验用品

仪器与材料:台秤、天平、烧杯、量筒、抽滤装置、容量瓶、蒸发皿、移液管、酸式滴定管、锥形瓶。

试剂:PAR 指示剂、$CuSO_4 \cdot 5H_2O$、$H_2C_2O_4 \cdot 2H_2O$、K_2CO_3、金属铜(基准物)、NaOH(2 mol/L)、HCl(2 mol/L)、H_2SO_4(3 mol/L)、氨水(1:1)、KMnO₄ 标准溶液(0.01 mol/L)、EDTA 标准溶液(0.02 mol/L)。

实验步骤

1. 二水合二草酸根合铜(Ⅱ)酸钾的合成

1)氧化铜的制备

称取 2.0 g $CuSO_4 \cdot 5H_2O$ 放入 100 mL 烧杯中,加入 40 mL 水溶解,在搅拌下加入 10 mL 2 mol/L NaOH 溶液,小火加热至沉淀变黑(生成 CuO),继续煮沸 20 min。稍冷后用双层滤纸抽滤,用少量去离子水洗涤沉淀 2 次。

2)草酸氢钾的制备

称取 3.0 g $H_2C_2O_4 \cdot 2H_2O$ 放入 250 mL 烧杯中,加入 40 mL 去离子水,微热溶解(温度不能超过 85℃,以避免 $H_2C_2O_4$ 分解)。稍冷后分数次加入 2.2 g 无水 K_2CO_3,溶解后生成 KHC_2O_4 和 $K_2C_2O_4$ 混合溶液。

3)二水合二草酸根合铜(Ⅱ)酸钾的制备

将 CuO 连同滤纸一起加入含 KHC_2O_4 和 $K_2C_2O_4$ 混合溶液的烧杯中,水浴加热,尽量使沉淀溶解(约 30 min)。趁热抽滤(若透滤应重新抽滤),用少量沸水洗涤沉淀,将滤液转入蒸发皿中。利用蒸汽浴加热将滤液浓缩到约原体积的 1/3,放置约 10 min 后用水彻底冷却。待大量晶体析出后抽滤,晶体用滤纸吸干,称量,计算产率,保存产品用于组成分析。

2. 产物的组成分析

1)试样溶液的制备

准确称取 0.95~1.05 g(准确到 0.0001 g)合成的晶体试样,置于 100 mL 烧杯中,加入 5 mL $NH_3 \cdot H_2O$ 使其溶解。再加入 10 mL 水,搅拌均匀后转移至 250 mL 容量瓶中,加水定容。

2) $C_2O_4^{2-}$ 含量的测定

取 25.00 mL 试样溶液于 250 mL 锥形瓶中，加入 10 mL 3 mol/L H_2SO_4 溶液，水浴加热至 75～85℃，在水浴中放置 3～4 min。趁热用 0.01 mol/L $KMnO_4$ 标准溶液滴定至淡粉色，30 s 不褪色为终点。记录消耗 $KMnO_4$ 标准溶液的体积。平行滴定 3 次。

3) Cu^{2+} 含量的测定

另取 25.00 mL 试样溶液，依次加入 1 mL 2 mol/L HCl 溶液、4 滴 PAR 指示剂和 10 mL pH = 7 的缓冲溶液，加热至近沸。趁热用 0.02 mol/L EDTA 标准溶液滴定至黄绿色，30 s 不褪色为终点。记录消耗 EDTA 标准溶液的体积。平行滴定 3 次。

数据记录与处理

(1) 产品记录。

形态和颜色：

质量：

理论产量：

产率：

(2) 滴定数据记录。

(3) 计算产物的组成。

计算 $C_2O_4^{2-}$ 的质量分数：
$$w_{C_2O_4^{2-}} = c_{KMnO_4} \times V_{KMnO_4} \times 88.02 \times 250 \times 5 \times 100\% / (m_{样} \times 1000 \times 25.00 \times 2)$$

计算 Cu^{2+} 的质量分数：
$$w_{Cu^{2+}} = c_{EDTA} \times V_{EDTA} \times 63.55 \times 250 \times 100\% / (m_{样} \times 1000 \times 25.00)$$

由 Cu^{2+} 和 $C_2O_4^{2-}$ 物质的量之比确定合成产物的组成。
$$物质的量比 = w_{Cu^{2+}} \times 88.02 \times 100\% / (63.55 \times w_{C_2O_4^{2-}})$$

注意事项

(1) 二水合二草酸根合铜(Ⅱ)酸钾在水中的溶解度很小，可加入适量氨水使 Cu^{2+} 形成铜氨离子而溶解，溶解时 pH 约为 10。溶剂也可采用 2 mol/L NH_4Cl 和 1 mol/L 氨水等体积混合组成的缓冲溶液。

(2) 指示剂本身在滴定条件下显黄色，而 Cu^{2+} 与 EDTA 显蓝色，终点为黄绿色。除 Cu^{2+} 外，PAR 在不同 pH 条件下能作铋、铝、锌、镉、铜、钇、钍、铊等元素的指示剂，终点由红色变黄色。

思考题

(1) 试设计由硫酸铜制备二水合二草酸根合铜(Ⅱ)酸钾晶体的其他方案。

(2) 实验中为什么不采用氢氧化钾与草酸反应生成草酸氢钾？

(3) $C_2O_4^{2-}$ 和 Cu^{2+} 分别测定的原理是什么？除本实验的方法外，还可以采用什么分析方法？

(4) 以 PAR 为指示剂滴定终点前后的颜色如何变化？

(5) 试样分析过程中，pH 过大或过小对分析有何影响？

参考文献

孙书静. 2004. 连续测定银电解液中的游离硝酸及铜银. 甘肃化工, 2: 47-48
魏士刚, 门瑞芝, 程新民, 等. 2003. 二草酸合铜酸钾中草酸根和铜离子测定方法的探讨. 广西师范大学学报(自然科学版), 21(4): 316-317
颜小敏. 2002. 二水合二草酸根合铜(Ⅱ)酸钾的制取及其组成测定. 西南民族学院学报, 28(1): 80-83
章慧. 2008. 配位化学原理与应用. 北京: 化学工业出版社

实验30 三草酸根合铁(Ⅲ)酸钾的合成及组成分析

实验目的

(1) 了解三草酸根合铁(Ⅲ)酸钾的合成方法。
(2) 掌握确定化合物化学式的基本原理和方法。
(3) 巩固无机合成、滴定分析和重量分析的基本操作。

实验原理

三草酸根合铁(Ⅲ)酸钾为亮绿色单斜晶体，易溶于水而难溶于乙醇、丙酮等有机溶剂。受热时，在110℃下可失去结晶水，230℃即分解。该配合物为光敏物质，光照下易分解。

本实验首先利用$(NH_4)_2Fe(SO_4)_2$与$H_2C_2O_4$反应制取FeC_2O_4：

$$(NH_4)_2Fe(SO_4)_2 + H_2C_2O_4 \longrightarrow FeC_2O_4\downarrow + (NH_4)_2SO_4 + H_2SO_4$$

在过量$K_2C_2O_4$存在下，用H_2O_2氧化FeC_2O_4即可制得产物：

$$6FeC_2O_4 + 3H_2O_2 + 6K_2C_2O_4 \longrightarrow 4K_3[Fe(C_2O_4)_3] + 2Fe(OH)_3\downarrow$$

在$H_2C_2O_4$和$K_2C_2O_4$存在下，反应的中间产物$Fe(OH)_3$也转化为产物：

$$2Fe(OH)_3 + 3H_2C_2O_4 + 3K_2C_2O_4 \longrightarrow 2K_3[Fe(C_2O_4)_3] + 6H_2O$$

利用如下的分析方法可测定该配合物各组分的含量，通过推算便可确定其化学式。

(1) 用重量分析法测定结晶水含量。

将一定量产物在110℃下干燥，根据失重的情况即可计算出结晶水的含量。

(2) 用高锰酸钾滴定法测定草酸根含量。

$C_2O_4^{2-}$在酸性介质中可被MnO_4^-定量氧化：

$$5C_2O_4^{2-} + 2MnO_4^- + 16H^+ \longrightarrow 2Mn^{2+} + 10CO_2 + 8H_2O$$

用已知浓度的$KMnO_4$标准溶液滴定$C_2O_4^{2-}$，由消耗$KMnO_4$的量便可计算出$C_2O_4^{2-}$的含量。

(3) 用高锰酸钾滴定法测定铁含量。

先用Zn粉将Fe^{3+}还原为Fe^{2+}，然后用$KMnO_4$标准溶液滴定Fe^{2+}：

$$5Fe^{2+} + MnO_4^- + 8H^+ \longrightarrow 5Fe^{3+} + Mn^{2+} + 4H_2O$$

由消耗$KMnO_4$的量便可计算出Fe^{3+}的含量。

(4) 确定钾含量。

配合物减去结晶水、$C_2O_4^{2-}$、Fe^{3+}的含量后即为K^+的含量。

实验用品

仪器与材料：分析天平、烘箱。

试剂：H_2SO_4(6 mol/L)、$H_2C_2O_4$(饱和)、$K_2C_2O_4$(饱和)、H_2O_2(w 为 0.05)、C_2H_5OH(w 为 0.95 和 0.5)、$KMnO_4$ 标准溶液(0.02 mol/L)、$(NH_4)_2Fe(SO_4)_2 \cdot 6H_2O$(s)、Fe 粉、丙酮、冰。

实验步骤

1. 三草酸根合铁(Ⅲ)酸钾的制备

将 5.0 g $(NH_4)_2Fe(SO_4)_2 \cdot 6H_2O$(s)溶于 20 mL 水中，加入 5 滴 6 mol/L H_2SO_4 酸化，加热使其溶解。在不断搅拌下再加入 25 mL $H_2C_2O_4$ 饱和溶液，然后加热至沸，静置。待黄色的 FeC_2O_4 沉淀完全沉降后，倾去上层清液，再用倾析法洗涤沉淀 2~3 次，每次用水约 15 mL。

在上述沉淀中加入 10 mL $K_2C_2O_4$ 饱和溶液，水浴加热至 40℃。用滴管缓慢地滴加 12 mL 质量分数为 0.05 的 H_2O_2 溶液，边加边搅并维持温度在 40℃左右，此时溶液中有棕色沉淀 $Fe(OH)_3$ 产生。加完 H_2O_2 后将溶液加热至沸，分两批共加入 8 mL $H_2C_2O_4$ 饱和溶液(先加入 5 mL，然后慢慢滴加 3 mL)。这时体系应变成亮绿色透明溶液(体积控制在 30 mL 左右)，如果体系浑浊可趁热过滤。在滤液中加入 10 mL 质量分数为 0.95 的乙醇，这时溶液如果浑浊，微热使其变清。放置暗处，让其冷却结晶。抽滤，用质量分数为 0.5 的乙醇溶液洗涤晶体，再用少量丙酮淋洗晶体两次，抽干，在空气中干燥。称量，计算产率。产物应避光保存。

2. 组成分析

1) 结晶水含量的测定

自行设计分析方案测定产物中结晶水含量。

2) 草酸根含量的测定

自行设计分析方案测定产物中 $C_2O_4^{2-}$ 含量。

3) 铁含量的测定

自行设计分析方案测定保留液中的铁含量。

4) 钾含量的确定

由测得 H_2O、$C_2O_4^{2-}$、Fe^{3+} 的含量计算出 K^+ 的含量，并由此确定配合物的化学式。

注意事项

(1) 产物在 110℃下烘 1 h，结晶水才能全部失去。

(2) 用高锰酸钾滴定 $C_2O_4^{2-}$ 时，为了加快反应速率需升温至 75~85℃，但不能超过 85℃，否则 $H_2C_2O_4$ 易分解。

(3) 滴定完成后保留滴定液，用来测量铁含量。

(4) 加入的还原剂铁粉需过量，反应体系需加热，以保证铁把 Fe^{3+} 完全还原成 Fe^{2+}。铁粉除与 Fe^{3+} 反应外，也与溶液中 H^+ 反应。因此，溶液必须保持足够的酸度，以免 Fe^{3+}、Fe^{2+} 等水解而析出。

(5) 滴定前过量的铁粉应过滤除去。过滤时要做到使 Fe^{2+} 定量地转移到滤液中，因此过滤后要对漏斗中的铁粉进行洗涤。洗涤液与滤液合并用来滴定，另外，洗涤不能用水而要用稀 H_2SO_4(为什么？)。

思考题

(1) 合成过程中，滴完 H_2O_2 后为什么还要煮沸溶液？

(2) 合成产物的最后一步，加入质量分数为 0.95 的乙醇，其作用是什么？能否用蒸干溶液的方法获得产物？为什么？

(3) 产物为什么要经过多次洗涤？洗涤不充分对其组成测定会产生什么影响？

(4) $K_3[Fe(C_2O_4)_3] \cdot 3H_2O$ 可用加热脱水法测定其结晶水含量，含结晶水的物质是否都可用这种方法进行测定？为什么？

实验 31 过氧化钙的制备及含量分析

实验目的

(1) 掌握制备过氧化钙的原理和方法。

(2) 掌握过氧化钙含量的分析方法。

(3) 巩固无机制备和化学分析的基本操作。

实验原理

过氧化钙为白色或淡黄色结晶粉末，室温下稳定，加热到 300℃可分解为氧化钙及氧，难溶于水，可溶于稀酸生成过氧化氢。它广泛用作杀菌剂、防腐剂、解酸剂、油类漂白剂、种子及谷物的无毒消毒剂，还用于食品、化妆品等作为添加剂。

过氧化钙可用氯化钙与过氧化氢及碱反应，或氢氧化钙、氯化铵与过氧化氢反应制备。在水溶液中析出的为 $CaO_2 \cdot 8H_2O$，再于 150℃左右脱水干燥，即得产品。

过氧化钙的含量可利用在酸性条件下过氧化钙与酸反应生成过氧化氢，用 $KMnO_4$ 标准溶液滴定而测得。

$$CaCl_2 + H_2O_2 + 2NH_3 \cdot H_2O + 6H_2O = CaO_2 \cdot 8H_2O + 2NH_4Cl$$

$$5CaO_2 + 2MnO_4^- + 16H^+ = 5Ca^{2+} + 2Mn^{2+} + 5O_2 \uparrow + 8H_2O$$

实验用品

仪器与材料：分析天平、酸式滴定管。

试剂：$CaCl_2 \cdot 2H_2O$(s)、H_2O_2(w=0.30)、$NH_3 \cdot H_2O$(浓)、HCl(2 mol/L)、$MnSO_4$(0.05 mol/L)、$KMnO_4$ 标准溶液(0.02 mol/L)，冰。

实验步骤

1. 过氧化钙的制备

将 7.5 g $CaCl_2 \cdot 2H_2O$ 溶解在 5 mL 水中，加入 25 mL w 为 0.30 的 H_2O_2 溶液，边搅拌边滴入由 5 mL 浓氨水和 20 mL 冷水配成的溶液。置于冰水中冷却 0.5 h，过滤，用少量冷水洗涤晶体 2～3 次。晶体抽干后，取出置于烘箱内在 150℃下烘 0.5～1 h。冷却后称量，计算产率。

2. 过氧化钙含量分析

用分析天平精确称取约 0.15 g 产物两份，分别置于 250 mL 锥形瓶中，各加入 50 mL 蒸馏水和 15 mL 2 mol/L HCl 溶液使其溶解，再加入 1 mL 0.05 mol/L $MnSO_4$ 溶液。用 0.02 mol/L $KMnO_4$ 标准溶液滴定至溶液呈微红色，30 s 内不褪色即为终点。计算 CaO_2 的质量分数。若测定值的相对平均偏差大于 0.2%，则需再滴定一次。

思考题

(1) 所得产物中的主要杂质是什么？如何提高产品的产率与纯度？

(2) $KMnO_4$ 是氧化还原滴定中最常用的氧化剂之一，该滴定通常在酸性溶液中进行，一般常用稀 H_2SO_4。本实验为什么不用稀 H_2SO_4？用稀 HCl 代替稀 H_2SO_4 对测定结果有无影响？如何证实？

实验 32 碱式碳酸铜的制备

实验目的

通过碱式碳酸铜制备条件的探索和生成物颜色、状态的分析，研究反应物的合理配比并确定制备反应合适的温度条件，培养独立设计实验的能力。

实验原理

碱式碳酸铜 $[Cu_2(OH)_2CO_3]$ 为天然孔雀石的主要成分，呈暗绿色或淡蓝绿色，加热至 200℃ 即分解，在水中的溶解度很小，新制备的试样在沸水中很易分解。

实验用品

学生自行列出所需仪器、试剂、材料的清单，经指导教师同意，即可进行实验。

实验内容

1. 反应物溶液的配制

配制 0.5 mol/L $CuSO_4$ 溶液和 0.5 mol/L Na_2CO_3 溶液各 100 mL。

2. 制备反应条件的探索

1) $CuSO_4$ 和 Na_2CO_3 溶液配比的探索

在四支试管中均加入 2.0 mL 0.5 mol/L $CuSO_4$ 溶液，再分别取 0.5 mol/L Na_2CO_3 溶液 1.6 mL、2.0 mL、2.4 mL 及 2.8 mL 依次加入另外四支编号的试管中。将八支试管放在 75℃ 恒温水浴中。几分钟后，依次将 $CuSO_4$ 溶液分别倒入不同体积的 Na_2CO_3 溶液中，振荡试管，比较各试管中沉淀生成的速度、沉淀的数量及颜色，得出两种反应物的最佳比例。

2) 反应温度的探索

在三支试管中各加入 2.0 mL 0.5 mol/L $CuSO_4$ 溶液，另取三支试管，各加入由上述实验得到的合适用量的 0.5 mol/L Na_2CO_3 溶液。从这两列试管中各取一支，将它们置于室温。几分钟后将 CuSO₄ 溶液倒入 Na₂CO₃ 溶液中,振荡试管并观察现象。按照同样操作过程试验在 50℃ 或 100℃ 恒温水浴中进行反应,由实验结果确定制备反应的合适温度。

3. 碱式碳酸铜的制备

取 60 mL 0.5 mol/L CuSO₄ 溶液,根据上面实验确定的反应物合适比例及适宜温度制备碱式碳酸铜。待沉淀完全后,减压过滤,用蒸馏水洗涤沉淀数次,直到沉淀中不含 SO_4^{2-} 为止,收集固体粉末并用滤纸吸干。将所得产品在烘箱中于 100℃ 烘干,待冷至室温后称量,并计算产率。

思考题

(1) 哪些铜盐适合制备碱式碳酸铜?写出硫酸铜溶液和碳酸钠溶液反应的化学反应式。
(2) 反应的条件,如反应温度、反应物浓度及反应物配比对反应产物是否有影响?
(3) 各试管中沉淀的颜色为什么会有差别?何种颜色产物的碱式碳酸铜含量最高?
(4) 若将 Na₂CO₃ 溶液倒入 CuSO₄ 溶液,其结果是否会有所不同?
(5) 反应在什么温度下进行会出现褐色产物?这种褐色物质是什么?
(6) 除反应物的配比和反应的温度对本实验的结果有影响外,反应物的种类、反应时间等因素是否也会对产物的质量有影响?
(7) 设计实验测定产物中铜及碳酸根的含量,从而分析所制得的碱式碳酸铜的质量。

实验 33　过氧化氢合钛(Ⅳ)配合物的组成和稳定常数的测定

实验目的

(1) 掌握分光光度法测定配合物浓度的原理和方法。
(2) 熟悉用等摩尔系列法确定配合物的组成和计算稳定常数的方法。
(3) 了解分光光度计的构造和使用方法。

实验原理

测定配合物的组成和稳定常数对于了解配合物的性质以及推断它的结构具有重要的作用。目前已经建立了各种测定配合物的组成和稳定常数的方法,如光度法、pH 电位滴定法、离子选择性电极法、极谱法和萃取法等。

光度法是最常用的测定方法之一。其优点是简单快速,特别适合于低浓度溶液,溶剂选择的范围也比较大。其原理是根据配合物在某一波长下对光有特征的最大吸收,而且配合物的溶液与原先的配体和金属离子的溶液对光有不同的吸收。溶液在某一波长下吸光度值 A 与溶液组成的关系在理想条件下符合吸收定律:

$$A = b \sum_{i=0}^{n} \varepsilon_i c_i$$

式中,b 为比色皿的光程长度;ε_i 为第 i 个质点在浓度为 c_i 时的摩尔吸光系数。体系中可能形成的各种配位比的配合物的摩尔吸光系数和浓度难以测得,因此不能从溶液的吸光度值直接求出配合物的平衡浓度来计算其稳定常数。如果目标配合物足够稳定且配体浓度足够高,就可以测得饱和的目标配合物的摩尔吸光系数。由于配合物在溶液中能以各种不同的形式存在,因此数据处理的方法也是多种多样的。

本实验用等摩尔系列法(又称为 Job's 法)测定配合物的组成和稳定常数。它是在固定溶液的体积且保持金属离子 M 和配体 L 的总摩尔数不变的情况下,依次改变 M 与 L 的摩尔比。随着这个摩尔比的不同,它所形成的饱和的目标配合物的量也就不同。配制一系列 M/L 的摩尔比不同的溶液组成系列待测液,分别测定它们的吸光度值,并画出溶液摩尔比-吸光度曲线。若生成的配合物很稳定,则曲线有明显的极大值[如图 5.1(a)所示],即可以求得配合物的组成。若生成的配合物不很稳定,即有一定的解离度,则曲线的极大值就不明显[如图 5.1(b)所示],这时可通过横坐标的两个端点向曲线作切线,由切线的交点可以确定曲线的最大值,再由极大值求得配合物的组成。

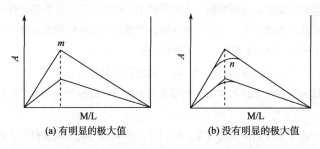

(a) 有明显的极大值　　　　(b) 没有明显的极大值

图 5.1　吸光度值 A 与 M/L 摩尔比的关系曲线

1. 配合物组成的确定

体系中存在配合物的配位反应为

$$\mathrm{M} + n\mathrm{L} \rightleftharpoons \mathrm{ML}_n \tag{5.1}$$

设 [M] + [L] = c,式中[M]、[L]分别为体系内金属离子、配体的起始浓度,实验中 c 为常数。

L 的摩尔分数为 x。平衡时金属离子的浓度为 c_M,则配体的浓度 c_L、配合物的浓度 y 分别为

$$c_\mathrm{M} = c(1-x) - y \tag{5.2}$$

$$c_\mathrm{L} = cx - ny \tag{5.3}$$

$$y = \beta c_\mathrm{M} c_\mathrm{L}^n \tag{5.4}$$

式中,β 为配合物的稳定常数。

微分上述各式得

$$\frac{\mathrm{d}c_\mathrm{M}}{\mathrm{d}x} = -c - \frac{\mathrm{d}y}{\mathrm{d}x} = 0 \tag{5.5}$$

$$\frac{\mathrm{d}c_\mathrm{L}}{\mathrm{d}x} = c - n\frac{\mathrm{d}y}{\mathrm{d}x} \tag{5.6}$$

$$\frac{\mathrm{d}y}{\mathrm{d}x} = \beta c_\mathrm{L}^n \frac{\mathrm{d}c_\mathrm{M}}{\mathrm{d}x} + n\beta c_\mathrm{M} c_\mathrm{L}^{n-1} \frac{\mathrm{d}c_\mathrm{L}}{\mathrm{d}x} \tag{5.7}$$

当配合物 ML_n 浓度极大时，$\dfrac{dy}{dx}=0$，则式(5.7)为

$$c_L \dfrac{dc_M}{dx} + nc_M \dfrac{dc_L}{dx} = 0 \tag{5.8}$$

将式(5.5)、(5.6)中的 $\dfrac{dc_M}{dx}$、$\dfrac{dc_L}{dx}$ 代入式(5.8)，整理得

$$c_L = nc_M$$

再将 c_L 代入式(5.3)得

$$nc_M = cx - ny \tag{5.9}$$

式(5.2)乘 n 得

$$nc_M = nc - ncx - ny \tag{5.10}$$

式(5.9)减式(5.10)，整理得

$$n = \dfrac{cx}{c(1-x)} = \dfrac{x}{1-x}$$

所以，通过测定系列溶液的吸光度值，作出溶液摩尔比-吸光度曲线得到最大吸光值时的 x，并计算 n(配位比)，由此确定配合物的组成 ML_n。

如果在不同的浓度范围内不生成其他配合物，则极大值的位置不变。浓度越小，配合物解离越明显，曲线越平。若配体和金属离子在同一波长也有吸收，则应从总吸光度值中减去配体和金属离子的吸光度值。这个方法不但可以求得配合物的组成，还可以计算配合物的稳定常数。

2. 配合物稳定常数的计算

体系中存在配合物的配位和解离平衡为

$$M + nL \rightleftharpoons ML_n$$

则配合物 ML_n 的稳定常数 β 为

$$\beta = \dfrac{[ML_n]}{[c_M][c_L]^n}$$

如果两个组成不同的溶液具有相同的吸光度值，则该两个溶液中配合物的浓度必然相同(如图 5.2 所示的 a 和 b，或 a' 和 b')。

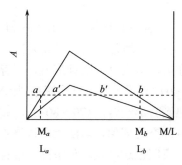

图 5.2　吸光度值 A 与 M/L 摩尔比的关系曲线

设这时配合物的浓度为 c_x，[M]和[L]分别为金属离子和配体的起始浓度，则

$$\beta = \frac{c_x}{(M_a - c_x)(L_a - nc_x)^n} = \frac{c_x}{(M_b - c_x)(L_b - nc_x)^n} \quad (5.11)$$

可任取同一条摩尔比-吸光度曲线上两个吸光度值相同的点，代入式(5.11)求出配合物浓度 c_x，并计算配合物的稳定常数 β。若配合物的组成为 ML(n=1)，就不能直接从一条摩尔比-吸光度曲线上的两点来计算 β，应从不同的摩尔比-吸光度曲线上找出吸光度值相同的两点来计算 β 值(如图 5.2 中的 a 和 a'，或 b 和 b')。

这个方法简单、迅速，结果也较可靠。但对于配合物稳定常数太大或太小，配位数太高的体系不能得到正确的结果。本实验用等摩尔系列法测定 Ti(Ⅳ)-H_2O_2 配合物的组成和稳定常数。

实验用品

仪器与材料：紫外-可见分光光度计、分析天平、电炉、滴定管架、锥形瓶、棕色试剂瓶、烧杯、容量瓶、酸式滴定管、刻度移液管、量筒。

试剂：草酸钛钾[$K_2TiO(C_2O_4)_2 \cdot 2H_2O$]、过氧化氢($H_2O_2$，30%)、高锰酸钾($KMnO_4$)、0.2 mol/L 硫酸锰溶液、草酸($H_2C_2O_4 \cdot 2H_2O$)、硫酸。

实验步骤

1. 钛(Ⅳ)溶液的配制

精确称取 0.3542 g 草酸钛钾，用 20 mL 1∶1(V/V) H_2SO_4 加热溶解，冷却后转入 100 mL 容量瓶中，用水稀释至刻度，摇匀，即得浓度为 1.00×10^{-2} mol/L 的 Ti(Ⅳ)溶液。

2. 过氧化氢溶液的配制和标定

(1) 用移液管吸取 1 mL 30% H_2O_2，用 2 mol/L H_2SO_4 溶液稀释至 100 mL。吸取上述 H_2O_2 溶液 10 mL 于锥形瓶中，用蒸馏水稀释至约 20 mL，加入 2~3 滴 0.2 mol/L 硫酸锰溶液，用高锰酸钾标准溶液进行标定。

(2) 配制浓度为 1.00×10^{-2} mol/L 的 H_2O_2 溶液：吸取一定量上述已知浓度的 H_2O_2 溶液于 100 mL 容量瓶中，用 2 mol/L H_2SO_4 溶液稀释至刻度，摇匀。

3. 高锰酸钾溶液的配制和标定

(1) 称取 1.70 g $KMnO_4$ 固体于 250 mL 烧杯中，用 200 mL 沸水溶解，将上层清液倒入棕色瓶中，再注入 300 mL 蒸馏水，摇匀。静置两天后用虹吸管将上层清液吸到 600 mL 烧杯中，弃掉瓶内剩余物。把棕色瓶洗净后，倒入高锰酸钾溶液，保存在暗处，浓度待标定。

(2) 精确称取 0.15 g $H_2C_2O_4 \cdot 2H_2O$，置于两个锥形瓶中，加入 25 mL 蒸馏水和 5 mL 2 mol/L H_2SO_4 溶液使其溶解，缓慢加热到有蒸汽冒出(70~80℃)。趁热用高锰酸钾溶液滴定，滴定时速度不能太快，滴定到终点时应充分摇匀，以防止超过终点，最后半滴高锰酸钾溶液在摇匀半分钟内仍保持微红色不褪，表明已达终点。记录高锰酸钾溶液的消耗体积。重复标定两三次。

(3) 计算高锰酸钾溶液的浓度。

4. Ti(Ⅳ)-H_2O_2 配合物的测定波长的选择

钛(Ⅳ)的硫酸溶液和过氧化氢溶液在可见光区没有吸收，可以选择 Ti(Ⅳ)-H_2O_2 配合物最大吸收时的波长为测定波长。用移液管吸取 2.5 mL 1.00×10^{-2} mol/L Ti(Ⅳ)溶液、2.5 mL 1.00×10^{-2} mol/L H_2O_2 溶液于 25 mL 容量瓶中，用 2 mol/L H_2SO_4 溶液稀释至刻度，摇匀，得待测液。以 2 mol/L H_2SO_4 溶液为参比液，用 1 cm 比色皿在 360~600 nm 波长范围内测定其吸光度值。以所测的吸光度值为纵坐标，波长为横坐标，绘制吸收曲线，由吸收曲线选择合适的测定波长。

5. Ti(Ⅳ)-H_2O_2 配合物的组成和稳定常数的测定

按等摩尔系列法，在 25 mL 容量瓶中，将 1.00×10^{-2} mol/L Ti(Ⅳ)溶液和 1.00×10^{-2} mol/L H_2O_2 溶液依照不同的体积比(摩尔比)配制一系列混合溶液，然后用 2 mol/L H_2SO_4 稀释至刻度，摇匀，测定其吸光度值。

数据记录与处理

1. Ti(Ⅳ)-H_2O_2 配合物的吸收曲线的绘制

测定不同波长下配合物的吸光度值，并记录于表 5.4。

表 5.4 不同波长下配合物的吸光度值

波长 λ/nm	360	380	400	410	420	440	460	480	500	520	560	600
吸光度 A												

以吸光度值 A 对波长 λ 作图，得出 Ti(Ⅳ)-H_2O_2 配合物的吸收曲线，再由吸收曲线确定配合物的测定波长。

2. Ti(Ⅳ)-H_2O_2 配合物的组成测定

测定不同摩尔比时的吸光度值，并记录于表 5.5。

表 5.5 不同摩尔比时的吸光度值

溶液编号	1	2	3	4	5	6	7	8	9	10
Ti(Ⅳ)溶液体积/mL										
H_2O_2 溶液体积/mL										
吸光度 A										

以 A 为纵坐标，Ti(Ⅳ)/H_2O_2 摩尔比为横坐标作图得出摩尔比-吸光度曲线，由曲线的极大值位置确定配合物的组成。

3. Ti(Ⅳ)-H_2O_2 配合物的稳定常数 β 的计算

在摩尔比-吸光度曲线上找出任一相同吸光度值的两点所对应的溶液组成，由式(5.11)求出配合物的浓度，由此计算配合物的稳定常数 β。

注意事项

(1) 过氧化氢溶液需临时配制。
(2) 必须在固定 M 和 L 总浓度不变的情况下，准确配制 M 与 L 摩尔比不同的溶液。
(3) 建议配制总体积为 5.0 mL，且摩尔比为 1∶9～9∶1 的 8～10 种溶液。

思考题

(1) 说明等摩尔系列法测定配合物稳定常数的适用范围及产生误差的原因。
(2) 计算配合物 ML_n 的稳定常数 β 时，对于 $n\neq1$ 型配合物可在同一摩尔比-吸光度值曲线上取吸光度值相等的两点，而对于 $n=1$ 型配合物必须取不同的摩尔比-吸光度值曲线上的两点来计算。为什么？

参考文献

南京大学配位化学研究所. 1984. 配位化学(无机化学丛书十二卷). 北京：科学出版社
王伯康, 钱文浙, 等. 1984. 中级无机化学实验. 北京：高等教育出版社
扬州大学, 徐州大学, 盐城师范大学等校. 2010. 新编大学化学实验(二). 北京：化学工业出版社

实验 34 钴铁氧体 $CoFe_2O_4$ 的制备及含量分析

实验目的

(1) 学习共沉淀法制备铁氧体及金属氧化物的原理。
(2) 设计利用共沉淀法制备 $CoFe_2O_4$ 磁性复合氧化物的方法。
(3) 熟悉用 XRD 表征样品物相，透射电子显微镜和扫描电子显微镜观察样品的相貌、晶粒尺寸及其分布，振动样品磁强计测定样品磁性参数或用磁铁石定性检验其磁性的原理和方法。

实验原理

铁氧体由于其高的立方磁晶各向异性，高的矫顽力和适中的饱和磁化强度被广泛地应用于永磁体、磁记录、磁性液体和抗电磁干扰以及电子设备等方面。铁氧体既是磁介质又是电介质，其内部存在共振吸收、散射吸收、单畴吸收和磁损耗、介电损耗等多重吸收机制和损耗机制，目前已成为较有发展前途的磁功能材料和电磁波吸收材料之一。磁性材料是浙江省的支柱型产业，具有优先发展的地位。因此，熟悉和掌握 $CoFe_2O_4$ 等磁性氧化物的结构、制备及其表征方法具有非常重要的战略意义。

在一定条件下，用一种沉淀剂使溶液中两种或两种以上的金属离子同时沉淀出来的方法称为共沉淀法。以制备 $CoFe_2O_4$ 为例，其制备原理如下：

$$Co^{2+} + 2Fe^{3+} + 8OH^- \longrightarrow Co(OH)_2 + 2Fe(OH)_3$$

$$Co(OH)_2 + 2Fe(OH)_3 \xrightarrow{热分解} CoFe_2O_4 + 4H_2O$$

$CoFe_2O_4$ 具有立方尖晶石结构，可看成是 Co^{2+} 取代了 Fe_3O_4 的 Fe^{2+}，其中两个 Fe^{3+} 分占尖晶石立方晶胞的四面体（A 位）和八面体（B 位），其单电子轨道自旋反平行排列，自旋磁矩相互抵消。故 $CoFe_2O_4$ 的磁性源于 Co^{2+}，属于亚铁磁性类。

实验用品

仪器与材料：天平、磁力搅拌器、温度计、烧杯、电炉、烘箱、马弗炉、X 射线粉末衍射仪、扫描电子显微镜、透射电子显微镜、振动样品磁强计等表征仪器。

试剂：六水合硝酸钴、九水合硝酸铁、氢氧化钠、聚乙二醇（或乙烯醇）等。

实验步骤

1. $CoFe_2O_4$ 的制备

按制备 0.01 mol $CoFe_2O_4$ 的化学计量比准确称取适量的各种金属硝酸盐，加入 40 mL 蒸馏水（滴加 2%～3%聚乙二醇溶液），在不断搅拌下使反应物完全溶解。将液体转移到三口烧瓶，并在 60℃水浴中和搅拌下滴加 1 mol/L NaOH 溶液，调节溶液 pH 至约 9。待溶液中出现大量棕色沉淀时停止滴加。将温度升至 80℃并搅拌反应 2 h。产物冷却后转移到烧杯，使其静置分层后用永磁铁把铁氧体沉淀在烧杯底部。弃去上层清液，用蒸馏水多次洗涤固体沉淀，直至上层清液呈中性，再用无水乙醇洗涤 2～3 次，得到 $CoFe_2O_4$ 的前驱物。将前驱物置于烘箱中 300℃煅烧 2 h，即得 $CoFe_2O_4$ 粉末（如欲得到晶形和磁性能更好的产品，可将粉末在 500℃下烧结 2 h）。

2. 微观结构和磁性表征

(1) 用 EDTA 法测定铁（水杨酸为指示剂）和钴（二甲酚橙为指示剂，六次甲基四胺，pH = 5～6）的含量。

(2) X 射线粉末衍射仪确定 $CoFe_2O_4$ 的晶相。

(3) 扫描电子显微镜观察 $CoFe_2O_4$ 的形貌。

(4) 透射电子显微镜观察 $CoFe_2O_4$ 晶粒的尺寸及其分布。

(5) 振动样品磁强计测定 $CoFe_2O_4$ 的磁性参数或用磁铁石定性检验其磁性。

3. 实验结果和讨论

(1) 计算 $CoFe_2O_4$ 的产率：

$$产率 = m_{CoFe_2O_4(实)} / m_{CoFe_2O_4(理)} \times 100\%$$

(2) 计算 $CoFe_2O_4$ 的含量：

$$CoFe_2O_4 的质量分数 = c_{EDTA} \times V_{EDTA} \times M_{CoFe_2O_4} / 2m \times 100\%$$

式中，m 为用于滴定的样品质量。

(3) 分析重量分析结果和容量分析结果产生差异的原因。

注意事项

(1) 制备 $CoFe_2O_4$ 的关键是 pH 和反应温度的调控。

(2) 如果制备的粉末样品 XRD 图谱中有杂质峰,表明样品中有氢氧化物未分解完全或 $CoFe_2O_4$ 结晶性能较差,可将样品在马弗炉中 400℃烧结 1 h 即可。

(3) XRD 图谱的特征峰为:$2\theta=18.37°$,30.34°,35.52°,43.17°,53.70°,57.14°和 62.76°(JCPDS-ICDD 22-1086)。

思考题

(1) 为什么要用共沉淀法制备 $CoFe_2O_4$?

(2) 制备 $CoFe_2O_4$ 时,为什么要调节溶液的 pH 约为 9?

实验 35　金(银)胶体的制备、光学吸收性质和稳定性

实验目的

(1) 了解贵金属纳米粒子的制备方法、性质和应用等。

(2) 掌握制备金、银胶体溶液的化学还原法和光化学法,学习光化学反应装置、紫外-可见分光光度计、激光粒度仪等的使用和产物表征数据的处理方法。

(3) 分析稳定剂种类对胶体溶液抵抗电解质聚沉作用的影响。

实验原理

1. 贵金属纳米粒子的性质概述

贵金属纳米粒子与相应的金属块体材料在很多方面表现出截然不同的性质。众所周知,块体金具有黄色金属光泽,而 3～20 nm 的金粒子受表面等离子共振(surface plasmon resonance,SPR)效应影响而呈现红色[1],且其颜色随粒子的尺寸大小和形貌改变而变化,呈现橙色、酒红色、紫色等多种颜色。SPR 吸收峰的位置还与粒子形状、溶液 pH、保护剂种类等多种因素有关。在催化活性方面,由于暴露出更多的高活性表面,一些特定形貌和结构的铂、金、铑等纳米粒子对很多化学反应的催化性能远高于微米或毫米尺度颗粒。基于高的化学稳定性、良好的生物相容性、特殊的表面化学性质、光学和电学性质以及催化性能等特点,金胶体和银胶体在光电子学、传感器、生物医学和工业催化等领域具有重要的应用价值[2]。

2. 贵金属纳米粒子的制备

制备贵金属纳米粒子的常用方法包括化学还原法、光化学法、辐射化学法、声化学法、浸渍法等,使用的贵金属前体化合物包括硝酸银、氯金酸、氯铂酸、氯化铑等化合物,在溶液或乳液状态下借助不同方法将其还原为单质原子,通过控制物料浓度和反应速率调控粒子的成核和生长进程,最终得到特定尺寸和形貌的粒子。稳定剂的使用能确保贵金属纳米粒子分散液具有一定的胶体稳定性。较大尺寸的离子、表面活性剂、高分子可以作为贵金属纳米粒子的稳定剂,柠檬酸根、聚乙烯醇、十六烷基三甲基溴化铵、硫醇等常用于金、银胶体的

制备。

以金为例，$[AuCl_4]^- + 3e^- \rightleftharpoons Au(s) + 4Cl^-$ 的电极电势为 0.93 V。在化学还原法中，使用强还原剂硼氢化钠常温下可以在 1 min 内完成反应，而使用还原性较弱的柠檬酸钠则需要 $HAuCl_4$ 溶液在煮沸状态下反应 15～30 min(Frens 法)。利用紫外光辐照也可以还原硝酸银或氯金酸制备银或金胶体[3-5]。乙醇、乙二醇或聚乙烯醇等醇的使用可以加速光化学还原反应的进程，进而影响金胶体的尺寸。紫外光辐照制备金纳米材料的机理参见文献[6]。Ag^+氧化性弱于 $Au(Ag^+ + e^- \rightleftharpoons Ag$ 的电极电势为 0.799 V)，采用上述合成方法同样可以制备银胶体溶液[7]。

3. 胶体溶液的稳定性

纳米粒子分散形成的胶体体系是热力学不稳定体系。一方面，粒子间倾向于相互聚结以降低表面能；另一方面，粒子间的静电斥力和表面修饰分子的位阻效应均有助于抑制粒子的团聚，保持胶体溶液的稳定。改变体系温度、电解质浓度可以破坏其稳定性，导致粒子聚沉。此时贵金属纳米粒子的 SPR 吸收峰的位置和强度均会发生变化。因此，可以通过溶胶的紫外-可见吸收光谱的检测获知胶体团聚的发生。此外，借助激光粒度仪提供的动态光散射技术可以测试贵金属纳米粒子在分散介质中的水力学粒径分布图，获得粒子的平均水力学粒径和多分散系数。当电解质导致粒子发生轻微团聚时，胶体溶液的紫外-可见吸收光谱可能无明显变化，但动态光散射可以准确地检测出胶体的平均水力学粒径和多分散系数增大，进而确定胶体溶液的稳定性。

实验用品

仪器与材料：激光粒度仪、紫外-可见分光光度计、紫外光化学反应装置(配 254 nm 紫外灯管)、磁力搅拌器、鼓风干燥箱、石英试管、比色皿、烧杯、试管、吸量管。

试剂：硝酸银溶液($AgNO_3$，2.5 mmol/L)、氯金酸溶液($HAuCl_4$，2.5 mmol/L)、柠檬酸钠溶液(25 mmol/L)、羧甲基壳聚糖溶液(CCT，1%)、硼氢化钠、聚乙二醇 2000 (PEG 2000)、十六烷基三甲基溴化铵(CTAB)、乙醇、盐酸、硝酸、氯化钠(1 mol/L)。

实验步骤

1. 金(或银)胶体溶液的制备

(1) 化学还原法：向 50 mL 烧杯中加入 2 mL 2.5 mmol/L $HAuCl_4$(或 $AgNO_3$)溶液、16 mL 水、2 mL 柠檬酸钠溶液和 1 粒磁力搅拌子。将烧杯放置在磁力搅拌器上搅拌，迅速加入 0.6 mL 用冰水现配的 1% $NaBH_4$ 溶液。观察溶液颜色稳定(约 30 s)后，停止搅拌，得到金(或银)胶体溶液。

(2) 光化学法：将 2 mL CCT 溶液、16 mL 水和 2 mL $HAuCl_4$(或 $AgNO_3$)溶液混合均匀后装入石英试管并固定在光化学反应装置内。开启紫外光源进行辐照，制备 CCT 稳定的金(或银)胶体溶液。在 15 min、30 min 和 45 min 时分别取样 3 mL，观测样品颜色，测定吸收光谱。用 2 mL 15 mg/mL PEG 2000 或 CTAB 溶液代替 CCT 溶液重复上述实验，可以得到 PEG 2000

或 CTAB 稳定的金(或银)胶体溶液。

2. 胶体的光学吸收性质和稳定性测试

(1)测定上述不同反应条件下制备的金(或银)胶体的紫外-可见吸收光谱。测试条件：以水为参比，波长扫描范围 300~800 nm。

(2)将三种用不同稳定剂制备的金(或银)胶体溶液各 2 mL 分别转移到三支干净试管中，在暗处用激光笔照射上述溶液，观察是否存在丁铎尔效应。将 1 mol/L NaCl 溶液分别逐滴加入上述胶体溶液中并轻微振荡混匀，观察溶液的颜色是否发生变化以及颜色变化时 NaCl 溶液的用量。比较胶体在高浓度电解质存在下的稳定性，扫描加入 NaCl 后胶体溶液的紫外-可见吸收光谱。

(3)选择步骤(2)中颜色变化最不明显的胶体样品，将原始胶体溶液和加入 NaCl 后的胶体溶液各 1.2 mL 分别装在两个样品池中。用激光粒度仪测试胶体的水力学粒径分布图，得出胶体的水力学粒径分布图、平均水力学粒径和多分散系数。

注意事项

(1)实验结束后，制备和测试胶体溶液所用的容器和比色皿需要用王水浸泡 20 min 以除去可能残余的胶体粒子，并用自来水、蒸馏水反复冲洗仪器后晾干。使用王水时需佩戴护目镜和防酸手套并在通风橱内进行。浸泡仪器后的王水应回收至指定容器。

(2)本实验所用的紫外光化学反应装置装配 8 W 低压汞灯，波长为 253.7 nm。使用光化学反应装置须严格遵照操作规程，避免紫外光照对人体的伤害。

(3)务必按量取用氯金酸和硝酸银溶液，避免浪费。用后尽快用锡箔纸包好瓶体并放回冰箱冷藏保存，避免长时间曝光发生光解。

思考题

(1)依据紫外-可见分光光度计测试结果的数据文件，使用 Excel 或 Origin 软件绘制胶体溶液的紫外-可见吸收光谱，分析和讨论影响胶体粒子表面等离子共振吸收峰位置的因素。

(2)说明稳定剂对胶体溶液抵抗电解质聚沉作用的能力差别。

(3)对激光粒度仪测试结果进行分析，结合紫外-可见分光光度计测试结果，评判该样品是否发生团聚并说明理由。

参考文献

[1] Daniel M C, Astruc D. 2004. Gold nanoparticles: assembly, supramolecular chemistry, quantum-size-related properties, and applications toward biology, catalysis, and nanotechnology. Chem Rev, 104: 293-346

[2] Dreaden E C, Alkilany A M, Huang X H, et al. 2012. The golden age: gold nanoparticles for biomedicine. Chem Soc Rev, 41(7): 2740-2779

[3] 吴泓橙, 董守安, 董颖男, 等. 2007. 金纳米粒子的阳光光化学合成和晶种媒介生长. 高等学校化学学报, 28(1): 10-15

[4] 陈延明, 王立岩, 李凤红. 2010. 聚乙烯吡咯烷酮为稳定剂制备银纳米粒子及表征. 高分子材料科学与工程, 26(10): 137-139

[5] 刘庆业, 覃爱苗, 蒋治良, 等. 2005. 聚乙二醇光化学法制备金纳米微粒及共振散射光谱研究. 光谱学与

光谱分析, 25(11): 1857-1860
[6] Eustis S, Hsu H Y, El-Sayed M A. 2005. Gold Nanoparticle formation from photochemical reduction of Au^{3+} by continuous excitation in colloidal solutions. A Proposed Molecular Mechanism. J. Phys. Chem. B, 109(11): 4811-4815
[7] 姚素薇, 曹艳蕊, 张卫国. 2006. 光还原法制备不同形貌银纳米粒子及其形成机理. 应用化学, 23(4): 438-440

实验36 金属离子与血清白蛋白的相互作用

实验目的

(1) 掌握紫外-可见分光光度计和荧光分光光度计的使用方法。
(2) 对荧光分光光度计相关数据进行初步计算。
(3) 检测生物体内常见金属离子对血清白蛋白的影响。

实验原理

金属离子在许多生命过程中发挥关键作用。研究金属离子与蛋白质的结合作用是生命科学的重要内容，是化学和生命科学研究的前沿领域。血清白蛋白是一种球蛋白，可以与无机离子、有机化合物以及小分子药物等结合，在生物体内发挥重要作用[1]。由于血清白蛋白在生理上的重要性和易于分离、提纯，从20世纪50年代(国内80年代末)开始，人们对血清白蛋白与金属离子(或药物分子等)的相互作用展开了大量研究，以期揭示生命过程的奥秘。

目前用于研究金属离子与血清白蛋白相互作用的研究方法主要有：①紫外-可见吸收光谱法；②荧光光谱法；③平衡透析法；④毛细管电泳法；⑤电泳法等。

紫外光谱和荧光光谱是研究生物大分子与小分子相互作用的重要方法[2,3]。通过研究小分子微扰的蛋白质光谱变化，可以推断生色基微环境及蛋白质结构的变化，并且可以推断小分子与蛋白质的结合情况。但是，紫外光谱通过电子跃迁产生，得到的是宽的谱带，难以分辨蛋白质中色氨酸残基、酪氨酸残基以及苯丙氨酸残基吸收谱带的振动结构。荧光测试中的发射峰特征、荧光偏振、能量转移及荧光寿命等指标可以对蛋白质分子中荧光生色基团的结构及其所处的微环境提供有用的信息。

下面对紫外-可见吸收光谱法和荧光分光光谱法进行介绍。

1. 紫外-可见光谱法

血清白蛋白通常有3个明显不同的紫外吸收带：①210 nm以下的吸收来自肽键以及许多构象因素；②210~250 nm的吸收来自多种因素，如芳香族和其他残基的吸收、某些氢键的吸收、与其他构象和螺旋的相互作用等；③250~290 nm附近的吸收为芳香族的残基，其中酪氨酸残基在278 nm(Tyr，260~290 nm)附近有强吸收，色氨酸(Trp)残基在290 nm附近有强吸收，而苯丙氨酸残基(Phe，250~260 nm)的吸收较弱。外界因素如溶剂极性及pH等也会影响吸收光谱。

当金属离子与蛋白质结合时，蛋白质或金属离子吸收光谱的强度或谱带位置会发生变化：①蛋白质微扰的金属离子光谱变化，可以推断金属离子的配位环境；②金属离子微扰的蛋白质光谱变化,可以推断生色基微环境及蛋白质结构的变化。通过对光谱的比较和计算,可以推断金属离子与蛋白质的结合情况。若蛋白质的吸收峰增强,则可认为小分子进入蛋白质的疏水腔,导致肽链伸展,疏水环境下降。在单纯的溶剂效应下,峰的蓝移表示原先深埋于蛋白质非极性区的生色基被暴露于极性溶剂,红移则表示生色基被翻转到极性较小的区域[4]。

2. 荧光光谱法

荧光光谱法是研究蛋白质分子构象的一种有效方法,具有灵敏度高、选择性强、用样量少、方法简便等优点。血清白蛋白中含有色氨酸、酪氨酸和苯丙氨酸残基,因此能发出天然荧光。在通常条件下,色氨酸、酪氨酸和苯丙氨酸的荧光强度比为 100∶9∶0.5,因此蛋白质的天然荧光主要来自色氨酸。当金属离子与血清白蛋白结合后,可引起蛋白质或少数金属离子荧光的改变,由此可进行定性和定量研究。常用荧光猝灭法测定金属离子与蛋白质的结合常数和结合位点数,利用热力学数据计算金属离子与蛋白质之间的相互作用力,用共振能量转移法测定金属离子与蛋白质分子中色氨酸残基之间的距离。

1) 荧光猝灭

荧光猝灭的原因是溶液中猝灭剂分子和荧光物质之间发生相互作用致使荧光效率降低或激发态寿命缩短。特异性的猝灭是荧光物质与猝灭物质之间发生了特异性的化学作用引起的。许多过程可引起荧光猝灭,如激发态反应、能量转移、配合物形成和碰撞猝灭。荧光猝灭的机制主要包括静态猝灭和动态猝灭。

动态猝灭又称为碰撞猝灭,是荧光体与猝灭体之间的扩散作用导致的。通常先假设猝灭为动态猝灭,则按照斯特恩-沃尔默方程[5]计算 K_q:

$$F_0/F = 1 + K_q\tau_0 [Q] = 1 + K_D [Q] \tag{5.12}$$

式中,F_0 为荧光体不存在猝灭的荧光强度;F 为加入猝灭体后的荧光强度;K_q 为双分子猝灭常数;$[Q]$ 为猝灭体浓度;τ_0 为荧光体不存在猝灭的荧光寿命;K_D 为斯特恩-沃尔默常数,$K_D = K_q\tau_0$。

通过荧光寿命计算猝灭常数 K_q。生物分子的荧光寿命约为 10 ns 数量级,各种猝灭剂对生物分子的最大扩散碰撞猝灭常数约为 $2.0×10^{10}$ L/(mol·s)[6]。当 K_q 大于 $2.0×10^{10}$ L/(mol·s) 时,则认为是静态猝灭。

2) 金属离子与蛋白质相互作用的结合常数 K_A 和结合位点数 n

假定小分子在蛋白质分子上有 n 个相同且独立的结合位点,根据文献[7]导出公式:

$$\lg(F_0 - F)/F = \lg K_A + n\lg[Q] \tag{5.13}$$

以 $\lg(F_0 - F)/F$ 对 $\lg[Q]$ 作图,由直线的截距和斜率分别求得结合常数 K_A 和结合位点数 n。

3) 蛋白质(给体)与金属离子(受体)之间的能量转移及结合距离

当蛋白分子的发射光谱与小分子的吸收光谱相互重叠时,通过分子间偶极-偶极的共振偶合可使能量从色氨酸(给体)转移到小分子(受体)。按照福斯特共振能量转移理论[5],可以求出小分子与蛋白质中色氨酸残基的距离 r,并由 r 的变化分析其他物质对血清白蛋白构象的影响。给体-受体间能量转移效率 E 与给体-受体间距离 r 和能量转移距离 R_0 存在如下关系:

$$E = 1 - F/F_0 = R_0^6/(R_0^6 + r^6) \tag{5.14}$$

$$R_0^6 = 8.8×10^{-25} K^2 N^{-4} \Phi J \tag{5.15}$$

式中，R_0 为转移效率为 50%时的临界距离；K^2 为偶极空间取向因子；N 为介质的折射指数；Φ 为受体的荧光量子产率。K^2 平均值取 2/3，N 取水(1.33)和有机物(1.39)的平均值 1.36，Φ 取 0.118[8]。J 为给体(蛋白质)的荧光发射光谱与受体的吸收光谱间的光谱重叠积分，可表示为

$$J = \frac{\sum F(\lambda) \cdot \varepsilon(\lambda) \cdot \lambda^4 \Delta\lambda}{\sum F(\lambda) \cdot \Delta\lambda} \tag{5.16}$$

式中，$F(\lambda)$ 为荧光给体在波长 λ 处的荧光强度；$\varepsilon(\lambda)$ 为受体在波长 λ 处的摩尔吸光系数。根据公式求得重叠积分 J，结合荧光数据求得 R_0、E 和 r。

4)金属离子与蛋白质之间作用力类型的测定

小分子和蛋白质之间的作用力包括氢键、范德华力、静电作用力、疏水作用力。根据反应前后热力学焓变 ΔH 和熵变 ΔS 的相对大小，可以判断小分子和蛋白质之间的主要作用力类型。$\Delta H > 0$，$\Delta S > 0$，通常为疏水作用力；$\Delta H < 0$，$\Delta S > 0$，是静电作用力；$\Delta H < 0$，$\Delta S < 0$，是氢键和范德华力[9]。当温度变化不大时，反应的焓变 ΔH 可看作是一常数。根据热力学公式：

$$\ln(K_2/K_1) = \Delta H (1/T_1 - 1/T_2) / R \tag{5.17}$$

$$\Delta G = -RT\ln K \tag{5.18}$$

$$\Delta S = (\Delta H - \Delta G) / T \tag{5.19}$$

根据不同温度下的结合常数 K 分别求得 ΔH 和 ΔS，进而判断金属离子与蛋白质之间的作用力类型。

5)金属离子对蛋白质构象的影响

固定波长的同步荧光光谱常用于判断蛋白质构象的改变。$\Delta\lambda = 15$ nm 时扫描显示酪氨酸残基的荧光，$\Delta\lambda = 60$ nm 时扫描显示色氨酸残基的荧光[5]。因为氨基酸残基的最大发射波长与其所处环境的疏水性有关，所以由发射波长的改变可判断蛋白质构象的变化。固定蛋白质的浓度，逐渐增加金属离子的浓度，扫描 $\Delta\lambda = 15$ nm 和 $\Delta\lambda = 60$ nm 的同步荧光光谱，由此判断金属离子对蛋白质构象的影响。

实验用品

仪器与材料：紫外-可见分光光度计、荧光分光光度计、生物制冰机、冰箱、自动三重纯水蒸馏器、干燥箱、水浴锅(6~8 孔)、pH 计、试管、微量注射器(50 μL)、分析天平、吸收池、烧杯、容量瓶、移液管、擦镜纸。

试剂：$Cu(Ac)_2 \cdot H_2O$、$FeCl_3$、$NaCl$、HCl(6 mol/L)、NaOH(6 mol/L)、牛血清白蛋白(BSA)、三(羟甲基)氨基甲烷(Tris-HCl)。

实验步骤

1. 溶液的配制

(1)配制 50 mmol/L Tris-HCl 并含 100 mmol/L NaCl 的缓冲溶液，维持体系的 pH(=7.0)和离子强度，并用此缓冲溶液配制 BSA 储备液，4℃存放。

(2) 准确称取 $Cu(Ac)_2 \cdot H_2O$ 和 $FeCl_3$，先在烧杯中用少量水溶解，然后转移到容量瓶中，稀释成 2 mmol/L 溶液，室温保存。

2. 金属离子与蛋白质相互作用的紫外光谱和荧光光谱测定

在一系列 5 mL 比色管中，依次加入牛血清白蛋白溶液和金属离子（Cu^{2+} 和 Fe^{3+}）溶液，用 Tris-HCl 缓冲溶液稀释至刻度，分别测定：①金属离子终浓度分别为 0、2.0、4.0、8.0、12.0、16.0、20.0（$\times 10^{-5}$ mol/L），310 K 恒温水浴中反应 30 min，以 Tris-HCl 缓冲溶液做参比，扫描 200~400 nm 的紫外吸收光谱；②金属离子终浓度分别为 0、2.0、4.0、8.0、12.0、16.0、20.0（$\times 10^{-6}$ mol/L），298 K 和 308 K 的荧光光谱和同步荧光光谱，固定牛血清白蛋白的浓度为 1.87×10^{-7} mol/L。

注意事项

(1) 实验要求：①4 人一组；②对紫外光谱进行简单分析；③对荧光光谱进行相关计算。
(2) 仪器扫描条件尽量保持一致。

思考题

(1) 测定为什么要在 37℃下进行？
(2) 缓冲溶液的 pH 为什么要保持在 7.0~7.4？缓冲溶液中 NaCl 的作用是什么？
(3) 如何区分动态猝灭和静态猝灭？

参考文献

[1] Babine R E, Bender S L. 1997. Molecular recognition of protein-ligand complexes: applied to drug design. Chem Rev, 97: 1359-1472
[2] 田伦富, 刘忠芳, 胡小莉, 等. 2012. [Hg(SCN)$_4$]$^{2-}$对蛋白质的荧光猝灭反应极其分析应用. 高等学校化学学报, 33(1): 59-65
[3] 王迎进, 张艳青, 李亚雄, 等. 2013. 光谱法研究根皮苷与人血清白蛋白相互作用. 分析测试学报, 32(2): 239-243
[4] Hu Y J, Liu Y, Pi Z B, et al. 2005. Interaction of cromolyn sodium with human serum albumin: a fluorescence quenching study. Bioorg Med Chem, 13(24): 6609-6614
[5] 陈国珍, 黄贤智, 许金钩. 1990. 荧光分析法. 2 版. 北京: 科学出版社
[6] 卢继新, 张贵珠, 赵鹏, 等. 1997. 阿霉素与血清白蛋白的作用及共存离子对反应影响的研究. 化学学报, 55: 915-920
[7] 谢孟峡, 徐晓云, 王英典, 等. 2005. 4′,5,7-三羟基二氢黄酮与人血清白蛋白相互作用的光谱学研究. 化学学报, 63(22): 2055-2062
[8] 杨曼意, 杨频, 张立伟. 1994. 荧光法研究咖啡酸类药物与白蛋白的作用. 科学通报, 39(1): 31-35
[9] 宋玉民, 吴锦绣. 2006. 稀土芦丁配合物的合成、表征及与血清白蛋白的相互作用. 无机化学学报, 22(12): 2165-2172

第6章 生活中的化学

实验37 硫代硫酸钠的制备和应用

实验目的

学习硫代硫酸钠的制备,了解其性质及应用技术,巩固抽滤、蒸发、结晶等操作。

实验原理

(1) $Na_2SO_3 + S + 5H_2O \xrightleftharpoons{\triangle} Na_2S_2O_3 \cdot 5H_2O$

(2) $2Na_2S + Na_2CO_3 + 4SO_2 \xrightleftharpoons{\triangle} 3Na_2S_2O_3 + CO_2\uparrow$

(3) 显影反应。

实验用品

仪器与材料:滤纸、表面皿、烧杯、蒸发皿、布氏漏斗、抽滤瓶、真空泵、石棉网、酒精灯。

试剂:$Na_2SO_3(s)$、$S(s)$、活性炭、2 mol/L H_2SO_4、1 mol/L KBr、0.1 mol/L Na_2S、碘水、0.5%淀粉溶液。

实验步骤

1. 硫代硫酸钠的制备

取一个100 mL烧杯,称取8.0 g无水亚硫酸钠(分子量126)和2 g硫粉,先用少量水(乙醇)润湿,再加入50 mL水,在不断搅拌下直接加热煮沸30 min,及时补充水。反应完毕后,在煮沸的溶液中加入1~2 g 活性炭(脱色剂,吸附过量的硫粉),在不断搅拌下,继续煮沸约10 min,趁热过滤,弃去杂质。将滤液置于蒸发皿中加热(微沸),待滤液浓缩到刚有结晶开始析出时停止蒸发,冷却,使硫代硫酸钠结晶析出,抽滤,即得白色 $Na_2S_2O_3 \cdot 5H_2O$(分子量248.2)结晶。将晶体放在烘箱中,40℃干燥40~60 min,称量,计算产率(理论产量15.75 g)。

2. 硫代硫酸钠的应用

1)硫代硫酸钠在洗相定影中的应用

a.硫代硫酸钠洗相定影的基本原理

在洗相过程中,相纸(感光材料)经过照相底版的感光,只能得到潜影。再经过显影液(如海德尔、米吐尔)显影以后,看不见的潜影才被显现成可见的影像。其主要反应如下:

$$HOC_6H_4OH + 2AgBr =\!=\!= OC_6H_4O + 2Ag + 2HBr$$

但相纸在乳剂层中还有大部分未感光的溴化银存在。由于它的存在,一方面得不到透明的影像；另一方面在保存过程中这些溴化银见光将继续发生变化，使影像不稳定。因此，显影后必须经过定影过程。

硫代硫酸钠(俗名海波、大苏打)的定影作用是由于它能与溴化银反应生成易溶于水的配合物。定影过程可用下列反应表示：

$$AgBr + 2Na_2S_2O_3 = Na_3[Ag(S_2O_3)_2] + NaBr$$

$$AgNO_3 + KBr = AgBr\downarrow(浅黄色) + KNO_3$$

注意：$AgNO_3$ 不能过量。

$$2Ag^+ + S_2O_3^{2-} = Ag_2S_2O_3\downarrow$$

b. 洗印照片

在暗室中，将相纸直接覆盖在感光箱的底片上进行感光。感光时间可根据底片情况进行选择。然后，将感过光的相纸放入显影液中进行显影。待影像基本清晰后，用镊子将相纸取出，放入水中清洗一下，紧接着再放入定影液中，定影 10~15 min。再把相纸取出放入水中，用水冲洗。然后，用上光机烘干上光，或贴在平板玻璃上自然晾干上光，最后把纸边剪齐。

2) 硫代硫酸钠定量分析中的应用(碘量法)

硫代硫酸钠标准溶液与单质碘定量反应，以淀粉为指示剂，滴定至溶液的蓝色刚好消失即为终点。反应式为

$$I_2 + 2S_2O_3^{2-} = 2I^- + S_4O_6^{2-}$$

根据消耗硫代硫酸钠标准溶液的体积和浓度计算碘的量。

注意事项

$Na_2S_2O_3·5H_2O$ 为无色透明或白色的单斜晶体，密度 1.685 g/cm³ (20℃)，易溶于水，水溶液显碱性，难溶于乙醇。在 33℃以上干燥空气中易风化，灼烧时分解为硫酸钠和硫化钠。在碱性或中性溶液中稳定，在酸性溶液中迅速分解。$S_2O_3^{2-}$中两个硫原子呈不同的氧化态：一个为+6，另一个为–2，是中等强度的还原剂。除上述应用外，与单质 Cl_2 反应，用于棉布漂白后的脱氯剂、鞣革、电镀等。

$$S_2O_3^{2-} + 2H^+ = S\downarrow + SO_2\uparrow + H_2O$$

可微热，硫逐渐析出。

$$Na_2S_2O_3 + 4Cl_2 + 5H_2O = 2H_2SO_4 + 2NaCl + 6HCl$$

显影液配方：

D-72	米吐尔	无水亚硫酸钠	对苯二酚	无水碳酸钠	溴化钾
	0.75 g	11.25 g	3 g	16.88 g	0.5 g

加冷水稀释至 250 mL。

定影液配方：

F-5	海波	无水亚硫酸钠	乙酸(28%)	硼酸	钾矾
	60 g	3.75 g	11.75 mL	1.88 g	3.75 g

加冷水稀释至 250 mL。

实验 38　自制植物酸碱指示剂

实验目的

(1) 强化有关酸碱指示剂知识，了解其变色的原理。

(2) 选择自己熟悉的植物，制备酸碱指示剂，并确定其变色范围。

实验原理

酸碱指示剂是指在酸性和碱性的溶液中显现出不同颜色的物质。许多植物的花、果、茎、叶中都含有色素，这些色素在酸性或碱性溶液中显示不同的颜色，可以作为酸碱指示剂。

实验用品

仪器与材料：试管、量筒、玻璃棒、研钵、滴管、点滴板、漏斗、纱布。

试剂：花瓣(如牵牛花)、叶子(如紫甘蓝)、萝卜(如胡萝卜、心里美萝卜)、乙醇溶液(乙醇与水的体积比为 1:1)、稀盐酸、稀 NaOH 溶液。

实验步骤

(1) 取一些花瓣、叶子、萝卜等，分别在研钵中捣烂后，各加入 5 mL 乙醇溶液，搅拌。再分别用 4 层纱布过滤，所得滤液分别是花瓣色素、叶子色素和萝卜色素等的乙醇溶液，将它们分装在 3 支试管中。

(2) 在白色点滴板的孔穴中分别滴入稀盐酸、稀 NaOH 溶液、蒸馏水，然后各滴入 3 滴花瓣色素的乙醇溶液。观察现象。

(3) 用叶子色素的乙醇溶液、萝卜色素的乙醇溶液等代替花瓣色素的乙醇溶液做上述实验，观察现象。

数据记录与处理

(1) 写出你知道的其他酸碱指示剂及其变色范围。

(2) 画坐标图(横坐标：乙酸的量；纵坐标：溶液 pH 的变化)描述向氢氧化钠溶液中缓慢滴加乙酸后溶液 pH 的变化。

(3) 带上 pH 试纸，回家后测定一种熟悉易得的植物作为指示剂的灵敏度(变色范围)，把它与酚酞的变色范围作对照，填充下表。

指示剂	酚酞	自制指示剂
变色范围	8~10	

附注

石蕊是常见的酸碱指示剂，是植物提取物。其他常见的酸碱指示剂还有：

百里酚蓝	甲基橙	甲基红	溴百里酚蓝	百里酚蓝	酚酞	百里酚酞
pH 1.2~2.8	3.1~4.4	4.4~6.2	6.2~7.6	8.0~9.6	8.0~9.6	9.4~10.6
红变黄	红变黄	红变黄	黄变蓝	黄变蓝	无色变红	无色变蓝

参考文献

郑长龙. 2009. 化学实验课程与教学论. 北京: 高等教育出版社

实验39 海带和食盐中碘的定性鉴定

实验目的

了解碘的主要氧化态化合物的性质，学习从海带中分离碘和鉴定含碘食盐的方法。

实验原理

1. 海带中碘的提取和鉴定

海带中碘含量较高，通过加热燃烧(炭化和灰化)可以把有机物氧化分解而烧成灰烬。经硝酸硝化和水溶液进一步浸取，可以获得含碘酸根的溶液，借助以下反应获得碘单质，利用碘单质遇淀粉溶液显蓝色的性质实现定性鉴定：

$$KIO_3 + 5KI + 3H_2SO_4 = 3I_2 + 3H_2O + 3K_2SO_4$$

2. 加碘食盐与无碘食盐的鉴别

市售碘盐大多含有碘(碘酸钾)，国家规定每千克食用碘盐中碘的含量必须为20~50 mg。用淀粉作指示剂，单质碘的检出限为1×10^{-7} g/mL。碘离子与碘酸根反应生成碘单质，据此可以简便地定性鉴别食盐中是否含碘。

实验用品

仪器与材料：试管、烧杯、研钵、抽滤瓶、布氏漏斗、蒸发皿、坩埚钳、酒精灯、表面皿、镊子、滤纸、pH试纸。

试剂：碘盐、氯化钠、干海带、2 mol/L H_2SO_4、3% H_2O_2、0.5 mol/L KIO_3、淀粉溶液、0.1 mol/L KI、四氯化碳。

实验步骤

(1) 自行设计实验方案提取海带中的碘并完成定性鉴定。

(2) 设计简单的实验鉴定加碘食盐与无碘食盐。

备选实验方案

1. 海带中碘的提取和鉴定

(1) 称取5~10 g 干海带点燃或焙烧，使海带完全灰化。

(2) 将海带灰倒在 50 mL 烧杯中，加入 3~5 mL 蒸馏水熬煮浸取 3~5 min，然后补加 3~5 mL 蒸馏水，重复上述浸取一次，过滤并收集滤液。总体积不宜超过 10 mL(注意：不要洗涤沉淀，以免浸取液总碘浓度太低)。

(3) 向一支试管中依次加入 1~2 mL 浸取液、10 滴 2 mol/L H_2SO_4 和 3% H_2O_2 溶液，振荡混匀后再加几滴淀粉溶液，变蓝色证明海带中存在碘元素。

对照实验：在一支试管中加入 1 mL 0.1 mol/L KI 溶液、10 滴 2 mol/L H_2SO_4 和 3% H_2O_2 溶液，振荡混匀，再加几滴淀粉溶液，观察现象。

(4) 往剩余滤液中加 2 mol/L H_2SO_4 小心酸化至 pH 呈中性。加入 1 mL 3% H_2O_2 溶液振荡(注意：海带灰中含有碳酸钾，酸化使其呈中性或弱酸性对下一步氧化析出碘有利。但硫酸加多了则易使碘化氢氧化出碘而损失)。向溶液中加入 2~3 mL 四氯化碳，振荡试管萃取水溶液中的碘，静置使溶液分层，观察四氯化碳的颜色。

2. 加碘食盐与无碘食盐的鉴别

取 2 支试管，分别加入少量加碘食盐和无碘食盐，加入 1 mL 2 mol/L H_2SO_4 和 10 滴 0.1 mol/L KI 溶液，搅拌溶解后再加入几滴淀粉溶液，观察现象。变蓝色的是加碘食盐，不变色的是无碘食盐。

对照实验：取 10 滴 0.5 mol/L KIO_3 溶液，依次加入 10 滴 2 mol/L H_2SO_4 和 0.1 mol/L KI 溶液，振荡混匀，再加入几滴淀粉溶液，观察现象。

实验 40　化学趣味实验(一)

实验目的

(1) 通过下列实验，使学生认识到生活中处处有化学，激发学生的学习兴趣，培养学生运用化学知识分析问题和解决问题的能力。

(2) 通过实验，掌握指纹鉴定、酒驾鉴定、壁画变色、隐形墨水、乙醇和甲醇的鉴别等实验原理和检验方法。

实验原理

1. 指纹鉴定(碘熏法)

碘受热时升华变成碘蒸气。碘蒸气能溶解在手指上的油脂等分泌物中，并形成棕色指纹印迹。每个人的手指上总含有油脂、矿物油和水，用手指按在纸面上时，指纹上的油、矿物油和汗水就会留在纸面上，只不过人的眼睛看不出来。当把隐藏有指印的纸放在盛有碘的试管口并加热时，碘开始升华，变成紫红色的蒸气(注意，碘蒸气有毒，不可吸入)。由于纸上指印中的油脂、矿物油都是有机溶剂，因此碘蒸气上升到试管口以后就会溶解在这些油类物质中，于是指纹就显示出来。

2. 酒驾鉴定

酒的主要成分乙醇有还原性，而重铬酸钾($K_2Cr_2O_7$，橙红色)具有强氧化性，在酸性条件下能将乙醇氧化为乙酸，同时本身被还原为三价铬（墨绿色），故可从溶液颜色的变化判断人呼出的气体或血液中是否含有乙醇。此反应非常灵敏，当乙醇含量大于 0.2% 即可检出。

$$2K_2Cr_2O_7 + 8H_2SO_4 + 3CH_3CH_2OH = 3CH_3COOH + 2Cr_2(SO_4)_3 + 11H_2O + 2K_2SO_4$$

3. 壁画变色

壁画上灰黑色（而非白色）的物质是硫化铅（PbS）。白色颜料为铅白[碱式碳酸铅，$2PbCO_3·Pb(OH)_2$]，具有很强的覆盖力，涂抹在壁画上时是雪白的。但由于空气中微量 H_2S 长期作用，发生如下反应：

$$2PbCO_3·Pb(OH)_2 + 3H_2S = 3PbS\downarrow + 2CO_2\uparrow + 4H_2O$$

若要使画面恢复，只需取一块软布蘸些过氧化氢在画面上轻擦，则可发生如下反应：

$$PbS + 4H_2O_2 = PbSO_4 + 4H_2O$$

4. 隐形墨水

(1) 用 $CoCl_2$ 溶液在白纸上写字，干后不会显现字迹，但若遇热（可以烘烤或用电吹风）即可显出蓝字；再遇水蒸气时（可放在盛放热水的杯口上移动），字迹又会再次隐没。发生如下反应：

$$CoCl_2·6H_2O(粉红色) \rightarrow CoCl_2·2H_2O(紫红色) \rightarrow CoCl_2·H_2O(蓝紫色) \rightarrow CoCl_2(蓝色)$$

当 $CoCl_2·6H_2O$ 溶液浓度较低时颜色浅，干后近乎无色，但遇热失水变成蓝色的 $CoCl_2$。用水蒸气烘时，上述变化反应进行，又变成浅色的 $CoCl_2·6H_2O$。

(2) 用 1% 酚酞的乙醇溶液在白纸上写字，干后不会显现字迹，但若在其表面涂上稀碱溶液则显出红色字迹，若再涂上乙酸，则红色褪为无色。这是因为酚酞是酸碱指示剂，pH < 8.2 时为无色，pH > 10 则为红色。

(3) 1% 淀粉水溶液或牛奶也可作隐形墨水。书写后无色，用碘溶液涂抹则显现蓝色，烘烤或微热后又褪色。这是因为碘遇淀粉显现蓝色，且碘单质容易升华，受热即可挥发，故又褪色。

5. 乙醇和甲醇的鉴别

甲醇和乙醇都是具有挥发性和特殊香味的无色透明液体，从外观上很难区分。但是甲醇及其代谢产物甲醛、甲酸对人体都有较强毒性，误饮会导致失明，甚至死亡。

甲醇和乙醇的鉴别方法是先将醇氧化成醛，再利用甲醛的特性反应鉴别。甲醇和乙醇在酸性条件下容易被高锰酸钾氧化成甲醛和乙醛，醛能与席夫试剂发生加成反应，生成带蓝影的紫红色加成物。甲醛所显示的颜色加硫酸后不消失，而其他醛如乙醛所显示的颜色则褪去。因此，这个反应可以作为区别甲醛与其他醛的方法。

综上所述，利用先将醇氧化为醛，再与席夫试剂反应（加硫酸）的方法可以鉴别乙醇和甲醇。甲醇的含量越高，则颜色越深。该方法鉴别甲醇的最低检出限为 0.2 g/L。

$$5CH_3OH + 2MnO_4^- + 6H^+ = 5HCHO + 2Mn^{2+} + 8H_2O$$
$$5CH_3CH_2OH + 2MnO_4^- + 6H^+ = 5CH_3CHO + 2Mn^{2+} + 8H_2O$$

实验用品

仪器与材料：白纸、烧杯、镊子、表面皿、量筒、试管、试管架、滴管、滤纸、玻璃棒、电吹风、酒精灯等。

试剂：0.1 mol/L $K_2Cr_2O_7$、H_2SO_4(1 mol/L)、碱式碳酸铅固体[$2PbCO_3·Pb(OH)_2$]、5% TAA(硫代乙酰胺)、3% H_2O_2、0.2 mol/L $CoCl_2$、1%酚酞、1%淀粉、0.1 mol/L 碘水、1 mol/L 乙酸、1 mol/L NaOH、碘(s)、乙醇、甲醇、掺有甲醇的乙醇、$KMnO_4$、$H_2C_2O_4$ 溶液、席夫试剂(品红磺酸盐，无色)。

实验步骤

1. 指纹鉴定(碘熏法)

取一小条白纸，用手指按下留下指纹。在小烧杯中放入 2~3 g 碘，盖上表面皿加热。碘升华成碘蒸气后，打开表面皿，将带有指纹的白纸放在烧杯口，有指纹的一面向下，慢慢地白纸上就会显示出棕色指纹。

由于碘挥发，实验要在通风橱中进行。

2. 酒驾鉴定

在试管中加入 1 滴 0.1 mol/L $K_2Cr_2O_7$ 溶液、2 滴 1 mol/L H_2SO_4，再加 1 滴乙醇，观察溶液颜色的变化。

3. 壁画变色

在滤纸上涂一层碱式碳酸铅，放置在盛有硫代乙酰胺溶液(加稀 H_2SO_4 酸化)的试管口，此时有 H_2S 产生，观察滤纸的变化(白→黑)。再用 3% H_2O_2 溶液滴在变黑处，观察颜色的变化。此实验应在通风橱中进行。

4. 隐形墨水

(1)用 0.2 mol/L $CoCl_2$ 溶液在白纸上写字，用电吹风吹干，有蓝色字迹出现。再把纸片放在水浴上接触水蒸气，蓝色字迹褪去。

(2)用 1%酚酞在白纸上写字，在字迹上涂 1 mol/L NaOH 溶液，即可呈现出红色字迹。再用 1 mol/L 乙酸溶液涂在红色字上，字迹隐去成无色。

(3)同样用 1%淀粉在白纸上写字，再用碘溶液涂在字迹上，即可呈现出蓝色字迹。将纸片烘热片刻，蓝色字迹褪去。此实验应在通风橱中进行。

5. 乙醇和甲醇的鉴别

取 3 支试管，分别加入 0.5 mL 乙醇、甲醇和掺有甲醇的乙醇，再各加入少量稀 H_2SO_4 溶液和 1~2 滴 $KMnO_4$ 溶液混匀。放置 10~20 s 后，再分别加入 1~2 滴 $H_2C_2O_4$ 溶液，振荡直至溶液褪色。再分别加入 3~4 滴席夫试剂，振荡后静置 5~10 min，观察现象。甲醇为紫红色(带蓝影)，乙醇为无色，掺有甲醇的乙醇为蓝色。

注意事项

碘蒸气有剧毒，需在通风橱中进行。

实验 41　化学趣味实验(二)

实验目的

(1) 通过下列实验，使学生认识到生活中处处有化学，并激发学生的学习兴趣，培养学生运用化学知识分析问题和解决问题的能力。

(2) 通过实验，掌握蓝瓶子、滴水升烟、水中芭蕾、溶洞奇观、水中黄金等实验原理及检验方法。

实验原理

1. 蓝瓶子实验

亚甲基蓝是一种暗绿色晶体，溶于水和乙醇。在碱性溶液中，蓝色的亚甲基蓝很容易被葡萄糖还原为无色的亚甲基白。振荡此无色溶液时，溶液与空气接触面积增大，溶液中氧气溶解量增多，氧气把亚甲基白氧化为亚甲基蓝，溶液又呈蓝色。静置此溶液时，有一部分溶解的氧气逸出，亚甲基蓝又被葡萄糖还原为亚甲基白。若重复振荡和静置溶液，其颜色交替出现"蓝色—无色—蓝色—无色"的现象，这称为亚甲基蓝的"化学振荡"。

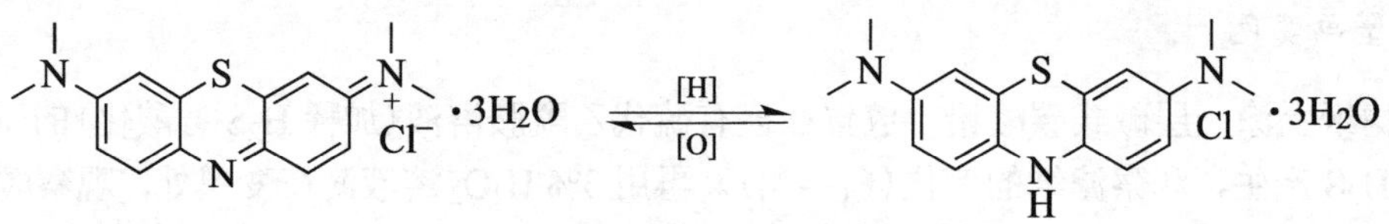

2. 滴水生烟

块状的金属铝表面有致密的氧化膜，使铝在常温下具有很高的稳定性，能广泛用于制造日用器皿。但是，当金属呈粉末状态时，其活性将大大提高。在常温下，以水作催化剂的条件下，铝粉能与碘单质反应生成碘化铝固体，并放出大量的热，使过量的碘升华成碘蒸气而呈现美丽的紫色烟雾。"滴水生烟"实验中的金属粉末不仅可以是铝，还可以是镁、锌、铁、铜等。

$$2Al + 3I_2 \xrightarrow{H_2O} 2AlI_3 + Q$$

3. 水中芭蕾

钠、钾在水中剧烈反应，产生氢气并放热，使金属熔化成小球，并在气体推动下在水面上快速旋转。钾甚至引起自燃，产生紫色火焰。生成的 MOH(M=Na、K)溶液呈碱性，使酚酞指示剂变红色。

$$2Na + 2H_2O == 2NaOH + H_2\uparrow$$

$$2K + 2H_2O == 2KOH + H_2\uparrow$$

4. 溶洞奇观

将各种能溶于水的金属盐颗粒放入硅酸钠溶液中，在表面生成难溶性的硅酸盐薄膜。由于这层薄膜具有半透膜性质，在渗透压的驱动下，膜外面的溶剂水分子不断向膜内渗透，导致薄膜膨胀。当达到一定压差时，薄膜破裂，金属盐溶液从裂口溢出，再与膜外的硅酸钠作用，又生成新的硅酸盐。这一过程不断重复，就像植物不断生长一样。由于不同金属的难溶性硅酸盐的颜色不同，所以可以看到一个五彩缤纷的"水中花园"。

$$M^{2+} + SiO_3^{2-} = MSiO_3\downarrow$$

5. 水中黄金

KI 与可溶性铅盐反应生成 PbI_2 沉淀，它是一种丝状有金黄色光泽的沉淀，其形貌酷似"黄金"。

$$2KI + Pb(NO_3)_2 = PbI_2\downarrow + 2KNO_3$$

因此，可利用上述反应在水中制备"黄金"。但是该盐易溶于沸水或过量的 KI 溶液中。

实验用品

仪器与材料：量筒、表面皿、烧瓶、橡皮塞、试管、研钵、蒸发皿、镊子、烧杯、漏斗、台秤、酒精灯。

试剂：1% 亚甲基蓝、20% Na_2SiO_3、KI(s)、I_2(s)、NaOH(s)、葡萄糖(s)、铝粉、Na(s)、K(s)、$CaCl_2$(s)、$Co(NO_3)_3$(s)、$CuSO_4$(s)、$NiSO_4$(s)、$MnSO_4$(s)、$FeSO_4$(s)、$FeCl_3$(s)、$Pb(NO_3)_2$(s)。

实验步骤

1. 蓝瓶子实验

在 250 mL 烧瓶中(塞子另配)加入 50 mL 水，分别称取 1 g NaOH 和 1 g 葡萄糖溶解在水中，然后加入 5 滴 1%亚甲基蓝水溶液。摇匀后，塞住瓶口，观察溶液颜色逐渐转为无色。再打开瓶塞摇动瓶子，观察溶液颜色又很快变成蓝色。再放置又转为无色。反复操作，观察颜色的变化。

2. 滴水生烟

取 10 颗碘粒置于研钵中研细，然后加 1 小匙铝粉(约为碘量的 1/10)，共同研磨，混合均匀。将混合物倒在蒸发皿中央，往混合物上滴 1～2 滴水，立即用大烧杯盖住蒸发皿(注意所用仪器和试剂必须是干燥的)。片刻后出现浓厚而美丽的烟雾，观察颜色。

3. 水中芭蕾

用镊子和小刀小心切取一小块(绿豆大小)金属钠和钾，并用滤纸吸干其表面煤油。把它投入已盛有半杯水的烧杯中，并立即用倒置漏斗覆盖在烧杯口上，以免金属钠和钾弹出烧杯而自燃。观察反应情况。反应完毕后，滴入 1～2 滴酚酞试剂，观察溶液颜色的变化。

4. 溶洞奇观

在 100 mL 烧杯中加入约 30 mL 20%硅酸钠溶液，然后各取一小粒氯化钙、硝酸钴、硫酸铜、硫酸镍、硫酸锰、硫酸亚铁、三氯化铁固体小心放到烧杯底部(注意各固体之间保持一定间隔)。记下各种金属盐的位置，静置一段时间后观察有何现象发生。

5. 水中黄金

取 2 个 50 mL 烧杯，各加入 20 mL 水，分别加入 3.5 g KI 和 3.5 g $Pb(NO_3)_2$，搅拌溶解，分别得 1 mol/L KI 溶液和 0.5 mol/L $Pb(NO_3)_2$ 溶液。将两者混合并加热，待即将沸腾时，将其骤冷，再放在冷水中，观察现象。

注意事项

(1) 碘蒸气有剧毒，实验需在通风橱中进行。

(2) 蓝瓶子实验的溶液一定要是碱性的，但碱不可过量。蓝瓶子实验的最佳温度是 25℃，所以冬天演示要略加热，但温度不可太高。

(3) 钠和钾不可多取，否则反应过于剧烈。多余的固体钠必须放回试剂瓶中，用过的滤纸也不可随处丢弃，以免残留固体发生自燃，可放入水中处理。

(4) 在溶洞奇观实验后，应及时将烧杯清洗干净，以免碱性硅酸盐久置腐蚀烧杯。

实验 42　食盐和亚硝酸钠的鉴别

实验目的

通过实验了解亚硝酸钠的性质，并对亚硝酸钠和食盐进行鉴别。

实验原理

亚硝酸钠为白色至淡黄色粉末或颗粒状，其外观和食盐相似，有微咸味，易溶于水。亚硝酸钠俗称"工业盐"，在工业、建筑业中广为使用，肉类制品中也允许作为发色剂限量使用。由亚硝酸钠引起食物中毒的概率较高，食入 0.3～0.5 g 亚硝酸钠即可引起中毒甚至死亡。食盐的市场价是每吨 2000 元左右，工业盐约是每吨 230 元，因此一些不法盐业制造商通过在食盐中添加亚硝酸钠获取高额利润。

可以通过物理性质和化学性质两个方面对亚硝酸钠和食盐进行鉴别，主要包括比色法、加热熔融法、水中溶解法、酸性高锰酸钾法、加酸法、KI-淀粉试纸法、硝酸银法、硫酸亚铁法、酚酞法等。

物理方法：

(1) 比色法。

氯化钠是立方晶体或细小的结晶粉末，呈白色；亚硝酸钠则是斜方晶体，略带有浅黄色。

(2) 加热熔融法。

氯化钠的熔点为 801℃，亚硝酸钠的熔点为 271℃。把亚硝酸钠和氯化钠置于同一小块金属片上,用酒精灯加热,先熔化的是亚硝酸钠,不熔化的是氯化钠。

(3) 水中溶解法。

氯化钠和亚硝酸钠虽然都溶解于水,但是两者溶解度却大不相同。温度对氯化钠溶解的影响很小;而亚硝酸钠极易溶于水,且溶解时吸热,在热水中溶解得更快。

化学方法:

(1) 亚硝酸钠的生成及不稳定性(加酸法):

$$NaNO_2 + H_2SO_4(稀) = NaHSO_4 + HNO_2$$
$$2HNO_2 = NO\uparrow + NO_2\uparrow(红棕色) + H_2O$$

食盐无类似反应和现象。

(2) 亚硝酸钠的氧化性(硫酸亚铁法和KI-淀粉试纸法):

$$2NaNO_2(无色) + 2FeSO_4 + 2H_2SO_4 = 2NO\uparrow + Fe_2(SO_4)_3(黄褐色) + Na_2SO_4 + 2H_2O$$
$$2NaNO_2(无色) + 2KI + 2H_2SO_4 = 2NO\uparrow + I_2(棕色) + 2H_2O + Na_2SO_4 + K_2SO_4$$

食盐无类似反应和现象。

(3) 亚硝酸钠的还原性(酸性高锰酸钾法):

$$2KMnO_4(紫红色) + 5NaNO_2 + 3H_2SO_4 = K_2SO_4 + 2MnSO_4(无色) + 5NaNO_3 + 3H_2O$$

食盐无类似反应和现象。

(4) 亚硝酸钠的鉴别(硝酸银法):

$$NaNO_2 + AgNO_3 = AgNO_2\downarrow(白色) + NaNO_3$$
$$NaCl + AgNO_3 = AgCl\downarrow(白色) + NaNO_3$$
$$AgNO_2 + HNO_3 = AgNO_3 + HNO_2(白色沉淀溶解)$$

AgCl与稀硝酸不反应。

(5) 亚硝酸钠溶液的酸碱性(酚酞法)。

$NaNO_2$属于强碱弱酸盐,其水溶液呈碱性,加热后随着水分的蒸发,碱性增强,因此可以使酚酞试纸变红。食盐属于强酸强碱盐,其水溶液呈中性,加热后酸碱性不改变,酚酞试纸颜色没有变化。

实验用品

仪器与材料:台秤、酒精灯、试管。

试剂:亚硝酸钠、氯化钠、酚酞指示剂、稀硫酸、0.1 mol/L $FeSO_4$、KSCN饱和溶液、0.1 mol/L KI、淀粉溶液、0.01 mol/L $KMnO_4$、0.1 mol/L $AgNO_3$、稀硝酸。

实验步骤

在室温时,分别称取5 g亚硝酸钠和氯化钠置于试管中,加10 mL水使其溶解,观察溶液温度的变化。两种溶液分别均分为6份于试管中,进行下面的实验。

(1) 在两种溶液中分别滴加酚酞指示剂,加热溶液,观察颜色的变化。

(2) 在两种溶液中加入冰冷的稀硫酸,静置片刻,振荡试管,观察溶液颜色变化及有无气体放出。

(3) 在两种溶液中分别加入 5 滴稀硫酸进行酸化，然后分别加入 5 滴 0.1 mol/L FeSO₄，振荡试管，静置观察溶液颜色变化。再加入 5 滴 KSCN 饱和溶液，观察颜色变化。

(4) 在两种溶液中分别加入 5 滴稀硫酸进行酸化，然后分别加入 5 滴 0.1 mol/L KI，振荡试管，静置观察溶液颜色变化。再加入 5 滴淀粉溶液，观察颜色变化。

(5) 在两种溶液中分别加入 5 滴稀硫酸进行酸化，然后分别加入 5 滴 0.01 mol/L KMnO₄，振荡试管，静置观察溶液颜色变化。

(6) 在两种溶液中分别加入 5 滴 0.1 mol/L AgNO₃，振荡试管，静置观察沉淀的生成。离心分离沉淀，在沉淀中加入稀硝酸，观察沉淀是否溶解。